算法设计与实现

（中英双语版）

王中生　沈 涵　冯孝周　编著

清华大学出版社

北京

内 容 简 介

世界一流大学和一流学科("双一流")的建设目标要求大学全面提升学生的综合素质,拓展国际视野,按照教育部关于"双一流"高校的培养目标,编写了这部《算法设计与实现(中英双语版)》教材。

本书主要内容包括算法概述、贪心算法、分治算法、动态规划算法、回溯算法、分支限界算法、图算法、概率分析和随机算法等。本书内容翔实,图文并茂,通俗易懂,适合普通高校工科类各年级本科生和研究生学习计算机算法及程序实现作为教材使用,也适合国际学院留学生作为学习算法与编程的教材或参考书。

Content Summary

The construction goal of world-class universities and first-class disciplines (the " double first-class") requires universities to enhance students' comprehensive quality and expand their international vision. In accordance with the Ministry of Education's training objectives for " double first-class" universities, this textbook, *Algorithm Design and Implementation* (*Chinese English Bilingual Edition*) has been compiled.

The main contents of this book include: Algorithm Overview, Greedy Algorithm, Divide-and-Conquer Algorithm, Dynamic Programming Algorithm, Backtracking Algorithm, Branch and Bound Algorithm, Graph Algorithm, Probability Analysis and Stochastic Algorithm. The content of this book is detailed, illustrated, and easy to understand. It is suitable for undergraduate and graduate students in engineering at various universities to learn computer algorithms and program implementation as a textbook, and it is also appropriate for international students to learn algorithms and programming as textbooks or reference books.

图书在版编目(CIP)数据

算法设计与实现:汉、英/王中生,沈涵,冯孝周编著. -- 北京:清华大学出版社,2025.9.
ISBN 978-7-302-70366-2

Ⅰ. TP301.6

中国国家版本馆 CIP 数据核字第 20255UF089 号

责任编辑:温明洁 薛 阳
封面设计:刘   键
责任校对:郝美丽
责任印制:沈   露

出版发行:清华大学出版社
　　　　网　　址:https://www.tup.com.cn,https://www.wqxuetang.com
　　　　地　　址:北京清华大学学研大厦 A 座　　　　　　邮　　编:100084
　　　　社 总 机:010-83470000　　　　　　　　　　　　邮　　购:010-62786544
　　　　投稿与读者服务:010-62776969, c-service@tup.tsinghua.edu.cn
　　　　质量反馈:010-62772015, zhiliang@tup.tsinghua.edu.cn
　　　　课件下载:https://www.tup.com.cn,010-83470236
印 装 者:三河市人民印务有限公司
经　　销:全国新华书店
开　　本:185mm×260mm　　　　印　　张:16.75　　　　字　　数:411 千字
版　　次:2025 年 9 月第 1 版　　　　　　　　　　印　　次:2025 年 9 月第 1 次印刷
印　　数:1~1500
定　　价:59.90 元

产品编号:109691-01

前 言

世界一流大学和一流学科（"双一流"）的建设目标要求大学全面提升学生的综合素质，拓展国际视野。教材是体现教学内容和教学要求的知识载体，是进行教学的基本工具，是教学质量的重要保证。双语教学已经成为"双一流"高校教学改革的重要内容，双语教学的教材更是一个非常关键的教学资源，教材的内容直接关系教学的质量和效果。实践证明，双语教材内容选取是影响双语教学发展的最重要因素，因此，按照教育部关于"双一流"高校建设培养目标，我们组织编写了这本中英文双语对照教材。

本书内容已经过西安工业大学计算机科学与工程学院计算机科学硕士留学生试用，通过和留学生 Hadush Rimon Yallow、Nouman Hameed、Nawaz Malik Saad、Jallah Willie Kamara、Mavhunga Tatenda Whitney and Emmanuel Tamanda Nhamo 等交流沟通，编写组老师对教材内容进行了多次商讨和取舍，最终确定本书的内容。本书主要内容包括算法概述、贪心算法、分治算法、动态规划算法、回溯算法、分支限界算法、图算法、概率分析和随机算法。

本书由王中生总体规划和统稿并编写第1章，沈涵编写了第2、7、8章并对程序代码进行了审核，牛宇欣、屈文静、常浩、骆仟收集了第3~6章的基础材料，并进行了代码编写和调试，冯孝周对全书进行了审核。由于本书采用英文和中文双语对照编写，限于篇幅，算法分析内容未能编入本书，如算法的时间复杂度分析和空间复杂度分析，有兴趣的读者可以通过邮件索取该部分内容。

本书在编写过程中参阅了大量中文教材及专著文献，在参考文献中业已标注，如果有未标明处，请与作者联系，作者对所有提供资料的老师和研究人员致以诚挚的感谢。由于作者水平有限，时间仓促，书中肯定存在一些疏漏和不足之处，欢迎广大读者提出建设性意见和建议，您的意见或建议将会对本书的优化与完善起到重要的促进作用，诚望不吝赐教。

本书配有完整的代码实现与教学课件资料（无偿提供），欢迎到清华大学出版社官网下载。

作 者
2025 年 6 月

Preface

The goal of building world-class universities and first-class disciplines (the "Double First-Class" initiative) requires universities to enhance students' comprehensive quality and international vision. A textbook is a knowledge carrier that embodies teaching content and requirements, serves as a fundamental tool for teaching, and is an important guarantee for teaching quality.

Bilingual teaching has become an important component of the teaching reform in "Double First-Class" universities, and bilingual teaching materials are a crucial teaching resource. The content of these materials is directly related to the quality and effectiveness of teaching. Practice has proved that the content selection is the most important factor affecting the development of bilingual teaching. Therefore, according to the goal of the Ministry of Education on the construction and training of "double first-class" universities, we have organized the compilation of this bilingual teaching material.

The content of this textbook has been tried out by international students with a master's degree in Computer Science in the School of Computer Science and Engineering of Xi'an Technological University. Through communication with international students, the teachers of the writing team discussed and selected the content of the textbook for many times, and finally determined the content of this book. The main contents of this book include: Algorithm Overview, Greedy Algorithm, Divide-and-Conquer Algorithm, Dynamic Programming Algorithm, Backtracking Algorithm, Branch and Bound Algorithm, Graph Algorithm, Probability Analysis and Stochastic Algorithm.

The book was planned and drafted by Wang Zhongsheng who wrote the first chapter. Shen Han wrote Chapters 2, 7 and 8 and reviewed the program code. Niu Yuxin, Qu Wenjing, Chang Hao and Luo Qian collected the basic materials from Chapter 3 to 6 and wrote and debuted the code. Feng Xiaozhou reviewed the whole book. Due to the bilingual nature of the book in English and Chinese, content on algorithm analysis, such as time complexity and spatial complexity analyses, could not be included in this volume due to space constraints. Readers can request this part by email.

During the preparation of this book, a large number of Chinese textbooks and monographs have been referred to, which have been marked in the references. If there is anything not marked, please contact the author. The author would like to thank all the teachers and researchers who have provided information. Due to the limited level of the author and the short time, there are certainly some shortcomings and inadequacies in this book. We welcome readers to put forward constructive comments and suggestions. Your comments or suggestions will play an important role in promoting the optimization and improvement of this book.

This book comes with a complete set of code implementations and teaching courseware, which are available free of charge. You are welcome to visit the official website of Tsinghua University Press to download these resources.

随书资源

目录 Contents

第1章

算法概述

Chapter 1　Algorithm Overview

算法是解决实际问题的方法和步骤,是人类文明出现以来计算工具及计算技术发展的基础。计算机则将计算步骤和方法程序化,以电子运行的方式让程序化的步骤自动地完成。本章将简要介绍算法设计的基础知识。

Algorithms represent the methods and steps designed to solve practical problems. They have been the foundation of computing tools and technology development since the dawn of human civilization. A computer operates by executing programmed instructions, which are sequences of calculations and methods encoded in electronic form, allowing for the automatic completion of predefined tasks. This chapter will offer a concise introduction to the fundamental knowledge of algorithms.

1.1　算法及算法描述(Algorithm and its Description)

1.1.1　算法概念(Algorithm Concept)

1. 算法的定义(Definition of Algorithm)

算法(Algorithm)是在有限步骤内求解某一问题所使用的一组准确而完整的定义明确的解题方案或规则,它是一系列解决问题的清晰指令。通俗地讲,就是计算机解题的过程,在这个过程中,无论是形成解题思路还是编写程序,都是在实施某种算法,所以可以理解为算法就是对特定问题求解步骤的一种描述,也就是解决问题的方法与步骤。

An algorithm is a set of accurate, complete, and well-defined solutions or rules used to solve a problem in a finite number of steps. It is a series of clear instructions that guide the problem-solving process. Whether the task is to conceive a solution idea or to write a program, it involves the implementation of an algorithm. Thus, an algorithm can be understood as a description of the steps to solve a specific problem, that is, the method and procedure to address the problem.

算法能够对一定规范的输入,在有限时间内获得所要求的输出。如果一个算法有缺陷,或不适合某个问题,执行这个算法将不会解决这个问题。不同的算法可能用不同的时间、空间或效率来完成同样的任务。

An algorithm can produce the required output in a finite time given a specific standard input. If an algorithm is flawed or inappropriate for a problem, executing it will not solve the problem. Different algorithms may perform the same task with varying amounts of time, space, or efficiency.

2. 算法的特点（Characters of Algorithm）

一个算法应该具有以下 5 个重要的特征。

An algorithm should possess the following five important characteristics.

（1）输入。一个算法有 0 个或多个输入，这些输入可以是具体的数值或其他信息，用于引导算法的计算过程。

Input. An algorithm may have zero or more inputs, which can consist of specific values or other data that guide the algorithm's computation process.

（2）输出。一个算法有一个或多个输出，以反映对输入数据加工后的结果，没有输出的算法是毫无意义的。

Output. An algorithm produces one or more outputs that reflect the results of processing the input data. An algorithm without outputs is considered meaningless.

（3）有穷性。一个算法必须保证在执行有限的步骤之后结束。也就是说，一个算法所包含的计算步骤是有限的。

Finiteness. An algorithm must be guaranteed to terminate after a finite number of steps. In other words, an algorithm comprises a finite number of computational steps.

（4）确切性。算法的每一步骤必须有确切的定义。即算法中所有有待执行的动作必须严格而不含糊地进行规定，不能有歧义性。

Definiteness. Each step of the algorithm must be well-defined. In other words, all actions to be performed in the algorithm must be specified clearly and unambiguously.

（5）可行性。算法原则上能够精确地运行，算法执行者甚至不需要掌握算法的含义即可根据该算法的每一步骤要求进行操作，并最终得出正确的结果。

Practicable. In principle, an algorithm can run precisely, and the executor of the algorithm does not even need to understand its meaning to follow the requirements of each step and ultimately obtain the correct result.

1.1.2　算法的描述方法（Description of Algorithm）

一个算法通常可以用自然语言、伪代码、流程图、N-S 图或者计算机语言来描述，介绍如下。

An algorithm is typically described using natural language, pseudocode, flowcharts, N-S diagrams, or a computer language, as detailed below.

1. 自然语言描述法（Natural Language Description）

自然语言描述法是用人们日常所用的语言来表示，如汉语、英语、法语等。使用这些语言不用专门训练，所描述的算法通俗易懂。但该方法文字冗长，易出现歧义，因此在实际应用中除了一些简单问题外，一般不使用该方法表示算法。

Natural language description involves the languages people use daily, such as

Chinese, English, French, and so on. Utilizing these languages requires no special training, and the algorithms described are easy to understand. However, the text produced by this method can be lengthy and prone to ambiguity, so it is generally not used to express algorithms, except for some simple problems in practical applications.

2. 伪代码描述法(Pseudocode Description)

伪代码(Pseudocode)是一种非正式的、类似于英语结构的、用于描述模块结构图的语言。人们在用不同的编程语言实现同一个算法时,实现的方式不同。伪代码是一种介于自然语言和计算机语言之间的、用文字和符号来描述算法的工具,它不用图形符号,因此,书写方便,格式紧凑,易于理解,便于向计算机程序设计语言过渡。

Pseudocode is an informal, English-like structure used to describe algorithms and module structures. When individuals implement the same algorithm using different programming languages, their approaches are different. Pseudocode serves as a bridge between natural language and computer language, describing algorithms using words and symbols without graphic symbols. It is convenient to write, has a compact format, is easy to understand, and facilitates the transition to computer programming languages.

3. 流程图描述法(Flowchart Description)

流程图描述法采用特定的符号,辅之以简要的文字或数字,以业务流程线加以连接,将业务的处理程序和内部控制过程反映出来。流程图绘制有直式流程图和横式流程图两种基本方式。流程图中符号的意义如表1.1所示。

The flowchart description method employs specific symbols, supplemented by brief text or numbers, connected by workflow lines, to depict the work process and its internal control mechanisms. There are two fundamental types of flowcharts: the vertical flowchart and the horizontal flowchart. The meanings of the symbols used in flowcharts are presented in Table 1.1.

表 1.1　流程图中符号的意义

Table 1.1　The meanings of symbols in flowchart

图形符号(Symbol)	名称(Name)	含义(Meaning)
	起止框(Start-stop frame)	算法的开始或结束(Rounded rectangles express the begin or end)
	处理框(Process frame)	处理或运算等功能(Rectangles express the functions of processing and operation)
	输入/输出(Input/Output)	进行输入/输出操作(Parallelograms express the input/output operations)
	判断框(Judgment box)	根据条件是否满足决定执行的路径(Rhombus express the different execution path according to the conditions are met or not)
	控制流(Control flow)	使用箭头指向算法执行的路径(Use arrows to point the path of algorithm is executed)

用流程图表示的算法有三种基本结构,介绍如下。

The algorithm depicted by the flowchart comprises three fundamental structures, described below.

（1）顺序结构（Sequential structure）。

顺序结构的程序设计是最简单的，只要按照解决问题的顺序写出相应的语句就可以，它的执行顺序是自上而下，依次执行。顺序结构流程图如图 1.1 所示。

The sequential structure in program design is the simplest. Simply write the corresponding statements in the order that solves the problem, and the execution order will be from top to bottom. The flowchart for the sequential structure is shown in Figure 1.1.

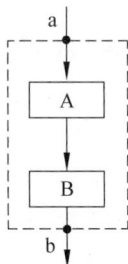

图 1.1　顺序结构流程图

Figure 1.1　Schematic diagram of sequential structure

（2）选择结构（Selection structure）。

选择结构根据给定的条件进行判断，根据判断结果来控制程序的流程。在选择结构中，必须有一个判断框。选择结构流程图如图 1.2 所示。

The selection structure operates by evaluating given conditions, and the program's flow is directed based on these evaluation outcomes. Within the selection structure, a decision box is mandatory. The flowchart for the selection structure is displayed in Figure 1.2.

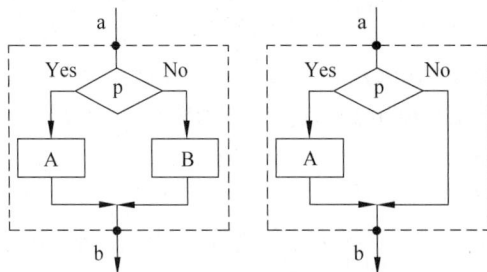

图 1.2　选择结构流程图

Figure 1.2　Schematic diagram of selection structure

（3）循环结构（Loop structure）。

循环结构是在程序中需要反复执行某个功能而设置的一种程序结构。它由循环体中的条件，判断是继续执行某个功能还是退出循环。

The loop structure refers to a program structure that needs to repeatedly execute a certain function in the program. It is determined by the conditions in the body of the loop to continue to perform a function or exit the loop.

循环结构包含三个要素：循环变量、循环体和循环终止条件。循环结构在程序框图中是利用判断框来表示的,判断框内写上条件,两个出口分别对应着条件成立和条件不成立时所执行的不同指令,其中一个要指向循环体,然后再从循环体回到判断框的入口处。

A loop structure comprises three elements: loop variables, loop body, and loop termination conditions. In program flow diagrams, the loop structure is represented by a decision box. This decision box contains conditions with two exits: one for when the condition is met and another for when it is not. One exit should point to the loop body, and from there, back to the entrance of the decision box.

根据判断条件,循环结构又可以细分为以下两种形式。

Based on the judgment conditions, the loop structure can be categorized into the following two forms.

① while 型循环(当型循环)。当型循环属于先判断后执行结构,即当满足条件时开始执行循环,不满足条件时退出循环,如图 1.3(a)所示。

While Loop. The while loop structure is evaluated first and then executed; that is, the loop begins execution when the conditions are satisfied and terminates if the conditions are not met, as illustrated in Figure 1.3(a).

② until 型循环。先执行循环,执行完后进行判断,当满足条件后退出循环体,如图 1.3(b)所示。

Until Loop. Execute the loop initially, then make a judgment; exit the loop body if the conditions are met. As shown in Figure 1.3(b).

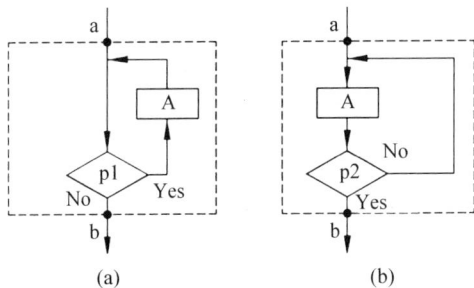

图 1.3　循环的两种形式示意图

Figure 1.3　Schematic diagram of two forms of loop

4. N-S 图(N-S Diagram)

N-S 图也被称为盒图或 NS 图(即 Nassi Shneiderman 图),是结构化编程中的一种可视化建模方法。该方法在流程图中完全去掉流程线,全部算法写在一个矩形阵内,在框内还可以包含其他框的流程图形式。即由一些基本的框组成一个大的框,这种流程图又称为 N-S 结构流程图。

N-S diagrams, also known as box diagrams or Nassi Shneiderman(NS) diagrams, are a visual modeling technique used in structured programming. This method eliminates the flow lines typically found in flowcharts, presenting all algorithms within a rectangular array, with the ability to incorporate the flowchart forms of other boxes within a single box. Composed of several basic boxes to form a larger structure, this type of flowchart is also referred to as the N-S structure flowchart.

（1）顺序结构 N-S 图（N-S Diagram of Sequential Structure）。

程序按照 A、B 顺序向下执行,如图 1.4 所示。

The program executes sequentially, with steps A and B, as depicted in Figure 1.4.

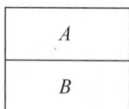

图 1.4　顺序结构 N-S 示意图

Figure 1.4　Schematic diagram of N-S for sequential structure

（2）选择结构 N-S 图（N-S Diagram of Selection Structure）。

当条件 P 满足时执行 A 程序,条件不满足时执行 B 程序,如图 1.5 所示。

When condition P is met, program A is executed. If condition P is not met, program B is executed, as shown in Figure 1.5.

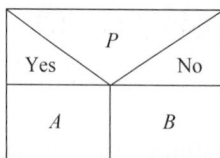

图 1.5　选择结构 N-S 示意图

Figure 1.5　Schematic diagram of N-S for selection structure

（3）循环结构 N-S 图（N-S Diagram of Loop Structure）。

该结构分为当型（while 型）和直到型（until 型）循环,如图 1.6 所示。

The loop structure is divided into two types: the while loop and the until loop, as shown in Figure 1.6.

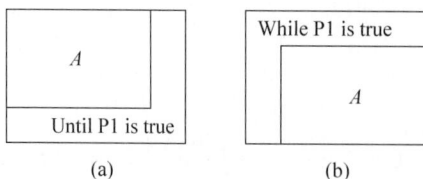

图 1.6　两种类型循环结构 N-S 示意图

Figure 1.6　Schematic diagram of N-S for two types of loop structure

5. 用计算机语言表示（Computer Language Description）

计算机无法识别流程图和伪代码,要让计算机实现某种算法,必须用计算机语言编写程序才能被计算机执行,因此在用流程图或伪代码描述出一个算法后,还要将它转换成计算机语言程序。

Computers cannot recognize flowcharts and pseudocode, hence to implement an algorithm, it must be written in a computer language for execution. Therefore, after an algorithm is described using a flowchart or pseudocode, it must be translated into a computer language program.

用计算机语言表示算法必须严格遵循所用语言的语法规则,这是和伪代码不同的。要完成一件工作,包括设计算法和实现算法两个部分。设计算法的目的是实现算法。

The representation of algorithms in a computer language must strictly adhere to the language's syntax rules, which differs from pseudocode. To complete a task, it involves two parts: the design and implementation of the algorithm. The purpose of designing an algorithm is to facilitate its implementation.

1.1.3　算法评价(Algorithm Evaluation)

算法对程序设计来说是非常重要的,因此一个良好的算法在整个设计开发中占有重要的地位,一个算法的优劣可以用以下5个因素来衡量。

Algorithms are crucial to program design; thus, a good algorithm holds a significant position in the entire design and development process. The merits and demerits of an algorithm can be assessed based on the following five factors.

1. 时间复杂度(Time Complexity)

算法的时间复杂度又称时间复杂性,是指运行该算法需要消耗的时间资源。它反映了算法在解决问题时所消耗的时间,也可以理解为算法的运行效率。算法的耗时和语句的执行次数成正比,即语句执行越多,耗时越多。

Time complexity refers to the amount of time resources required to execute an algorithm. It reflects the time an algorithm consumes in solving a problem and can also be understood as a measure of the algorithm's efficiency. The execution time of an algorithm is proportional to the number of statements executed; that is, the more statements executed, the more time is spent.

2. 空间复杂度(Space Complexity)

算法的空间复杂度是指算法需要消耗的内存空间大小。其计算和表示方法与时间复杂度类似,一般都用复杂度的渐近性来表示。同时间复杂度相比,空间复杂度的分析要简单得多。

Space complexity refers to the amount of memory space required by an algorithm. Its calculation and representation are analogous to those of time complexity, typically expressed in terms of asymptotic complexity. Analyzing space complexity is generally simpler than analyzing time complexity.

3. 正确性(Correctness)

算法的正确性是评价一个算法优劣的最重要的标准。算法应能正确地实现预定的功能,达到满足具体问题的需要。处理数据使用的算法是否得当,决定了能不能得到预想的结果。

Correctness is the most important criterion for evaluating the quality of an algorithm；it must correctly implement the predetermined function and satisfy the requirements of specific problems. The appropriateness of the algorithms used to process data and their ability to yield desired results are also critical considerations.

4. 可读性（Readability）

算法的可读性是指一个算法可供人们阅读的容易程度。算法应易于阅读、理解和交流，便于调试、修改和扩充。写出的算法如果通俗易懂，在系统调试、修改或者扩充功能时，使系统维护更为便捷，就是一个好的算法。

Readability refers to the ease with which people can read an algorithm. An algorithm should be easy to read，understand，and communicate，as well as easy to debug，modify，and expand. A good algorithm is one that is easy to comprehend and facilitates system maintenance during debugging，modification，or expansion.

5. 健壮性（Robustness）

健壮性也称容错性，是指一个算法对不合理数据输入的反应能力和处理能力。输入非法数据时，算法也应能适当地做出反应后进行处理，不会产生预料不到的运行结果。数据的形式多种多样，算法可能接收各种各样的数据，当算法接收到不适合算法处理的数据时，算法本身该如何处理呢？如果算法能够处理异常数据，处理能力越强，健壮性越好。

Robustness，also known as fault tolerance，refers to an algorithm's capacity to respond to and manage unreasonable data inputs effectively. If illicit data is entered，the algorithm should be able to react appropriately and process it without yielding unexpected outcomes. Given the multitude of data forms，an algorithm may encounter the need to handle diverse data types. In cases where the algorithm encounters data unsuitable for processing，it is crucial to address how it should manage such situations. The ability of an algorithm to handle anomalous data is directly related to its processing strength；the greater the capability，the higher the robustness.

1.2　基本数据结构（Basic Data Structure）

数据结构和算法是程序设计的基础，多数算法的设计都是基于良好的数据结构，因此，数据结构在一定程度上影响着算法的设计和性能。在数据结构中，顺序表与链表、栈与队列、图与树是算法设计中经常使用的，了解这些基本知识，对算法设计是相当有益的。

Data structures and algorithms are the foundations of program design. Most algorithms are designed based on good data structures. Consequently，data structures significantly influence the design and performance of algorithms to a certain extent. Within data structures，sequences，linked lists，stacks，queues，graphs，and trees are frequently utilized in algorithm design. Understanding this foundational knowledge is quite beneficial for algorithm design.

1.2.1　顺序表与链表（Sequential List and Linked List）

线性表是最简单、最常用的一种数据结构，一个线性表是 $n(n \geqslant 0)$ 个数据元素的有限序列。顺序表和链表是常用的两种线性数据结构,也是线性结构中最简单的实例。两者的区别是:顺序表是顺序随机存取,而链表是顺序存取。

A linear table is the simplest and most commonly used type of data structure, consisting of a finite sequence of n data elements, where $n \geq 0$. The sequential list and the linked list are two commonly used linear data structures and are among the simplest examples of linear structures. The primary difference between them is that a sequential list allows sequential and random access, whereas a linked list allows only sequential access.

1. 顺序表(Sequential List)

把线性表的元素按照逻辑次序依次存放在一组地址连续的存储单元,用这种方式存储的线性表简称为顺序表。

The elements of a linear table are stored in logical sequence within a contiguous block of storage units, each having consecutive addresses. Such a linear table is commonly referred to as a sequential table.

顺序表是在计算机内存中以数组的形式保存的线性表,线性表的顺序存储是指用一组地址连续的存储单元依次存储线性表中的各个元素,使得线性表中在逻辑结构上相邻的数据元素存储在相邻的物理存储单元中,即通过数据元素物理存储的相邻关系来反映数据元素之间逻辑上的相邻关系,采用顺序存储结构的线性表通常称为顺序表。

A sequential list is a type of linear table stored in computer memory as an array. Sequential storage of a linear table implies that a series of storage units with consecutive addresses are used to store the elements of the linear table sequentially. Consequently, the data elements in the linear table are stored in adjacent physical storage units, maintaining their logical order. In other words, the logical adjacency among data elements is mirrored by their physical adjacency in storage. Linear tables that have a sequential storage structure are commonly referred to as sequential tables.

顺序表是将表中的节点依次存放在计算机内存中一组地址连续的存储单元中,如图1.7所示。

A sequential list stores nodes in a sequence within a contiguous block of storage units in computer memory. This arrangement is depicted in Figure 1.7.

2. 链表(Linked List)

链表是一种物理存储单元上非连续、非顺序的存储结构,数据元素的逻辑顺序是通过链表中的指针链接实现的。链表由一系列节点(链表中每一个元素称为节点)组成,节点可以在运行时动态生成。

A linked list is a type of storage structure that is neither continuous nor sequential on physical storage media. The logical sequence of data elements is maintained through the use of pointers within the linked list. A linked list is composed of a series of nodes, with each element in the list referred to as a node, and these nodes can be dynamically allocated at runtime.

每个节点包括两个部分:一部分是存储数据元素的数据域,另一部分是存储下一个节

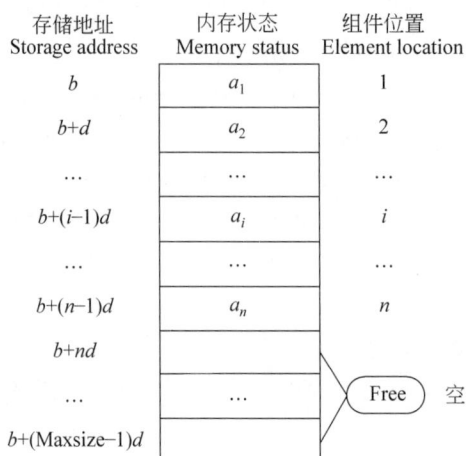

图 1.7　顺序表存储示意图

Figure 1.7　Schematic diagram of the storage structure of sequence table

点地址的指针域。

Each node comprises two components: a data field for storing the data element and a pointer field for the next node's address.

链表的特点如下：数据元素不是连续存储的，而是分散存储在物理存储单元上；数据元素的逻辑顺序是通过链表中的指针链接次序实现的；每个数据元素包含两部分：一部分是实际存储的数据，另一部分是指向下一个数据元素的指针。链表支持灵活的插入和删除操作，不需要移动大量元素，只需要更新相关指针即可。

The characteristics of linked lists are as follows: Data elements are not stored contiguously but are scattered across the physical storage medium; the logical sequence of data elements is determined by the order of pointer linkages within the linked list. Each data element comprises two parts: one is the actual data being stored, and the other is a pointer to the next data element. Linked lists facilitate flexible insertion and deletion operations, which do not require moving a large number of elements; instead, only the relevant pointers need to be updated.

链表有很多种不同的类型：单向链表、双向链表以及循环链表。链表可以在多种编程语言中实现。线性链表逻辑结构示意图如图 1.8 所示。

There are several types of linked lists: single linked lists, circular linked lists, and bidirectional linked lists. Linked lists can be implemented in various programming languages. The logical structure of a linear linked list is depicted in Figure 1.8.

综上所述，线性表提供了一致且快速的顺序访问性能，但不适合于频繁的插入和删除操作；而链表提供了高效的插入和删除能力，但牺牲了一些顺序访问的性能。在实际应用中，选择哪种数据结构取决于具体的需求和性能权衡。

In summary, linear lists offer consistent and rapid sequential access but are less suitable for frequent insertions and deletions. Conversely, linked lists facilitate efficient insertions and deletions, albeit with some sacrifice in sequential access performance. In

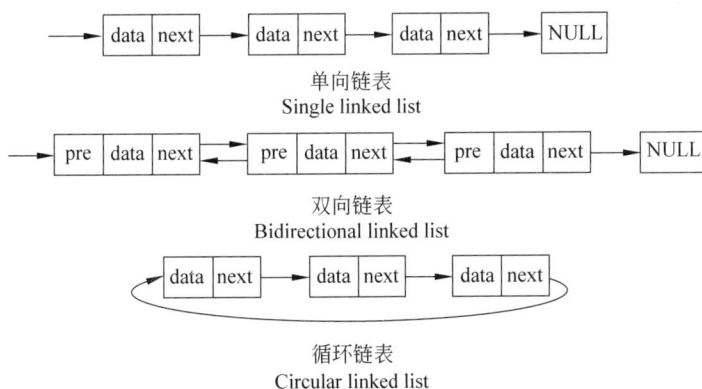

单向链表
Single linked list

双向链表
Bidirectional linked list

循环链表
Circular linked list

图 1.8　线性链表逻辑结构示意图

Figure 1.8　Schematic diagram of linear linked list logic structure

practice, the choice of data structure hinges on specific requirements and performance tradeoffs.

1.2.2　栈与队列(Stack and Queue)

栈和队列是两种常见的线性数据结构,分别适用于不同的数据类型。

Stacks and queues are two common linear data structures used to handle various data types.

1. 栈(Stack)

栈又名堆栈,是一种特殊的线性表,仅在表尾进行插入和删除操作,通常把这一端称为栈顶;相对地,把另一端称为栈底。向一个栈中插入新元素又称作进栈、入栈或压栈,它是把新元素放到栈顶元素的上面,使之成为新的栈顶元素;从一个栈中删除元素又称作出栈或退栈,它是把栈顶元素删除掉,使其相邻的元素成为新的栈顶元素。

A stack is a specialized linear data structure where insertion and deletion operations occur only at one end of the structure, known as the top of the stack. The opposite end is referred to as the bottom of the stack. Inserting a new element into a stack is called pushing, and it involves placing the new element at the top, making it the new top element. Removing an element from a stack is called popping, and it involves removing the top element, allowing the adjacent element to become the new top.

栈作为一种数据结构,是一种只能在一端进行插入和删除操作的特殊线性表。它按照后进先出的原则存储数据,先进入的数据被压入栈底,最后进入的数据在栈顶。需要读数据时,从栈顶开始弹出数据(最后一个进入的数据被第一个读出来)。栈具有记忆作用,对栈的插入与删除操作不需要改变栈底指针。出栈、入栈结构如图 1.9 所示。

As a data structure, a stack is a specialized linear table where insertion and deletion occur exclusively at one end. It adheres to the last-in, first-out (LIFO) principle: the earliest data is placed at the bottom of the stack, while the most recent data resides at the top. When data needs to be read (i.e., the last data to enter is the first to be read out), it is ejected from the top of the stack. A stack exhibits memory retention, and its insertion and deletion operations do not require altering the bottom pointer. The structures of out-stack and in-stack are depicted in Figure 1.9.

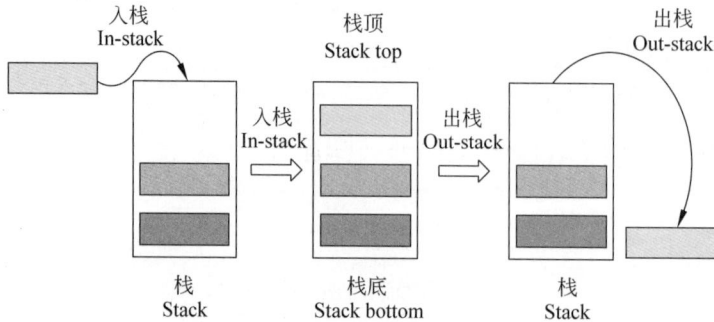

图 1.9　出栈、入栈结构示意图

Figure 1.9　Schematic diagram of out and in stack structure

栈可以用来在函数调用时存储断点，做递归时要用到栈。堆栈是一种存储部件，即数据的写入和读出不需要提供地址，而是根据写入的顺序决定读出的顺序。

The stack can be used to store breakpoints during function calls and is essential for recursion. It serves as a storage unit, meaning that data writing and reading do not require providing an address; instead, the order of writing determines the order of reading.

2. 队列(Queue)

队列是一种特殊的线性表，特殊之处在于它只允许在表的前端进行删除操作，而在表的后端进行插入操作。和栈一样，队列是一种操作受限制的线性表，进行插入操作的端称为队尾，进行删除操作的端称为队头。队列中没有元素时，称为空队列。队列结构示意图如图 1.10 所示。

A queue is a specialized type of linear table where deletion operations occur only at the front of the queue, and insertion operations occur only at the rear of the queue. It is a linear list with restricted operations: the insertion point is referred to as the ' tail' of the queue, and the deletion point is referred to as the ' head'. When a queue contains no elements, it is referred to as an empty queue. Figure 1.10 illustrates the structure of a queue.

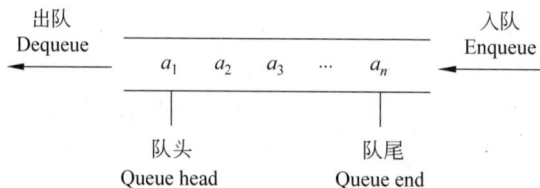

图 1.10　队列结构示意图

Figure 1.10　Schematic diagram of queue structure

队列的数据元素又称为队列元素。在队列中插入一个队列元素称为入队，从队列中删除一个队列元素称为出队。因为队列只允许在一端插入，在另一端删除，所以只有最早进入队列的元素才能最先从队列中删除，故队列又称为先进先出(First In First Out,FIFO)线性表。

The data within a queue are also referred to as queue elements. The process of inserting data into a queue is termed 'enqueuing', and the process of removing data from a queue is termed 'dequeuing'. Since a queue permits insertion only at one end and removal only at the other, the elements that enter the queue the earliest are the first to be removed, which is why the queue is also known as a first-in-first-out (FIFO) linear list.

1.2.3 树与图(Structure of Tree and Diagram)

1. 树结构(Tree Structure)

树结构是一种以分支关系定义的层状数据结构。它是一种非线性结构,是由 $n(n \geq 0)$ 个有限节点组成一个层次关系的集合。计算机系统中的文件和信息的管理方式采用该结构。

A tree structure is a type of hierarchical data structure defined by branch relationships. It is a nonlinear structure composed of n finite nodes ($n \geq 0$) arranged in a set of hierarchical relationships. This structure is used for managing files and information within computer systems.

树结构指的是数据元素之间存在着"一对多"的关系。在树结构中,树根节点没有前驱节点,其余每个节点有且只有一个前驱节点。叶子节点没有后续节点,其余每个节点的后续节点数可以是一个也可以是多个。

A tree structure refers to a hierarchical relationship where each element has a one-to-many relationship with its descendants. In a tree structure, the root node has no parent node, while every other node has exactly one parent node. A leaf node has no child nodes, whereas the number of child nodes for each non-leaf node can range from one to many.

树是由一个集合以及在该集合上定义的一种关系构成的。集合中的元素称为树的节点,所定义的关系称为父子关系。父子关系在树的节点之间建立了一个层次结构,在这种层次结构中有一个节点具有特殊的地位,这个节点称为该树的根节点,或称为树根。树结构示意图如图 1.11 所示。

A tree is composed of a set of elements and a relation that defines the relationships among those elements. The elements in this set are referred to as nodes of the tree, and the defined relationships are known as parent-child relationships. These parent-child relationships establish a hierarchy within the tree, with one node holding a special status—this node is referred to as the root node of the tree. The diagram of the tree structure is depicted in Figure 1.11.

树结构中的重要术语如下:根节点、父节点、子节点、兄弟节点、叶子节点。树结构中有一种十分重要的二叉树结构,该结构中每个节点至多只有两个子节点,且有左右之分,其次序有时不能任意颠倒。树结构用途广泛,如哈夫曼树、最优二叉查找树及解空间树等,在后面章节将有详细介绍。

The key terms in tree structures are as follows: root node, parent node, child node, sibling node, and leaf node. A particularly important type of tree structure is the binary tree, where each node has at most two child nodes, referred to as the left and

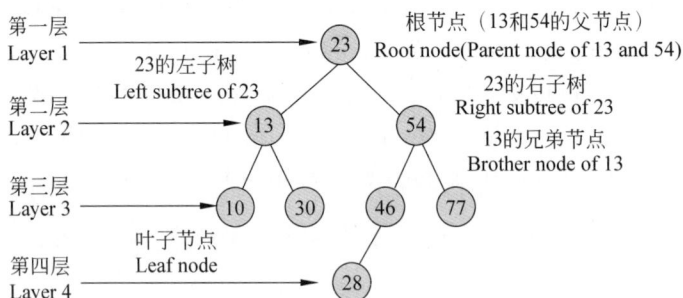

图 1.11　树结构示意图

Figure 1.11　Schematic diagram of tree structure

right children. The order of these children cannot be arbitrarily reversed at times. Tree structures have numerous applications, including the Huffman tree, optimal binary search trees, and solution space trees, which will be discussed in detail in subsequent chapters.

2. 图结构(Graph Structure)

（1）图结构概述(Overview)。

图结构是一种比树结构更复杂的非线性结构。在树结构中,节点间具有分支层次关系,每一层上的顶点只能和上一层中的至多一个顶点相关,但可能和下一层的多个顶点相关。而在图结构中,任意两个顶点之间都可能相关,即顶点之间的邻接关系可以是任意的。

A graph structure is a non-linear structure that is more complex than a tree structure. In a tree structure, vertexes exhibit a branching hierarchical relationship where each vertex on a given layer is connected to at least one vertex in the preceding layer and may be connected to multiple vertexes in the subsequent layer. In a graph structure, any two vertexes may have a relationship, meaning the adjacency between vertexes can be arbitrary.

图是由顶点和边所组成的集合,顶点通常用圆圈来表示,边就是这些圆圈之间的连线。

A graph consists of a set of vertexes and edges; vertexes are typically represented by circles, and edges are the lines connecting these circles.

图是一种数据结构,其中,顶点可以具有零个或多个相邻元素,两个顶点的连接称为边,节点在图结构中也被称为顶点,一个顶点到另一个顶点经过的线路称为路径。图结构有三种类型：无向图、有向图、带权图,如图 1.12 所示。

A graph is a data structure consisting of vertexes, where each vertex can have zero or more adjacent vertexes. The connection between two vertexes is called an edge, and a vertex is also referred to as a vertex within the structure. A path is defined as a line segment connecting one vertex to another. Graph structures come in three types: undirected graphs, directed graphs, and weighted graphs, as illustrated in Figure 1.12.

① 无向图。顶点 A 与顶点 B 之间的边是无方向的,可以从 A 到 B,也可以从 B 到 A。

Undirected graph. An undirected graph is characterized by edges that have no direction, allowing traversal from vertex A to vertex B or vice versa.

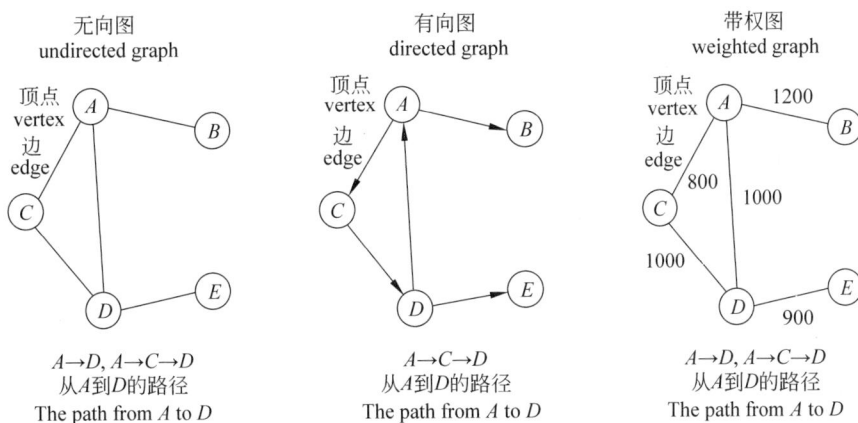

图 1.12　图结构三种类型示意图

Figure 1.12　Schematic diagram of three types of graph structure

② 有向图。顶点 A 与顶点 B 之间的边是有方向的，可以从 A 到 B，但不可以从 B 到 A。

Directed graph. A directed graph features edges that are oriented, allowing travel from vertex A to vertex B, but not vice versa.

③ 带权图。顶点 A 与顶点 B 之间的边是带有属性的，如 A 到 B 的距离。

Weighted graph. A weighted graph is a type of graph where the edges between vertex A and vertex B, have associated properties, such as the distance from A to B.

（2）图的表达方式（Representation of Graph）。

图的表达方式有两种：邻接矩阵（使用二维数组）和邻接表（使用数组+链表）。

A graph can be represented in two ways: using adjacency matrices, which are two-dimensional arrays, and using adjacency lists, which combine arrays with linked lists.

① 邻接矩阵。邻接矩阵是表示图形中各顶点之间的关系，矩阵的行和列对应各顶点，坐标位置上的值对于它们之间的关系，1 为连接，0 为没有连接，在程序中用二维数组来实现，如图 1.13 所示。

Adjacency Matrices. The adjacency matrix represents the relationships between vertexes in a graph, where the rows and columns correspond to each vertex. The values at the intersection of the rows and columns indicate the connection between them: 1 for connected and 0 for not connected. This structure is implemented in programs using two-dimensional arrays, as illustrated in Figure 1.13.

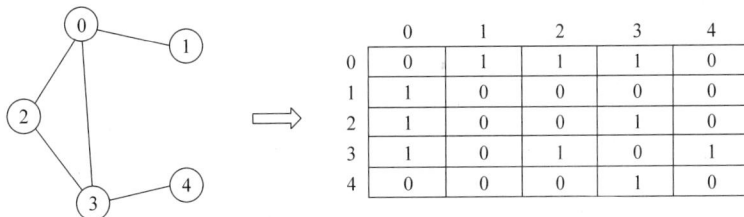

图 1.13　邻接矩阵图形中各顶点关系图

Figure 1.13　Graph of vertexes relationships in adjacency matrix

② 邻接表。邻接表只关心存在的边，不需要为不存在的边分配空间，同邻接矩阵相比，避免了不必要的空间浪费。在程序中用数组+链表的形式实现数组存储对应的顶点，链表存储该顶点连接的所有顶点，如图 1.14 所示。

Adjacency List. An adjacency list focuses only on existing edges, eliminating the need to allocate space for non-existent edges. This approach avoids unnecessary space waste compared to the adjacency matrix. In programs, adjacency lists are implemented using a combination of arrays and linked lists: the array stores the corresponding vertexes, and the linked list stores all vertexes connected to each vertex. As illustrated in Figure 1.14.

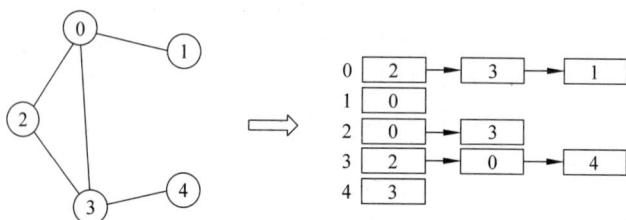

图 1.14　邻接表图形中各顶点连接图

Figure 1.14　Graph of adjacency list in vertex connection

（3）图结构属性(Graphic Structure Attribute)。

图结构被用于描述各种复杂的数据对象，在自然科学、社会科学和人文科学等许多领域有着非常广泛的应用。图结构在计算机科学、人工智能、电子线路分析、最短路径寻找、遗传学、控制论等方面均有不同程度的应用，可以这样说，图结构在所有数据结构中应用最为广泛。

Graph structures are used to describe a variety of complex data objects and have a wide range of applications across many fields, including natural science, social science, and the humanities. These structures find varying degrees of application in computer science, artificial intelligence, electronic circuit analysis, shortest pathfinding, genetics, cybernetics, and other areas. It can be said that the graph structure is the most widely used of all data structures.

图结构的属性如下。

The properties of a graph structure are outlined below.

① 顶点。图中的每个节点就是顶点。

Vertex. Every node within a graph is called a vertex.

② 边。图中两个顶点之间的线叫作边。

Edge. The line connecting two vertexes in the diagram is called an edge.

③ 路径。路径是从某个顶点到另一个顶点所要经过的顶点序列。

Path. A path is a sequence of vertexes that connects one vertex to another.

④ 路径长度。路径长度指的是一条路径上经过的边的数量。

Path length. Path length refers to the number of segments or edges that a path traverses.

⑤ 回路。一条路径的起点和终点为同一个顶点。

Loop. A loop is defined by a path where the starting and ending vertexes are the same.

⑥ 度。在无向图中，顶点的度指与该点相连的边的数量；在有向图中，分为出度和入度，指向该点的边数称为入度；反之，则称为出度。有向图某点度的大小等于该点出度和入度之和。

Degree. In an undirected graph, the degree of a vertex refers to the number of edges connected to that vertex. In a directed graph, it is divided into out-degree and in-degree. The number of edges pointing to a vertex is called the in-degree, while the number of edges pointing away from it is called the out-degree. The total degree of a vertex in a directed graph is equal to the sum of the out-degree and in-degree.

(4) 图结构算法(Graph Structure Algorithm)。

图结构算法有很多，下面介绍两种常见的算法，这些算法的实现在第7章中将详细介绍。

There are many graph structure algorithms, and now let's introduce two common ones. The detailed implementation of these algorithms is covered in Chapter 7.

① 深度优先搜索(Depth First Search,DFS)。深度优先搜索属于图算法的一种，其过程是对每一个可能的分支路径深入到不能再深入为止，而且每个节点只能访问一次。这样的访问策略是优先往纵向进行深入挖掘，而不是对一个顶点的所有邻接顶点进行横向访问。

Depth First Search (DFS). Depth First Search is a type of graph algorithm. It involves exploring as deeply as possible along each branch before backtracking. Each vertex is visited only once. This strategy prioritizes vertical exploration over horizontal access to all adjacent vertexes.

从当前顶点选一个与之连接但未被访问过的顶点，将当前顶点往该邻接顶点移动，如果所有邻接顶点都访问过，则回溯到上一个顶点位置，继续该步骤，直到所有顶点都访问过，当所有顶点都被访问过时，退出算法。

Select an unvisited vertex from the list of adjacent vertexes to the current vertex and move the current vertex to that vertex. If all adjacent vertexes have been visited, return to the previous vertex, and repeat this process until all vertexes have been visited. When all vertexes have been visited, terminate the algorithm.

② 广度优先搜索(Breadth First Search,BFS)。广度优先搜索是最简便的图结构算法之一，这一算法也是很多重要的图的算法的原型。Dijkstra 单源最短路径算法和 Prim 最小生成树算法都采用了广度优先搜索的思想。

Breadth First Search (BFS). Breadth first search is one of the simplest graph traversal algorithms, and it serves as a foundation for many important graph algorithms. Algorithms such as Dijkstra's single-source shortest path algorithm and Prim's minimum spanning tree algorithm are inspired by BFS.

广度优先搜索算法类似于一个分层搜索的过程，该算法需要一个队列以保持访问顶点的顺序，以便按这个顺序来访问这些顶点的邻接顶点。

The BFS algorithm resembles a hierarchical search process, requiring a queue to maintain the order in which vertexes are visited. This ensures that the adjacent vertexes of each vertex are explored in the correct sequence.

依次访问当前顶点的邻接顶点，并按访问顺序将这些邻接顶点存储在队列中，当前顶点的所有邻接顶点都被访问后，从队列中弹出一个顶点，以该顶点为当前顶点继续该步骤，直到所有顶点都被访问过。

Access the adjacent vertexes of the current vertex one by one, and store them in the queue in the order they are accessed. Once all adjacent vertexes of the current vertex have been accessed, pop a vertex from the queue, and continue this process with that vertex as the new current vertex until all vertexes have been accessed.

1.3　算法设计（Algorithm Design）

1.3.1　算法与程序（Algorithms and Programs）

算法是解决问题的步骤，程序是算法的代码实现，算法要依靠程序来完成功能，程序要以算法作为灵魂。程序是结果，算法是手段，完成一个同样功能的程序，可以使用不同的算法，从而使程序的大小和执行效率差别很大，因此算法是编写程序的关键所在。

An algorithm is a set of steps to solve a problem, while a program is the code implementation of that algorithm. An algorithm relies on a program to execute its functions, and a program should consider the algorithm as its core. The program is the outcome, the algorithm is the method. To create a program with the same functionality, different algorithms can be used, which can result in very different program sizes and execution efficiencies. Therefore, the choice of algorithm is crucial in writing a program.

算法的含义与程序十分相似，但又有区别。一个程序不一定满足有穷性。例如，操作系统，只要整个系统不遭到破坏，它将永远不会停止，即使没有作业需要处理，它仍处于动态等待中。因此，操作系统不是一个算法。另外，程序中的指令必须是机器可执行的，而算法中的指令则无此限制。算法代表了对问题的解，而程序则是算法在计算机上的特定的实现。一个算法若用程序设计语言来描述，则它就是一个程序。

The meaning of an algorithm is similar to that of a program, but there are key differences. A program is not necessarily finite; for example, an operating system continues to run as long as the entire system is functional, even if there are no tasks to process, as it remains in a state of dynamic waiting. Therefore, an operating system is not considered an algorithm. Besides, the instructions in a program must be machine-executable, whereas the instructions in an algorithm do not have such a requirement. An algorithm represents a solution to a problem, whereas a program is a specific implementation of that algorithm on a computer. An algorithm becomes a program when it is described in a programming language.

算法与数据结构是相辅相成的。解决某一特定类型问题的算法可以选定不同的数据结构,而且选择得恰当与否直接影响算法的效率。反之,一种数据结构的优劣由各种算法的执行来体现。

Algorithms and data structures are complementary. An algorithm solving a particular type of problem can choose from different data structures, and the efficiency of the algorithm is directly affected by the appropriate choice of data structure. Conversely, the merits of a data structure are reflected in the execution of various algorithms.

算法与程序又是互相联系的两个方面。甚至还可以说,问题的算法往往取决于选定的数据结构。所以 *N.Wirth* 教授认为:程序设计=算法+数据结构。

Algorithms and programs are two interrelated aspects. It might even be said that the algorithm for solving a problem often depends on the selected data structure. Therefore, Professor *N. Wirth* posits that programming = algorithm + data structure.

1.3.2 算法设计原则(Algorithm Design Principle)

随着计算机的出现,算法被广泛地应用于计算机的问题求解中,被认为是程序设计的精髓。对算法的认识主要包括 5 个方面:算法设计、算法表示、算法确认、算法分析、算法验证。

With the emergence of computers, algorithms have become widely used in computer problem solving and are considered the essence of program design. Understanding algorithms mainly involves five aspects: algorithm design, algorithm representation, algorithm correctness proof, algorithm analysis, and algorithm verification.

对于一个特定问题的算法在多数情况下都不是唯一的。也就是说,同一个问题,可以有多种解决问题的算法,而对于特定的问题、特定的约束条件,相对好的算法还是存在的。因此,选择合适的算法,对解决问题有很大的帮助。算法设计原则主要包括以下 7 个方面。

The algorithm for a particular problem is not unique in most cases. In other words, there can be a variety of algorithms to solve the same problem, and for specific problems with specific constraints, relatively good algorithms still exist. Therefore, choosing the right algorithm can be of great help in solving the problem. Algorithm design principles mainly include the following seven aspects.

1. 正确性(Correctness)

算法应当满足以特定的"规则说明"方式给出的需求。对算法是否"正确"的理解可以从几个层次来看:程序语法是否存在错误,对于一组输入数据能否得出满足需求的输出,是否能对精心挑选的典型苛刻的输入数据给出正确的结果,以及是否对所有合法输入数据都能得到满足要求的输出。

The algorithm should satisfy the requirements given in the form of a specific ' rule specification.' The understanding of whether an algorithm is ' correct' can be seen at several levels: whether there are errors in the program's syntax, whether it can produce a desired output for a set of inputs, whether it can give the correct result for a carefully selected set of typically demanding inputs, and whether it can produce a desired output for all legitimate inputs.

2. 确定性（Definiteness）

对于每种情况下算法应该执行的操作，在算法中都应该有明确的规定，使算法的执行者或阅读者能够明确其含义及如何执行。并且在任何条件下，算法只能有一条执行路径。也就是说，对于任意一组给定的合法输入值，算法要执行的操作是确定的。

For each case, the operations performed by the algorithm should be clearly specified within the algorithm. This clarity ensures that both the operator and the reader understand what is meant and how to execute each step. Moreover, in any case, an algorithm should have only one execution path. That is, for any given set of valid input values, the operations to be performed by the algorithm are deterministic.

3. 可行性（Practicable）

算法中的所有操作都必须足够基本，都可以通过已实现的基本操作运算有限次数实现。这意味着算法在执行时必须有明确的执行路径，确保每次运行都能得到相同的输出。

All operations within an algorithm must be fundamental enough to be executed with a finite number of steps. This means that the algorithm must have a clear execution path to ensure that it produces the same output every time it is run.

4. 健壮性（Robustness）

当输入的数据非法时，算法应当恰当地做出反应或进行相应处理，而不是产生莫名其妙的输出结果。并且，处理出错的方法不应是中断程序执行，而是应当返回一个表示错误或错误性质的值，以便在更高的抽象层次上进行处理。

When the input data is invalid, the algorithm should react appropriately or process the data accordingly, rather than producing inexplicable output. Moreover, the approach to handling an error should not be to interrupt program execution, but to return a value that indicates the error or the nature of the error, allowing it to be managed at a higher level of abstraction.

5. 可读性（Readability）

算法是为了人类的阅读与交流设计的，因此在设计和编写时需要考虑到易理解和可维护性，晦涩难懂的程序易于隐藏错误而难以调试。

Algorithms are designed for human reading and communication, so they need to be designed and written with ease of understanding and maintainability in mind. Programs that are obscure tend to conceal errors and are difficult to debug.

6. 高效性（Efficiency）

算法的设计应考虑在不同情况下的执行效率，包括执行时间和所需的存储空间，高效的算法能够在有限的时间内完成计算任务，同时占用较少的内存。

The design of an algorithm should consider its execution efficiency under various conditions, including execution time and required storage space. Efficient algorithms can complete tasks within a limited time frame while using minimal memory resources.

7. 简洁性（Simplicity）

算法应该是简洁的，避免不必要的复杂性，使得理解和实施更为容易。

Algorithms should be concise, avoid unnecessary complexity, and facilitate understanding and implementation.

以上 7 个方面构成了算法设计的主要原则,它们共同保证了算法的有效性和实用性。

The aforementioned seven aspects constitute the main principles of algorithm design, which together ensure the effectiveness and practicality of the algorithm.

习题 1(Exercises One)

1. 算法的基本定义是什么? 算法具有什么特点?

What is the basic definition of an algorithm? What are the features of the algorithm?

2. 算法的表示方法有哪些? 各有什么特点?

What are the representation methods of the algorithm? What are the features of each method?

3. 简单叙述流程图表示的算法中各种符号的意义。

What is the meaning of symbols inan algorithm represented by a flowchart?

4. 循环有哪几类? 分别用线性流程图表示。

How many types of loops(cycles)are there? How are each represented by linear flowcharts?

5. 什么是 N-S 图? 它具有什么特点?

What is an N-S graph and what are its features?

6. 顺序表与链表有什么相同点和不同点?

What are the similarities and differences between a sequential list and a linked list?

7. 堆栈与队列的数据进出顺序是什么?

What is the order of data entry and exit in a stack and queue?

8. 什么是无向图? 什么是有向图?

What is the definition of an undirected graph? What is the definition of a directed graph?

9. 图的结构类型有哪些? 分别简述。

What are the structure types of a graph? Briefly describe each.

10. 邻接矩阵和邻接表各有什么特点?

What are the characteristics of an adjacency matrix and an adjacency list?

11. 图结构具有什么属性? 分别叙述。

What properties does a graph structure have? Describe them separately.

12. 什么是深度优先搜索算法? 什么是广度优先搜索算法?

What is the depth-first search algorithm? What is the breadth-first search algorithm?

第 2 章 贪心算法

Chapter 2 Greedy Algorithm

贪心算法(*Greedy Algorithm*)是把一个复杂问题分解为一系列较为简单的局部最优选择,每一步选择都是对当前解的一个扩展,直到获得问题的完整解。贪心算法的典型应用是求解最优化问题,而且对许多问题都能得到整体最优解,即使不能得到整体最优解,通常也能得到最优解的近似值。

A *Greedy Algorithm* decomposes a complex problem into a series of relatively simple local optimal choices, where each step of the selection is an extension of the current solution until the complete solution is obtained. The typical application of the Greedy Algorithm is to solve optimization problems. For many problems, the global optimal solution can be obtained, and even if it cannot, the Greedy Algorithm usually provides a good approximation of the optimal solution.

2.1 概述(Overview)

贪心算法可以称为最接近人们日常思维的一种解题策略,具有简单、直接和高效的特点。对范围相当广泛的许多实际问题它通常都能产生整体最优解,在一些情况下,即使采用贪心法不能得到整体最优解,但其最终结果却是最优解的很好的近似解。基于此,它在对*NP*(*Non-deterministic Polynomial*,多项式复杂程度的非确定性问题)完全问题的求解中发挥着越来越重要的作用。

A greedy algorithm can be considered a problem-solving strategy that is closest to people's daily thinking; it is simple, direct, and efficient. For a wide range of practical problems, it can often produce the global optimal solution. In some cases, even if the greedy method cannot achieve the global optimal solution, the final result is still a good approximation of the optimal solution. Because of this, it plays an increasingly important role in solving *NP* (*Non-deterministic Polynomial*) complete problems.

2.1.1 算法思想(Algorithm Idea)

从问题的某一个初始解出发,在每一个阶段都根据贪心策略来做出当前最优的决策,逐步逼近给定的目标,尽可能快地求得更好的解。当达到算法中的某一步不能再继续前进时,算法终止。贪心算法可以理解为以逐步的局部最优达到最终的全局最优。

The algorithm begins with an initial solution to the problem, makes the current optimal decision according to the greedy strategy at each stage, gradually approaches the given goal, and aims to obtain a better solution as quickly as possible. When a certain step in the algorithm is reached and no further progress can be made, the algorithm terminates. A greedy algorithm can be understood as achieving the final global optimum by taking local optimal steps sequentially.

从算法的思想中,很容易得出以下特点。

From the concept of the algorithm, the following characteristics can be easily deduced.

(1)贪心算法在每个阶段面临选择时,都做出对眼前最有利的选择,不考虑该选择对将来是否有不良影响。

When a greedy algorithm faces a choice at each stage, it makes the most favorable choice for the present, without considering whether the choice will have adverse effects on the future.

(2)每个阶段的决策一旦做出,就不可更改,该算法不允许回溯。

Once a decision is made at each stage, it cannot be changed, and the algorithm does not allow backtracking.

(3)贪心算法根据贪心策略来逐步构造问题的解。如果所选的贪心策略不同,则得到的贪心算法就不同,贪心解的质量当然也不同。

A greedy algorithm constructs a solution to the problem step by step according to the greedy strategy. If the choices are different, the resulting greedy algorithm will yield different solutions, and the quality of these solutions will, of course, vary.

(4)贪心算法具有高效性和不稳定性,因为它可以非常迅速地获得一个解,但这个解不一定是最优解。即便不是最优解,也一定是最优解的近似解。

The greedy algorithm is efficient and unstable because it can obtain a solution very quickly, but this solution is not necessarily the optimal one. Even if it's not the optimal solution, it must be an approximate solution to the optimal solution.

贪心算法的本质就是从眼前某个初始解出发,在每一个阶段都做出当前最优的解决策略,即贪心策略,逐步逼近给定的目标,尽可能快地求得更好的解。贪心算法可以理解为以逐步的局部最优,达到最终的全局最优的方法。

A greedy algorithm starts from an initial solution at the beginning, makes the current optimal choice at each stage according to the greedy strategy, gradually approaches the given goal, and aims to obtain a better solution as quickly as possible. A greedy algorithm can be understood as a method that achieves the final global optimization by taking local optimization steps sequentially.

2.1.2　算法的设计与描述（Algorithm Design and Description）

1. 算法描述（Algorithm Description）

一个问题的最优解一定包含子问题的最优解,贪心算法求解问题的流程就是依序研究每个子问题,其最优解组合成原问题的最优解,所以最优子结构性质是能够采用贪心算法求解问题的关键,只有具有最优子结构性质才能保证贪心算法得到的解是最优解。

An optimal solution to a problem must include the optimal solutions to its subproblems. The process of solving a problem using a greedy algorithm involves studying each subproblem in sequence, and synthesizing their optimal solutions to form the optimal solution to the original problem. Therefore, the property of optimal substructure is key to using the algorithm to solve the problem, and it is the only property that can ensure the final optimal solution.

所求问题的整体最优解可以通过一系列局部最优的选择获得,即通过一系列的逐步局部最优选择使得最终的选择方案是全局最优的。其中每次所做的选择,可以依据以前的选择,但不依赖于将来的选择。每次选择面对的子问题都是独立的,不依赖于其他子问题,子问题间有严格的先后顺序。

The global optimal solution to a problem can be obtained through a series of local optimal choices, that is, by making step-by-step local optimal decisions, the final selection scheme becomes the global optimal. Each of these choices can be based on previous choices, but not on future ones. The subproblems faced by each choice are independent of other subproblems, and there is a strict order among them.

2. 算法设计（Algorithm Design）

算法的设计分为分解问题、解决子问题、合并问题和证明算法正确性4个步骤。

The design of the algorithm is divided into four steps: decomposing the problem, solving subproblems, merging solutions, and demonstrating the correctness of the algorithm.

1) 分解（Decomposition）

分解是将原问题分解为若干个相互独立的阶段。

Decomposition of the problem involves breaking down the original problem into several independent subproblems.

2) 解决（Solving）

解决是指对于每个阶段依据贪心策略进行贪心选择,求出局部的最优解。

Solving the problem involves making greedy selections at each stage according to the greedy strategy and finding the local optimal solution.

3) 合并问题（Merging）

合并问题是将各个阶段的解合并为原问题的一个可行解。

Merging the problem involves combining the solutions from various stages into a single feasible solution for the original problem.

4）证明（Proof）

证明算法的正确性，即证明最优子结构和贪心选择性质。

To demonstrate a proof of the algorithm's correctness involves showing that it has both optimal substructure and the greedy selection property.

2.2 活动安排问题（Event Scheduling Problem）

2.2.1 问题描述与分析（Problem Description and Analysis）

1. 问题描述（Problem Description）

活动安排问题要求高效地安排一系列争用某一公共资源的活动。贪心算法提供了一个简单漂亮的方法，使尽可能多的活动能兼容地使用公共资源。

Event scheduling problems require efficiently arranging a series of activities that compete for a common resource. Greedy algorithms offer a simple and elegant method to maximize the compatibility of as many activities as possible with the use of shared resources.

设有 n 个活动的集合 $E = \{1, 2, \cdots, n\}$，其中每个活动都要求使用同一资源如演讲会场等，而在同一时间内只有一个活动能使用这一资源。每个活动 i 都有一个要求使用该资源的起始时间 s_i 和一个结束时间 f_i，且 $s_i < f_i$。

A set with n activities, denoted as $E = \{1, 2, \cdots, n\}$ Where each event requires the use consists of events where each requires the use of the same resource, such as a lecture hall, and only one event can use this resource at any given time. Each activity i has a start time s_i and an end time f_i that require the use of the resource, with $s_i < f_i$.

如果选择了活动 i，则它在半开时间区间 $[s_i, f_i)$ 内占用资源。若区间 $[s_i, f_i)$ 与区间 $[s_j, f_j)$ 不相交，则称活动 i 与活动 j 是相容的。也就是说，当 $s_i \geq f_i$ 或 $s_j \geq f_j$ 时，活动 i 与活动 j 相容。活动安排问题就是要在所给的活动集中选出最大的相容活动子集合。

If activity i is selected, it occupies resources in the half-open interval $[s_i, f_i)$. Activity i is said to be compatible with activity j if the intervals $[s_i, f_i)$ and $[s_j, f_j)$ do not overlap. That is, activity i is compatible with activity j when $s_i \geq f_i$ or $s_j \geq f_j$. The problem of event scheduling is to select the largest set of compatible activities from the given set of activities.

2. 问题分析（Problem Analysis）

根据问题能够分析出以下的约束信息。

Based on the problem, the following constraint information can be analyzed.

（1）n 个活动的集合 $E = \{1, 2, \cdots, n\}$，是由活动编号组成的集合。

The set of n activities, denoted as $E = \{1, 2, \cdots, n\}$, is a collection of activity numbers.

（2）活动 $i(i = 1, 2, \cdots, n)$ 的开始时间表示为 s_i。

The start time of activity $i(i=1,2,\cdots,n)$ is expressed as s_i.

（3）活动 $i(i=1,2,\cdots,n)$ 的结束时间表示为 f_i。

The end time of activity $i(i=1,2,\cdots,n)$ is expressed as f_i.

（4）活动 $i(i=1,2,\cdots,n)$ 的资源使用时间为 f_i-s_i。

The resource usage time for activity $i(i=1,2,\cdots,n)$ is given by the difference f_i-s_i.

（5）活动 i 和活动 j 相容的资源使用条件为 $s_i\geqslant f_j$ 或 $s_j\geqslant f_i$，如图 2.1 所示。

The resource usage conditions compatible that make activities i and j compatible are $s_i\geqslant f_j$ or $s_j\geqslant f_i$, as shown in Figure 2.1.

图 2.1　活动安排问题关系示意图

Figure 2.1　Relationship between event scheduling problems

（6）满足相容条件的活动子集都是活动安排问题的解，含有活动个数最多的子集就是最优解。

The activity subsets that satisfy the compatibility condition constitute all the feasible solutions to the event scheduling problem, and the subset with the largest number of activities is considered the optimal solution.

要实现上述条件约束下的设计目标，可以考虑三种贪心策略：开始最早先安排、结束最早先安排、活动时间最短先安排。事实上，结束时间越早，则后续空余的时间段越多，有更多安排的可能，也能够通过实例分析。因此，在活动安排问题中，应当使用活动结束时间最早的优先安排的贪心策略。

To achieve the design goal under the aforementioned conditions, three greedy strategies can be considered: the earliest start time, the earliest end time, and the shortest activity duration. In fact, the earlier the end time, the more free time becomes available later, which increases the potential for scheduling. This can also be analyzed through examples. Therefore, in the activity scheduling problem, the greedy strategy of prioritizing activities with the earliest end times should be employed.

2.2.2　算法的设计（Algorithm Design）

1. 设计思想（Design Idea）

活动安排问题需要输入活动集合、活动的开始时间和结束时间，输出活动安排的结果。

The activity scheduling problem requires inputs such as the activity set, the start and end times of each activity, and outputs the result of the scheduling.

用 n 维向量 $\boldsymbol{x}$ 表示问题的输出，$x_i=1$ 表示 i 号活动被选择，$x_i=0$ 表示 i 号活动未

被选择,解向量要满足值为 1 的分量所表示的活动相容且个数最多。问题的建模过程如下。

The output of the problem is represented by the n-dimensional vector x, where $x_i = 1$ indicates that activity i is selected, and $x_i = 0$ indicates that activity i is not selected. The solution vector must ensure that the activity represented by components with a value of 1 that are compatible with this set is maximized. The modeling process for the problem is as follows.

(1) 输入:活动集合 $E = \{1, 2, \cdots, n\}$,每个活动的开始时间 s_i,结束时间 f_i,$i = 1, 2, \cdots, n$,$f_1 \leqslant f_2 \leqslant \cdots \leqslant f_i$。

Input. Activity set $E = \{1, 2, \cdots, n\}$, the start time of each activity s_i, the end time f_i, where $i = 1, 2, \cdots, n$, and $f_1 \leqslant f_2 \leqslant \cdots \leqslant f_i$.

(2) 输出(Output):解向量(Solution vector) $(x_1, x_2, \cdots, x_n)$。

(3) 目标函数(Objective function):$\max \sum\limits_{i=1}^{n} x_i$。

(4) 约束条件(Constraint condition):$\begin{cases} (x_i = 1) \text{和} (x_j = 1), & (s_j \geqslant f_i) \text{或} (s_i \geqslant f_j) \\ x_i \in \{0, 1\} & i = 1, 2, \cdots, n \end{cases}$。

(5) 贪心策略:贪心策略按照活动结束时间从小到大排列,然后依次在相容的条件限制下安排活动。

Greedy strategy. Arrange the activities in order of their end times from smallest to largest, and then schedule the activities sequentially under the compatibility conditions.

2. 算法伪代码(Algorithmic Pseudocode)

算法 2.1 活动安排的贪心策略选择(Algorithm 2.1 Greedy strategic choice of event scheduling)

输入:活动集合 $E = \{1, 2, \cdots, n\}$,每个活动的开始时间 s_i 和结束时间 f_i,$i = 1, 2, \cdots, n$,$f_1 \leqslant f_2 \leqslant \cdots \leqslant f_i$。

Input:The activity set $E = \{1, 2, \cdots, n\}$, the start time s_i and the end time f_i for each activity, where $i = 1, 2, \cdots, n$, and $f_1 \leqslant f_2 \leqslant \cdots \leqslant f_i$.

输出:问题解向量 $(x_1, x_2, \cdots, x_n)$。

Output:Problem solution vector $(x_1, x_2, \cdots, x_n)$.

```
1.  n←|E|        //The total number of all events (mathematically expressed as the module of the set of events E),
                 //in which all events have been sorted in order of end time from smallest to largest
2.  x1←1         //The first event can surely be arranged
3.  j←1          //The j variable is used to record the number of the currently arranged event
4.  for i←2 to n:   //Search for events in E in order
5.     if si≥fj:  //If the event can be scheduled (the start time is greater than or equal to the end time of all
                  //currently scheduled events) (The program compares the end time of the last scheduled event
                  //with the start time of the current search event)
6.        xi←1    //The event is scheduled
7.        j←i     //Update the number of the latest scheduled event
8.     else:      //Corresponding to line 5 if, if not satisfied, the event is not scheduled
9.        xi←0
10.    end if     //Mark the end of the judgment segment
11.  end for      //Marks the end of the loop segment
12. return x      //The program ends and the result is returned
```

2.2.3 算法实现（Algorithm Implementation）

1. 实例分析（Instance Analysis）

假如有 10 个活动等待安排，这些活动的开始时间和结束时间如表 2.1 所示，用贪心算法找出满足目标要求的活动集合。

If there are 10 activities waiting to be arranged, their start and end times are shown in Table 2.1, and the greedy algorithm is used to find the set of activities that meet the target requirements.

表 2.1　活动编号、开始时间和结束时间

Table 2.1　Activity number, start time and end time

i	1	2	3	4	5	6	7	8	9	10
s_i	3	1	5	2	5	3	8	6	8	12
f_i	6	4	7	5	9	8	11	10	12	14

按照结束时间从早到晚排序，如表 2.2 所示。

Reorder by activity end time from earliest to latest, as shown in Table 2.2.

表 2.2　按照结束时间从早到晚排序

Table 2.2　Reorder by activity end time from earliest to latest

i	2	4	1	3	6	5	8	7	9	10
s_i	1	2	3	5	3	5	6	8	8	12
f_i	4	5	6	7	8	9	10	11	12	14

首先选择表格中的第一个活动，编号为 2，结束时间 4；然后向后寻找首个开始时间大于或等于 4 的活动，即编号 3，其开始时间为 5，结束时间为 7；向后寻找结束时间大于或等于 7 的活动，即编号 7，其开始时间为 8，结束时间为 11；继续向后寻找开始时间大于或等于 11 的活动，找到编号 10，开始时间为 12，结束时间为 14。全部活动搜索结束，活动编号 2、3、7、10 能够被安排。

First, select the first activity in the table, number 2, with an end time of 4. Then, look backward for the first activity whose start time is greater than or equal to 4, which is number 3, with a start time of 5 and an end time of 7. Next, find the activity whose end time is greater than or equal to 7, which is number 7, with a start time of 8 and an end time of 11. Continue looking backward for activities with a start time greater than or equal to 11, and find number 10, with a start time of 12 and an end time of 14. All activity searches are completed, and activities numbered 2, 3, 7, and 10 can be arranged.

2. 数据结构设计（Design of Data Structure）

算法的程序实现中，选用元组 *tuple* 存储单个活动的编号、开始时间和结束时间，然后选用列表 *list* 将 n 个活动对应的元组组织起来。问题的解用列表 *result* 表示 n 维向量 **x**。

In the program implementation of the algorithm, a *tuple* is used to store the number, start time, and end time of a single activity. Then, a *list* is used to organize these tuples corresponding to the n activities. The solution to the problem is represented by the list *result*, which corresponds to the n-dimensional vector x.

3. 代码实现(Code Implementation)

Python 程序的算法实现过程中不需要调用依赖包。

首先定义一个函数 *meetings_Greedy_Select*(*meetings*),用于接收活动数据并输出选择结果,也是算法的核心策略实现。然后定义 *Python* 的入口 *main*( ) 函数,准备数据集,调用选择函数并打印结果输出到显示器中。程序的输出结果如图 2.2 所示。

Python programs do not need to call dependency packages during the algorithm implementation.

First, define a function *meetings _ Greedy _ Select*(*meetings*), which is used to receive data and output a choice as a result; this function also implements a core strategy of the algorithm. Then, define *Python's* entry *main*( ) function, prepare the dataset, call the selection function, and print the resulting output to the display. The output of the program is shown in Figure 2.2.

程序代码如下(The program code is as follows)。

```
def meetings_Greedy_Select(meetings):
    length=len(meetings)
    meetings.sort(key=lambda x:x[2])
    result=[False for i in range(length)]
    j=0                                    #当前选中的活动(The currently selected activity)
    result[j]=True                         #选中第一个活动(Select the first activity)
    for i in range(1,length):
        if meetings[i][1]>=meetings[j][2]:
            j=i
            result[j]=True
    return result
if __name__=='__main__':
    meetings=[(4,1,4),(2,3,5),(1,0,6),(5,5,7),(7,3,8),(8,5,9),(9,6,10),(3,8,11),(10,8,12),(11,2,
13),(6,12,14)]
#每个活动的编号,开始时间,结束时间(Number of each activity, start time, end time)
    result=meetings_Greedy_Select(meetings)
    length=len(result)
    print('安排的活动编号为:')(The scheduled activities are numbered)
    count=0
    for i in range(length):
        if result[i]:
            print('第',meetings[i][0],'号活动')('Number ',meetings[i][0],' activity')
            count+=1
    print('\n 共计',count,'个活动')('\n Total ',count,' one activity ')
```

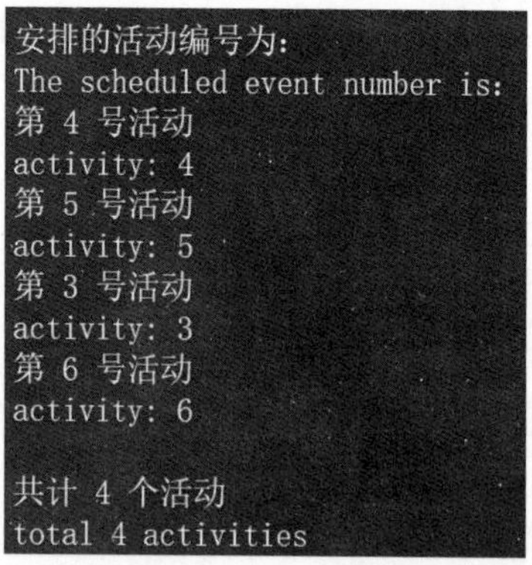

```
安排的活动编号为:
The scheduled event number is:
第 4 号活动
activity: 4
第 5 号活动
activity: 5
第 3 号活动
activity: 3
第 6 号活动
activity: 6

共计 4 个活动
total 4 activities
```

图 2.2　贪心算法活动安排问题程序结果

Figure 2.2　Program result of greedy algorithm activity scheduling

2.3　最短路径问题（Shortest Path Problem）

2.3.1　基本思想（Basic Idea）

给定一个有向带权图 $G=(V,E)$，其中每条边的权是一个非负实数。另外，给定 V 中的一个顶点，称为源点。现在要计算从源点到其他所有各个顶点的最短路径长度。

Given a directed weighted graph $G=(V,E)$ where the weight of each edge is a non-negative real number, and suppose a vertex V called the source vertex, Now calculate the shortest path length from the source vertex to all the other individual vertex.

该问题的算法由 $Dijkstra$（迪杰斯特拉）给出。他提出按照各个顶点与源点之间路径长度的递增次序，生成源点到各个顶点的最短路径的方法，即先求出长度最短的一条路径，再参照它求出长度次短的一条路径，以此类推，直到从源点到其他各个顶点的最短路径全部求出为止，该算法称为 $Dijkstra$ 算法。

The algorithm for this problem was developed by $Dijkstra$. He proposed a method to generate the shortest paths from the source point to each vertex in increasing order of path length between each vertex and the source. That is, one first finds the path with the shortest length, then uses that path to find the next shortest path, and so on, guiding the determination of the shortest path from the source vertex to all other vertexes until all shortest paths are found. This algorithm is known as $Dijkstra's$ algorithm.

其中的主要约定条件如下。

The main agreed conditions are as follows.

（1）源点（Source vertex）。算法首先从图中选定一个点，相当于出发点。

The algorithm initially selects avertex from the graph, which serves as the starting vertex.

（2）S 集合（S-set）。已经确定到源点最短路径的点构成的集合。

A set of vertexes whose shortest paths to the source vertex have been determined.

（3）V-S 集合（V-S set）。尚未确定到源点最短路径的顶点构成的集合。

A set of vertexes whose shortest paths to the source vertex have not yet been determined.

（4）特殊路径（Special path）。从源点出发，只经过 S 中的点，到达 V-S 中的点的路径。

The path from the source vertex to a vertex in V-S, passing only through vertexes in S.

最初，源点到自身的路径已经确定，其长度为 0，故 S 集合中只有源点；源点到其他顶点的路径尚未确定，故集合 V-S 是除源点之外的其他所有点组成的集合。

Initially, the path from the source vertex to itself has been determined, and its length is 0, so there is only the source vertex in the S set. The paths from the source vertex to the other vertexes have not been determined, so the set V-S consists of all vertexes except the source vertex.

该算法的贪心策略为，选择特殊路径长度最短的，将其相连的 V-S 中的顶点加入集合 S 中，检查新增加的特殊路径是否优于原来找到的特殊路径，若新的特殊路径最优，则优化。

The greedy strategy of the algorithm is to select the vertexes in V-S that are connected by the shortest special paths and add them to the set S. This process checks whether the newly added special paths are better than the original ones. If a new special path is found to be optimal, it will replace the previous one.

2.3.2 算法的设计与描述（Algorithm Design and Description）

1. 设计思想（Design Idea）

输入：有向带权图 $G=(V,E)$，$V=\{1,2,\cdots,n\}$，源点 $s=1$。

Input：Directed weighted graph $G=(V, E)$, $V=\{1, 2,\cdots,n\}$, the source vertex $s=1$.

输出：从 S 到每个顶点的最短路径。

Output：The shortest path from the source vertex S to each vertex.

初始时，$S=\{1\}$，然后计算特殊路径长度，对于 $i\in V$-S，计算 1 到 i 的最短特殊路径长度，记为 $dist[i]$。选择 V-S 中的 $dist$ 值最小的 $dist[j]$，将相连的 V-S 中的 j 点加入 S 中，优化 V-S 中顶点的 $dist$ 值。循环操作，直到 $S=V$ 为止。

Initially，$S=\{1\}$. Then, calculate the special path lengths, for $i\in V$-S, the shortest special path length from 1 to i, denoted as $dist[i]$. $dist[j]$ in V-S with the smallest $dist$ value is selected, and the vertex j is added to S to optimize $dist$ value of vertexes in V-S. This loop operation continues until $S=V$.

2. 算法伪代码（Algorithmic Pseudocode）

算法 2.2　贪心算法解决最短路径问题（Algorithm 2.2　Greedy algorithm to solve the shortest path problem）

输入：有向带权图 $G=(V,E)$，$V=\{1,2,\cdots,n\}$，源点 $s=1$。

Input：Directed weighted graph $G=(V, E)$, $V=\{1, 2,\cdots,n\}$, the source point $s=1$.

输出：从 s 到每个顶点的最短路径 pre。

Output：The shortest path from s to each vertex pre.

1.　S[1]←1　//Have found the path

续

2.   dist[1]←0   //The length of the path from the source point to itself is 0
3.   pre[1]←0   //The leading point of the source point is 0
4.   for i←2 to n： //Path length record array initialization
5.   dist[i]←w(s,i)   //In the initial state, the direct weights of other points to the source point are recorded,
                      //and those that are not directly connected are recorded as infinity
6.   end for
7.   while V-S≠Φ： //Pull the vertexes one by one from the V-S and determine their shortest path until empty
8.   Take vertex j of the shortest path of dist[j] from V-S   //Take the shortest path and its vertex j from the
                      //remaining solution space
9.   S[j]←1   //Mark
10.   for i←2 to n： //If there is a shorter path to another point through point j, the information about point i
                      //is updated
11.   if s[i]==0 and dist[j]+w(j,i)<dist[i]：
12.   dist[i]←dist[j]+w(j,i)
13.   end if
14.   pre[i]←j
15.   end while   //loop end
16.   return pre   //return result

2.3.3 算法实现（Algorithm Implementation）

1. 实例分析（Instance Analysis）

在如图 2.3 所示的有向带权图中，求源点 1 到其余顶点的最短路径以及最短路径长度。

In the directed weighted graph depicted in Figure 2.3, find the shortest path from the source vertex 1 to all other vertexes, as well as the lengths of these shortest paths.

1 号顶点作为原点，寻找到其他各个点的最短路径。

Vertex 1 serves as the origin, and the goal is to find the shortest path to each of the other vertexes.

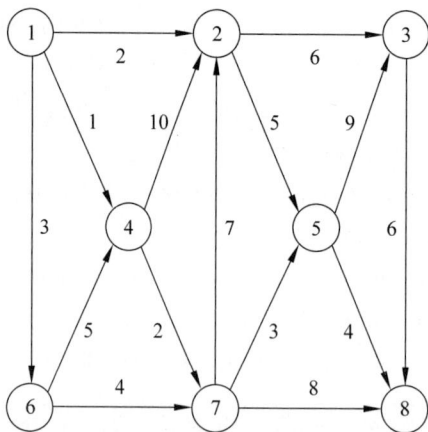

图 2.3 最短路径问题实例的有向带权图

Figure 2.3   Directed weighted graph of the shortest path problem

首先以 1 为顶点，可以直接到达点 2、4、6，各条路径的权重为 2、1、3，前置节点都为 1。其他点暂时不能直接从 1 号点到达，在记录表中路径权重记为无穷大，前置节点记为 0。1

号节点此时已经被使用过,相关量标记为 1,其他节点标记为 0,如表 2.3 所示。

First, with vertex 1 as the source and can directly reach vertexes 2, 4, and 6, with path weights of 2, 1, and 3 respectively, and vertex 1 as the preceding vertex. Other vertexes cannot be reached directly from vertex 1 at this time, so their path weights are marked as infinity in the record table, and their leading vertex is marked as 0. vertex 1 has already been processed at this point, with the relevant quantity marked as 1, while other vertexes are marked as 0. As shown in Table 2.3.

表 2.3 从顶点 1 开始第一步选择

Table 2.3 Start at vertex 1 and select the first step

vertex	1	2	3	4	5	6	7	8
S	1	0	0	0	0	0	0	0
dist	0	2	∞	1	∞	3	∞	∞
pre	0	1	0	1	0	1	0	0

path graph	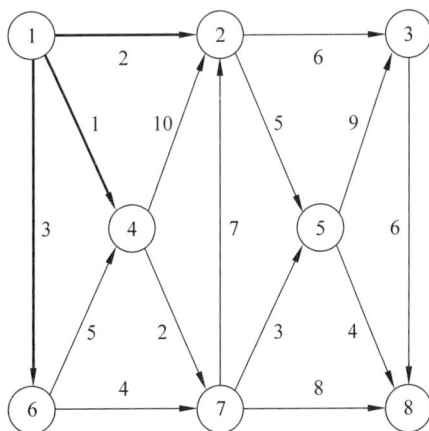

然后在当前未被使用过的顶点中,选择权重记录最小的作为中转点,继续向外寻找路径。选择 4 号顶点,可以到达 2 号顶点和 7 号顶点。2 号顶点的新路径权重为 11,原路径权重为 2,原路径更短,因此不更新 2 号顶点的信息。7 号顶点的新路径权重为 3,比原路径记录的无穷大更小,因此更新 7 号顶点的权重和前置点信息。4 号顶点已经被使用,更改相应标记,如表 2.4 所示。

Then, among the currently unused vertex, select the one with the smallest weight as the transit point and continue to find the path. Select vertex 4 to reach vertexes 2 and 7. The weight of the new path to vertex 2 is 11, while the weight of the original path is 2; since the original path is shorter, the information for vertex 2 is not updated. The weight of the new path to vertex 7 is 3, which is smaller than the infinity recorded for the original path; therefore, the weight and the preceding information for vertex 7 are updated. vertex 4 is now in use. Change the corresponding flag. As shown in Table 2.4.

表 2.4　第二步以顶点 4 为中转点计算

Table 2.4　The second step takes vertex 4 as the transfer point

vertex	1	2	3	4	5	6	7	8
S	1	0	0	1	0	0	0	0
dist	0	2	∞	1	∞	3	3	∞
pre	0	1	0	1	0	1	4	0
path graph	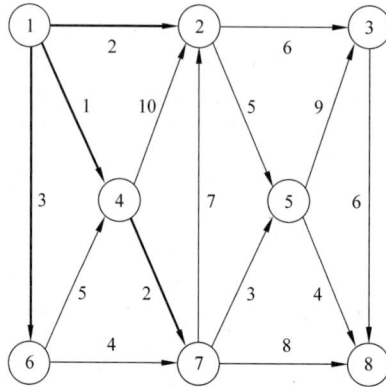							

按照规则，下一个被考虑的顶点是 2 号，能够到达 3 号顶点和 5 号顶点，比较新路径的权重值和原权重值，选择路径更短的进行记录，并更新 2 号顶点的标记，如表 2.5 所示。

According to the rules, the next vertex to be considered is vertex 2, which can reach vertexes 3 and 5. Compare the weight values of the new paths with the original weight values, select the shorter path for recording, and update the mark for vertex 2. As shown in Table 2.5.

表 2.5　第三步以节点 2 为中转点计算

Table 2.5　The third step takes vertex 2 as the transfer point

vertex	1	2	3	4	5	6	7	8
S	1	1	0	1	0	0	0	0
dist	0	2	8	1	7	3	3	∞
pre	0	1	2	1	2	1	4	0
path graph	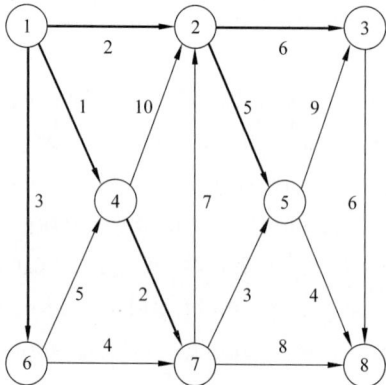							

下一个被考虑的顶点是 6 号(虽然 6 号与 7 号顶点的权重是一样的,但是按照顺序搜索,6 号顶点的权重会被认为是最小的),能够直接到达 4 号和 7 号顶点,但是其新路径的权重并没有更小,因此不更新,只对 6 号顶点做标记,表示已经测算过,如表 2.6 所示。

The next vertex to be considered is vertex 6 (although the weights of vertexes 6 and 7 are the same, according to the search order, the weight of vertex 6 is considered the smallest). Vertex 6 can directly reach vertexes 4 and 7, but the weight of its new paths is not smaller, so no updates are made; only vertex 6 is marked, indicating that it has been evaluated. As shown in Table 2.6.

表 2.6   第四步以顶点 6 为中转点计算

Table 2.6   The fourth step takes vertex 6 as the transfer point

vertex	1	2	3	4	5	6	7	8
S	1	1	0	1	0	1	0	0
dist	0	2	8	1	7	3	3	∞
pre	0	1	2	1	2	1	4	0
path graph								

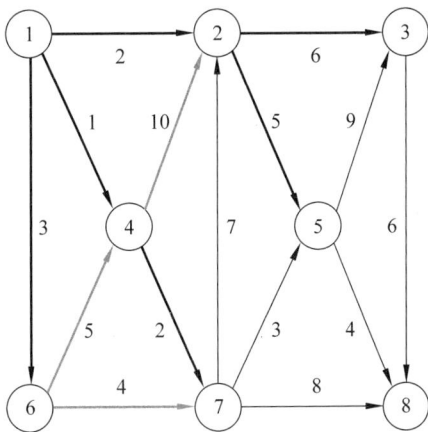

下一个被考虑的是顶点 7,能够到达顶点 5 和 8,对比已经存储的信息并更新,如表 2.7 所示。

The next vertex to be considered is vertex 7, which can reach vertexes 5 and 8. Compare and update the information already stored for these vertexes, as shown in Table 2.7.

下一个被考虑的节点是顶点 5,更新信息后如表 2.8 所示。

The next vertex to be considered is vertex 5, as shown in Table 2.8 after the information is updated.

表 2.7 第五步以顶点 7 为中转点计算

Table 2.7 The fifth step takes vertex 7 as the transfer point

vertex	1	2	3	4	5	6	7	8
S	1	1	0	1	0	1	1	0
dist	0	2	8	1	6	3	3	11
pre	0	1	2	1	7	1	4	7

path graph

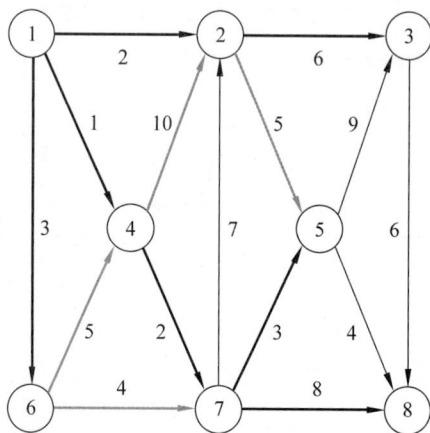

表 2.8 第六步以顶点 5 为中转点计算

Table 2.8 The sixth step takes vertex 5 as the transfer point

vertex	1	2	3	4	5	6	7	8
S	1	1	0	1	1	1	1	0
dist	0	2	8	1	6	3	3	10
pre	0	1	2	1	7	1	4	5

path graph

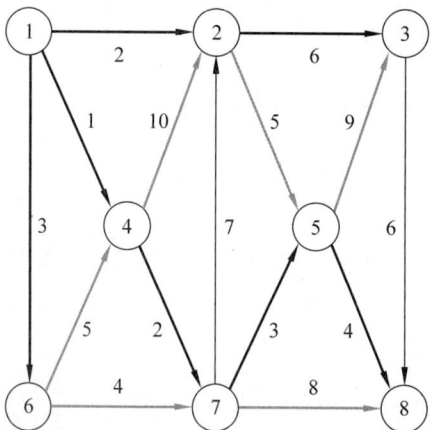

下一个是顶点 3，如表 2.9 所示。

The next calculation is forvertex 3, as shown in Table 2.9.

表 2.9 以顶点 3 为中转点计算

Table 2.9 Vertex 3 is used as the transit point

vertex	1	2	3	4	5	6	7	8
S	1	1	1	1	1	1	1	0
dist	0	2	8	1	6	3	3	10
pre	0	1	2	1	7	1	4	5
path graph	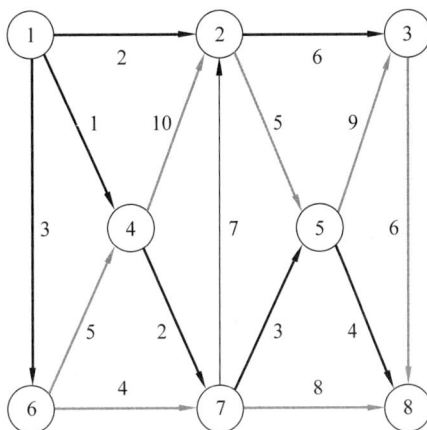							

下一个顶点是 8,更新后如表 2.10 所示。

The next calculation isvertex 8, as shown in Table 2.10.

表 2.10 以顶点 8 为中转点计算

Table 2.10 Vertex 8 is used as the transit vertex

vertex	1	2	3	4	5	6	7	8
S	1	1	1	1	1	1	1	1
dist	0	2	8	1	6	3	3	10
pre	0	1	2	1	7	1	4	5
path graph	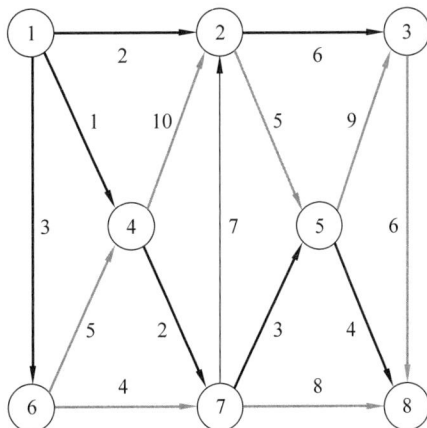							

所有节点都已经被标记，没有剩下的需要被考虑的节点，因此计算结束。由表格结果可见，每个节点的路径，可以通过前置节点的标记数据，逐次向前对应获取。

All vertexes have been marked, and there is no vertex left to consider, indicating that the calculation is complete. As can be seen from the results in the table, the path to each vertex can be determined by the marked data of the preceding vertexes.

2. 数据结构设计（Design of Data Structure）

给定的有向带权图用图的邻接表进行存储，选用 *Python* 中的 *list* 存储邻接表，*list* 的下标表示弧尾，其元素是由共同弧尾的 *list* 组成。即在下标为 0 的位置，存储的数据都是以 0 为弧尾的有向边的顺序。选用 *list* 存储集合 S、当前最短路径长度 *list* 和顶点的前驱数据 *pre*。

A given directed weighted graph is stored using the adjacency list representation of the graph, and *Python's list* data structure is selected to store this adjacen*cy list*. The index of the *list* represents the tail of an arc, and its elements are composed of pairs *list*, each representing an edge with a common tail. Specifically, at index 0, the data stored corresponds to directed edges with 0 as the tail. A *list* is also used to store set S, *list* for the current shortest path lengths, and *pre* for the predecessor data for each vertex.

3. 代码实现（Code Implementation）

首先定义算法的核心函数 *dijkstra*(*start_point*, *graph*)，接收指定的源点和有向带权图 G 的邻接表 *graph*，输出最短路径长度 *dist* 和前驱 *pre*。然后定义入口函数，准备数据集，调用核心程序函数并打印结果输出到显示器中。程序的输出结果如图 2.4 所示，代码如下。

Firstly, the core function *dijkstra*(*start_point*, *graph*) of the algorithm is defined. It receives the specified source vertex and the adjacency list representing the directed weighted *graph G*, and outputs the shortest path lengths *dist* and the predecessor *pre*. Then, define the entry function, prepare the dataset, call the core function, and print the result to the display. Figure 2.4 shows the program's output. The code is as follows.

```python
def dijkstra(start_point, graph):
    n = len(graph)
    #初始化各项数据,把 dist[start]初始化为 0,其他为无穷大(Initializes each item of data, initializing dist[start] to
    #0 and everything else to infinity)
    dist = [99999999 for _ in range(n)]  #路径长度(Path length)
    pre = [-1 for _ in range(n)]  #前驱 pre(precursor pre)
    s = [False for _ in range(n)]  #集合 S(Set S)
    dist[start_point] = 0
    for i in range(n):
        minLength = 99999999  #相当于设置无穷大,为不引入依赖包(This is equivalent to setting infinity so that no
                              #dependency packages are introduced)
        minVertex = -1
        for j in range(n):  #线性时间找最小(Linear time minimization)
            if not s[j] and dist[j] < minLength:
                minLength = dist[j]
```

续

```
                minVertex = j
            s[minVertex] = True
            #从这个顶点出发,遍历与它相邻的顶点的边,计算特殊路径长度,更新 dist 和 pre
#(From this vertex, we traverse the edges of the vertexes adjacent to it, compute the special path length, and update
#dist and pre)
            for edge in graph[minVertex]:
                if not s[edge[0]] and minLength + edge[1] < dist[edge[0]]:
                    dist[edge[0]] = minLength + edge[1]
                    pre[edge[0]] = minVertex
    return dist, pre
if __name__ == "__main__":
    data = [[1,2,2],[1,4,1],[1,6,3],[2,3,6],[2,5,5],[3,8,6],[4,2,10],[4,7,2],[5,3,9],[5,8,4],
[6,4,5],[6,7,4],[7,5,3],[7,8,8]]  #边集合(顶点,顶点,边权)(Edge sets (vertexes, vertexes, edge weights))
    n = 8#图的顶点数 n
    graph = [[] for _ in range(n+1)]#图的邻接表
    #根据输入的图构建图的邻接表
    for edge in data:
        graph[edge[0]].append([edge[1], edge[2]])
        #graph[edge[1]].append([edge[0], edge[2]])
    #print(graph)
    dist,pre = dijkstra(1,graph)
    print("dist=",dist[-n:])
    print("pre=",pre[-n:])
    print("start node: 1")
    for nnode in range(2,n+1):
        rodelist=[]
        print("Path to node",nnode,":",end="")
        prenode = pre[nnode]
        rodelist.append(prenode)
        while prenode !=1:
            prenode = pre[prenode]
            rodelist.append(prenode)
        rodelist=rodelist[::-1]
        for ii in range(len(rodelist)):
            print('-->',rodelist[ii],end="")
        print(",",end=" ")
        print("path length:",dist[nnode])
```

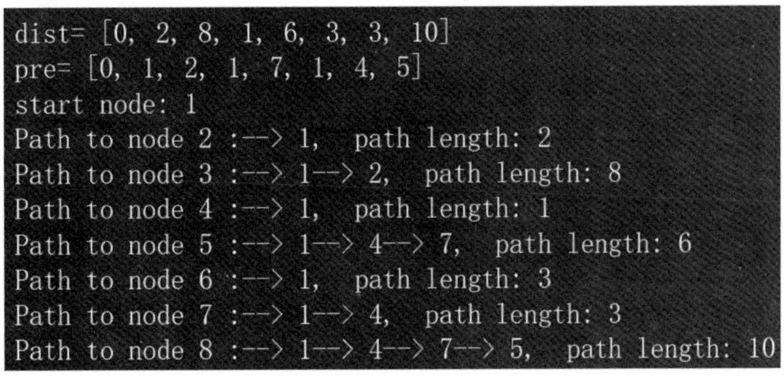

```
dist= [0, 2, 8, 1, 6, 3, 3, 10]
pre= [0, 1, 2, 1, 7, 1, 4, 5]
start node: 1
Path to node 2 :--> 1,  path length: 2
Path to node 3 :--> 1--> 2,  path length: 8
Path to node 4 :--> 1,  path length: 1
Path to node 5 :--> 1--> 4--> 7,  path length: 6
Path to node 6 :--> 1,  path length: 3
Path to node 7 :--> 1--> 4,  path length: 3
Path to node 8 :--> 1--> 4--> 7--> 5,  path length: 10
```

图 2.4 贪心算法最短路径问题程序结果

Figure 2.4 Program result of algorithm shortest path problem

2.4 哈夫曼编码（Huffman Coding）

2.4.1 基本思想（Basic Idea）

1. 压缩编码概述（Compressed Coding Overview）

当需要解决远距离通信以及大容量存储问题时，通常涉及字符的编码和信息的压缩问题。一般来说，较短的编码能够提高通信的效率并节省存储空间。常见的编码方式有固定长度编码和不等长度编码两种。

When solving problems related to long-distance communication and large-capacity storage, it usually involves character encoding and information compression. Generally, shorter codes can improve communication efficiency and save storage space. There are two common coding methods: fixed-length coding and variable-length coding.

如果每个字符的使用频率相等，那么等长码是空间效率最高的方法。但是在信息的实际处理过程中，每个字符的使用频率有很大的差异，此时再用等长码，就会导致空间效率降低。而不等长编码则能够大大压缩编码长度。

If each character is used with equal frequency, then equal-length codes are the most space-efficient method. However, in the actual process of information processing, the frequency of use of each character varies greatly, and using equal-length codes will lead to reduced space efficiency. Unequal length encoding can greatly compress the encoding length.

不等长编码的思想是利用字符的使用频率来编码，使经常使用的字符编码较短，不常使用的字符编码较长，既能够节约磁盘空间，又能够提高运算和通信效率。

The idea behind unequal length encoding is to use the frequency of characters to determine their encoding lengths, such that frequently used characters have shorter encodings, and infrequently used characters have longer encodings. This approach can save disk space and improve the efficiency of operations and communication.

变长编码方案必须满足每个字符的编码不能是其他字符编码的前缀，此为前缀码性质，否则会在译码的时候引起歧义，利用二叉树结构表示的变长码是满足前缀码性质的。

The variable-length coding scheme must satisfy the requirement that no character encoding can be a prefix of another character's encoding; this is known as the prefix property. Otherwise, it would cause ambiguity during decoding. The variable-length code represented by a binary tree structure adheres to the prefix property.

哈夫曼编码是一种变长码编码方式，是数学家哈夫曼于 1952 年提出的，其完全根据字符出现频率来构造平均长度最短的码字。哈夫曼编码算法根据字符出现的频率来建立一个用 0、1 串表示各字符的最优表示方式，有时称为最佳编码。

Huffman coding is a variable-length coding method proposed by mathematician *David Huffman* in 1952. It constructs the shortest average length code words entirely based on the frequency of character occurrence. The Huffman coding algorithm uses the frequency of character occurrence to establish an optimal representation with 0s and 1s to represent each character, sometimes referred to as optimal coding.

例如,有一列字母 $\{A,B,C,D,E,F,G\}$,其各自出现的频率权重为 $\{5,24,7,17,34,5,13\}$,考虑最短的等长编码和哈夫曼编码的情况,结果汇总在表 2.11 中。

For example, consider a list of letters $\{A,B,C,D,E,F,G\}$ with their corresponding frequency weights $\{5, 24, 7, 17, 34, 5, 13\}$. We will compare the case of the shortest isometric-length encoding with Huffman encoding, and the results are summarized in Table 2.11.

表 2.11 等长编码与不等长编码的结果对比

Table 2.11 Comparison of the results of equal length encoding and unequal length encoding

Symbol		A	B	C	D	E	F	G	Sum
Weight		5	24	7	17	34	5	13	
Equal Length Coding	Coding	000	001	010	011	100	101	110	325
	Total weight	15	72	21	51	102	15	39	
Huffman Coding	Coding	10110	01	1010	00	11	10111	100	267
	Total weight	25	48	28	34	68	25	39	

使用频率较高的字母使用较少位数的编码,而使用频率较低的字母使用较多位数的编码,可以使总共的编码长度在一定程度上减少。

Letters with higher frequencies use fewer digits of code, while letters with lower frequencies use more digits of code, thereby reducing the total encoding length to some extent.

2. 问题分析(Problem Analysis)

哈夫曼编码是使平均码长最短的编码方式,以字符的使用频率作权值构建一棵哈夫曼树,然后对字符进行编码,核心思想是频率越大的字符离树根越近。因此,算法的贪心策略是,从树的集合中选取两个频率最低的字符,使其作为左右子树构造一棵新树,父节点的频率为左右节点频率之和,然后将新树插入树的集合中。

Huffman coding is a coding method designed to minimize the average code length. It constructs a Huffman tree using the frequency of characters as weights and then encodes the characters accordingly. The core idea is that characters with higher frequencies are placed closer to the root of the tree. Therefore, the greedy strategy of the algorithm is to select the two characters with the lowest frequencies from the set, make them the left and right subtrees to construct a new tree, with the frequency of the parent node being the sum of the frequencies of its left and right child nodes. The new tree is then inserted back into the set of trees.

2.4.2 算法的设计与描述(Algorithm Design and Description)

1. 设计思想(Design Idea)

输入:字符集 $C=\{c_1,c_2,\cdots,c_n\}$ 及字符出现的频率 $f(c_i),i=1,2,\cdots,n$。

Input. Character set $C = \{c_1, c_2, \cdots, c_n\}$ and frequency $f(c_i)$, $i = 1, 2, \cdots, n$ of character occurrence.

输出：哈夫曼树 Q。

Output. Huffman tree Q.

首先将字符集中的每一个字符看作一棵只含有根节点的树，构造一个 n 棵树构成的树的集合 Q；然后做 $n-1$ 次贪心选择，每次都选择两个出现频率最小的节点，让其作为左右子树构造一棵新树，将新树插到树的集合 Q 中，直到 Q 中只含有一棵树为止。

Firstly, each character in the character set is regarded as a tree containing only a root node, and a set Q of n trees is constructed. Then, perform $n-1$ greedy selections. Each time, choose the two nodes with the least frequency, make them the left and right subtrees to construct a new tree, and insert the new tree into the set Q. Continue this process until Q contains only one tree.

从树根深度优先遍历，左子树输出 0，右子树输出 1，搜索到叶子节点就得到了叶子字符的编码。

Starting from the root, perform a depth-first traversal: output ' 0' for the left subtree and ' 1' for the right subtree, and search to the leafnode to obtain the leaf character's encoding.

2. 算法伪代码（Algorithmic Pseudocode）

算法 2.3　贪心算法解决哈夫曼编码（Algorithm 2.3　Greedy algorithm solves of Huffman coding）

输入：字符集 C 及每个字符出现的频率 f(i)，i ∈ C

Input：character set C and frequency f(i) for each character, i ∈ C

输出：Q（Output：Q）

1. n←|C|
2. Q←sort(C)//频率由小到大排序（Frequency is ordered from small to large）
3. for i←1 to n-1
4. 　　构造节点 z（Construct node z）
5. 　　z.left←Q 中频率第一小字符 x（The first small character x of frequency in z.left ←Q）
6. 　　z.right←Q 中频率第二小字符 y（The second smallest character y in frequency z.right ←Q）
7. 　　f(z)←f(x)+f(y)
8. 　　 insert(Q,z)
9. end for
10. 　return Q

2.4.3　算法实现（Algorithm Implementation）

1. 实例分析（Instance Analysis）

已知某系统在通信联络中只可能出现 5 种字符，分别为 A、B、C、D、E，其使用频率分别为 15、7、6、6、5，设计哈夫曼编码。

It is known that only 5 symbols can appear in the communication system：A, B, C,

D, E. Their use frequencies are 15, 7, 6, 6, 5, respectively. Now, it is needed to design a Huffman code to minimize the average code length.

首先将每个字符视为一棵树,根据字符出现的频率从小到大排序,规定最小的字符为新树的左枝,第二小的为新树的右枝,构成一棵新树,新树的频率为两棵子树之和。新树与剩下的其他树再在一起排序,选取频率最小的两棵树从左到右排列构成新的树。重复上述过程,直到最后生成一棵哈夫曼树。最后,每个节点的左枝编码为0,右枝编码为1,逐层编码,所有字符的编码从根节点开始往下依次书写。过程和结果如图2.5和表2.12所示。

First of all, each character is regarded as a node in a tree. According to the frequency of characters from smallest to largest, the least frequent character is assigned as the left branch of the new tree, and the second least frequent as the right branch, forming a new tree. The frequency of the new tree is the sum of the frequencies of the two subtrees. This new tree is then sorted together with the remaining trees, and the two trees with the lowest frequencies are selected to form a new tree, arranged from left to right. Repeat the process until it ultimately has a single Huffman tree. Finally, assign code 0 to the left branch of each node and code 1 to the right branch, coding layer by layer. The encoding of all characters is then written successively from the root node. The process and results are shown in Figure 2.5 and Table 2.12.

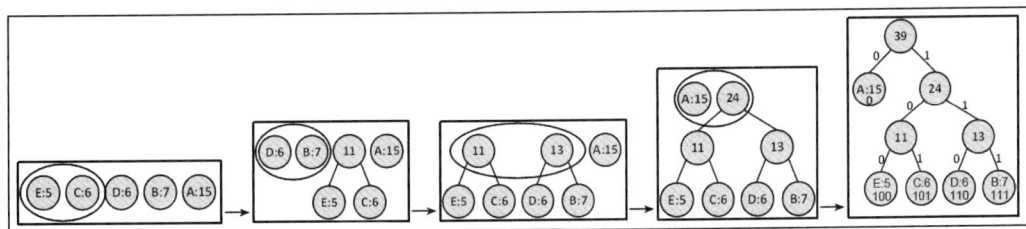

图 2.5　哈夫曼树计算过程

Figure 2.5　Huffman tree computation procedure

表 2.12　字符出现的次数与概率

Table 2.12　Frequency and probability of character occurrence

符号(symbol)	A	B	C	D	E
Frequency	15	7	6	6	5
Code	0	111	101	110	100

2. 数据结构设计(Data Structure Design)

采用 *list* 数据结构存储哈夫曼树构造过程中树的集合 Q。树中有 n 个字符节点,n-1 个中间节点。每个节点包括字符、频率、左孩子、右孩子、父亲5个属性,在生成字符编码时,左孩子为0,右孩子为1,所以每个节点还包括一个判断自己是否为左孩子节点的过程。由此,定义一个节点类将节点的属性和操作方法封装到一起。

The *list* data structure is used to store the set Q of trees in Huffman tree construction. There are n character nodes and n-1 intermediate nodes in the tree. Each node includes five attributes：character, frequency, left child, right child, and parent. In the generation of character encoding, the left child is represented by 0 and the right child by 1. Therefore, each node also includes a process to determine whether it is a left child node. From this, a node class is defined to encapsulate the attributes and operation methods of the node.

3. 代码实现（Code Implementation）

Python 实现中需要引入 *copy* 依赖包，即 *import copy*。

Python implementations need to import the copy module, namely import copy.

节点类 *node* 定义了 *char*、*freq*、*left*、*right*、*father* 5 个字段，分别对应字符、频率、左孩子、右孩子、父亲；定义函数 *create_prim_nodes*（*data_set*, *labels*），用于根据字符集和频率构造 n 个节点的树集合；定义函数 *create_HF_tree*（*nodes*）用于构造哈夫曼树，该函数做 n−1 次，每次从树的集合中选择两个频率最小的节点，让其作为左右子树构造一棵新树，新树的根节点的频率为左右子树根节点频率之和，将新树插入树的集合中；定义哈夫曼编码的函数 *get_huffman_code*（*root*, *nodes*），从叶子节点顺着父亲节点找到根节点，得到该叶子节点的编码；最后定义入口函数 *main*（），提供字符集和字符的频率，先构造 n 棵单个节点的树集合，然后构造哈夫曼树，获取哈夫曼编码，最后将结果打印显示。

The node class defines five attributes：*char*, *freq*, *left*, *right*, and *parent*, corresponding to character, frequency, left child, right child, and parent, respectively. Define the function *create_prim_nodes*（data_set, labels）to construct a set of n-node trees based on the character set and frequency. The function *create_HF_tree*（*nodes*）is defined to construct a Huffman tree. This function performs n−1 iterations, each time selecting two nodes with the lowest frequency from the set of trees to construct a new tree with them as the left and right subtrees. The frequency of the root node of the new tree is the sum of the frequencies of the left and right subtree nodes, and then inserts the new tree into the set of trees. Define the Huffman coding function *get_huffman_code*（*root*, *nodes*）to find the path from the leaf node to the root node along the parent nodes and obtain the code for the leaf node. Finally, the entry *main*（）function is defined to provide the character set and the frequency of the characters. First, a set of n single nodes is constructed, and then the Huffman tree is built to obtain the Huffman codes. Finally, the results are printed and displayed.

程序输出结果如图 2.6 所示。代码如下。

The program output is shown in Figure 2.6. The code is as follows.

```
import copy
classnode：                              #定义节点类（Define node class）
    def _ _init_ _(self, name, weight)：
        self.name = name                 #节点名（node name）
        self. freq = freq                #节点频率（node frequency）
        self.left = None                 #节点左孩子（node left child）
```

续

```python
        self.right = None                      #节点右孩子(node right child)
        self.father = None                     #节点父节点(node parent)
    #判断是否是左孩子(Determine if it's a left child)
    def is_left_child(self):
        return self.father.left == self
def create_prim_nodes(data_set, labels):      #构造单节点树集合(Construct a single node tree set)
    if(len(frequency) != len(char_set)):
        raise Exception('数据和标签不匹配!')(Data and labels do not match)
    nodes = []
    for i in range(len(char_set)):
        nodes.append(node(char_set[i], frequency[i]))
    return nodes
def create_HF_tree(nodes):                     #构造哈夫曼树(Construct Huffman tree)
    #此处注意,copy()属于浅拷贝,只复制最外层元素,内层嵌套元素则通过引用,而不是独立分配内存(Note that the
    #copy() is a shallow copy that copies only the outermost elements, while the inner nested elements are referenced
    #instead of allocated memory independently)
    tree_nodes = nodes.copy()
    tree_nodes.sort(key= lambda node: node.freq)  #升序排列(Ascending arrangement)
    while len(tree_nodes) >1:                   #只剩根节点时,退出循环(Exit the loop when only the root node remains)
        new_left = tree_nodes.pop(0)
        new_right = tree_nodes.pop(0)
        new_node = node(None, (new_left.weight +new_right.weight))
        new_node.left = new_left
        new_node.right = new_right
        new_left.father = new_right.father = new_node
        j = len(tree_nodes)
        for i in range(len(tree_nodes)):
            if new_node.freq <= tree_nodes[i].freq:
                j = i
                break
        tree_nodes.insert(j,new_node)
    tree_nodes[0].father = None                #根节点父亲为None(The father of the root node is None)
    return tree_nodes[0]                        #返回根节点(Return to root node)
def get_huffman_code(root,nodes):              #构造哈夫曼编码树(Construct Huffman code tree)
    codes = {}
    for node in nodes:
        code=''
        name = node.name
        while node.father !=None:
            if node.is_left_child():
                code = '0' +code
            else:
                code = '1' +code
            node = node.father
        codes[char] = code
    return codes
if __name__ =='__main__':                      #主函数入口(Main function entry)
    data_set = ['A','B','C','D','E']
    frequency = [15,7,6,6,5]
    nodes = create_tree_nodes(frequency,data_set)  #创建初始叶子节点(Create the initial leaf node)
    root = create_HF_tree(nodes)               #创建哈夫曼树(Create Huffman tree)
    codes = get_huffman_code(root,nodes)       #获取哈夫曼编码(Get the Huffman code)
```

续

```
#打印哈夫曼码(Print Huffman code)
for key in codes.keys( ) :
    print( key,' : ',codes[ key])
```

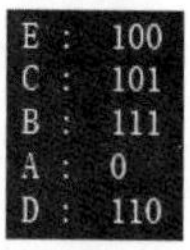

```
E :  100
C :  101
B :  111
A :  0
D :  110
```

图 2.6　哈夫曼算法编码运行结果

Figure 2.6　Result of Huffman algorithm program coding

2.5　最小生成树（Minimum Spanning Tree,MST）

2.5.1　基本思想（Basic Idea）

假设要在 n 个城市之间建立通信网络,则连通 n 个城市至少需要 $n-1$ 条线路。n 个城市之间,最多可能设置 $n(n-1)/2$ 条线路。这时,自然会考虑一个问题：如何在这些可能的线路中选择 $n-1$ 条,以便在最节省费用的前提下建立该通信网络？最小生成树就是在这个题设下进行分析的。

Ii is supposed to build a communication network between n cities; it needs at least $n-1$ lines to connect them. Between n cities, a maximum of $n(n-1)/2$ lines may be installed. At this point, it is natural to consider a problem：how to select $n-1$ of these possible lines to establish the most cost-effective communication network. This problem is analyzed under the concept of the minimum spanning tree.

设 $G=(V,E)$ 是无向连通带权图。E 中每条边 (i,j) 的权为 $w(i,j)$。如果 G 的子图 G' 是一棵包含 G 的所有顶点的树,则称 G' 为 G 的生成树。图 G 的生成树并不唯一,其中两棵生成树如图 2.7 所示。

Let $G=(V,E)$, is a connected, weighted graph. The weight of each edge (i,j) in E is $w(i,j)$, if the subgraph G' of G is a tree containing all vertexes of G, then G' is called a spanning tree of G. The spanning trees in Figure G are not unique；two of them are shown in Figure 2.7.

图的生成树上各边权的总和称为树的耗费,因此图 2.7 中的两棵生成树的耗费分别为 16 和 23,最小生成树是指耗费最小的生成树。

The sum of the weights of the edges in a spanning tree of the graph is referred to as the tree's cost. Therefore, the costs of the two spanning trees in the above figure are 16 and 23, respectively. The minimum spanning tree refers to the spanning tree with the least cost.

因此,用无向连通带全图 $G=(V,E)$ 来表示城市通信网络,图的顶点表示城市,顶点之间的边表示城市之间的通信线路,边的权值表示线路的费用。对于 n 个顶点的连通网可以

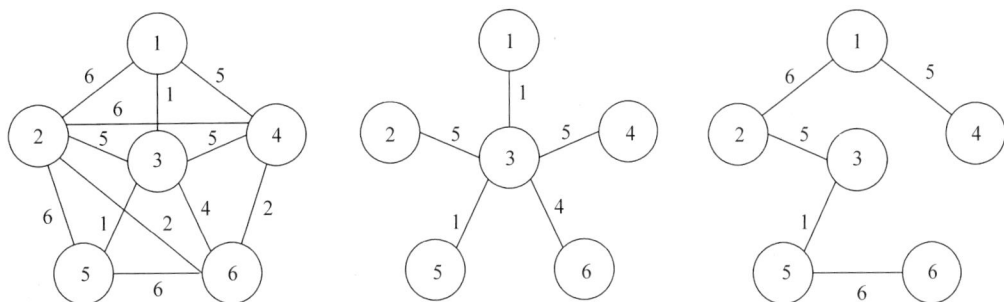

图2.7 图 G 及其两棵生成树示意图

Figure 2.7 Diagram of figure G and its two spanning trees

建立许多包含 $n-1$ 条通信线路且各个城市互相连通的通信网,则每个通信网即为图 G 的一棵生成树。因此,问题变成了寻找一棵总耗费最少的生成树,即构造连通网的最小生成树问题。

Therefore, the city communication network is represented by the undirected, connected, total graph $G=(V,E)$, where the vertexes of the graph represent the cities, the edges between the vertexes represent the communication lines between the cities, and the weights of the edges represent the costs of the lines. For an interconnection network with n vertexes, many communication networks can be established, each containing $n-1$ lines and ensuring that each city is connected to every other city. Thus, each communication network is a spanning tree of graph G. Consequently, the problem becomes one of finding a spanning tree with the least total cost, which is, constructing the minimum spanning tree of the connected network.

在解决寻找无向连通带权图的最小生成树问题中,*Prim* 算法和 *Kruskal* 算法是两个常用的算法。

Prim's and *Kruskal's* algorithms are two commonly used methods for finding the minimum spanning tree of undirected, connected, weighted graphs.

2.5.2 问题分析(Problem Analysis)

最小生成树(*MST*)原则,设 $G=(V,E)$ 是一个无向连通带权图,U 是顶点集 V 的一个非空真子集。若 (u,v) 是 G 中一条一个顶点在 U 中 $(u \in U)$,另一个顶点不在 U 中的边 $(v \in V-U)$,且 (u,v) 具有最小权值,则一定存在 G 的一棵最小生成树包括此边 (u,v),即将顶点集 V 划分为两个互不相交的真子集 U 和 $V-U$,连接两个真子集 U 和 $V-U$ 的边中,权最小的边一定在一棵最小生成树中。

Minimum spanning tree (*MST*) property: Let $G=(V,E)$ be an undirected connected weighted graph, and let U be a non-empty proper subset of vertexes V. If (u,v) is one endpoint of G in $U(u \in U)$, and the other endpoint is no longer an edge in $U(v \in V-U)$, and if (u,v) has a minimum weight among all such edges, then there must be a minimum spanning tree of G that includes this edge (u,v). That is, the vertex set V is divided into two non-intersecting proper subsets U and $V-U$. Among the edges connecting the two proper subsets U and $V-U$, the edge with the least weight must be in a minimal spanning tree.

2.5.3　Prim 算法（Prim Algorithm）

Prim 算法也称为 *Prim-Jarnik* 算法，最早是捷克人 *V.Jarnik* 于 1930 年在文章中提出的，而 *R.C.Prim* 在 1957 年提出了完整的算法。*Prim* 算法从图的顶点出发，紧扣最小生成树的性质，把顶点集分成两个互不相交的集合，贪心选择连接两个集合最小权的边加入最小生成树中。

Prim's algorithm, also known as the *Prim-Jarnik* algorithm, was first proposed by Czech *V. Jarnik* in 1930, and the complete algorithm was proposed by *R. C. Prim* in 1957. The *Prim* algorithm starts from a vertex in the graph, closely follows the properties of the minimum spanning tree, divides the vertex set into two disjoint sets, and greedily selects the edge with the minimum weight connecting the two sets to join the minimum spanning tree.

Prim 算法的贪心策略是，选取连接两个集合的最小权边，将其 $V-U$ 中的端点加入 U 中。

The greedy strategy of *Prim's* algorithm is to take the edge with the least weight connecting two sets and add its endpoints in $V-U$ to U.

1. 设计思想（Design Idea）

输入：无向连通带权图 $G=(V,E)$。

Input. undirected connected weighted graph $G=(V,E)$.

输出：最小生成树边集 *TE*。

Output. Minimum generated tree edge set *TE*.

初始时，从顶点集 V 中选取一个顶点 1，加入 U 集合中，即 $U=\{1\}$，最小生成树 *TE* 为空集，即 $TE=\{\}$。

In the initial step, vertex 1 is selected from the vertex set V and added to the set U, so $U=\{1\}$. The minimum spanning tree *TE* is initially the empty set, so $TE=\{\}$.

n 个顶点的图，需要 $n-1$ 步贪心选择，选取满足的条件 $i\in U, j\in V-U$，且边 (i,j) 是连接 U 和 $V-U$ 的所有边中的最短边，将顶点 j 加入集合 U，边 (i,j) 加入集合 *TE*。算法一直进行到 $U=V$ 为止，此时，选择到的所有边恰好构成 G 的一棵最小生成树 T。

For a graph with n vertexes, the greedy selection requires $n-1$ steps, where the condition $i\in U, j\in V-U$ is satisfied, and the edge (i,j) is the shortest of all the edges connecting U and $V-U$. The vertex j is added to the set U, and the edge (i,j) is added to the set *TE*. The algorithm continues until $U=V$, at which point all selected edges form exactly a minimum spanning tree T of G.

采用穷举的方法选取满足条件的边 (i,j)，扫描 U 中的每一个点连接 $V-U$ 中的顶点的所有边，找出权最小的边 (i,j)，所耗时间显然是 $O(n^2)$。做 $n-1$ 次贪心选择，算法的时间复杂度会达到 $O(n^3)$。

The exhaustive method is used to select the edge (i,j) that meets the conditions:

scan every vertex in U to connect all edges of vertexes in $V-U$, and find the edge with the least weight (i,j). The time complexity is obviously $O(n^2)$. By making $n-1$ greedy choices, the time complexity of the algorithm will reach $O(n^3)$.

Prim 算法借助两个 n 存储单元的辅助空间 *closest* 和 *lowcost*，*closest*[j]用于存储对 $V-U$ 中的每个点 j，U 中哪个点离 j 最近，*lowcost*[*closest*[j], j]用于记录最小的权值。

The *Prim* algorithm uses the auxiliary space of two n storage units *closest* and *lowcost*. *closest*[j] is used to store each point j in $V-U$, indicating which point in U is closest to j, *lowcost*[*closest*[j], j] is used to record the smallest weight.

2. 算法伪代码(Algorithmic Pseudocode)

算法 2.4　贪心策略解决最小生成树 Prim 算法(Algorithm 2.4　Greedy strategy solves the minimum spanning tree Prim)

输入：无向带权图 G(Input：undirected weighted graph G)
输出：最小生成树 T(Output：minimum spanning tree T)

```
1.  S←{1},T←{}
2.  for i←1 to n do
3.     if G[1][i]<∞
4.        closest[i]←1
5.        lowcost←G[1][i]
6.     end if
7.  end for
8.  while V-S≠Φ do
9.     V-S 中选择最小的 lowcost[j](Select the smallest lowcost[j] in V-S)
10.    S←SU{j}
11.    T←TU{(closest[j],j)}
12.    for i←1 to n do
13.       if S[i]=0 and G[j][i]<lowcost[i]
14.          closest[i]←j
15.          lowcost[i]←G[j][i]
16.       end if
17.    end for
18. end while
19. return T
```

3. 算法实例(Algorithm Instance)

按 Prim 算法对一个无向连通带权图构造一棵最小生成树。假设初始为顶点 1，即设定最小生成树 T 的顶点集合 $U=\{1\}$，$V-U=\{2,3,4,5,6\}$，如图 2.8 所示。

A minimum spanning tree is constructed for an undirected, connected, weighted graph using the Prim algorithm. Suppose we start withvertex 1, that is set the vertex set $U=\{1\}$, and $V-U=\{2,3,4,5,6\}$ for the minimum spanning tree T. As shown in Figure 2.8.

从原顶点 1 出发，其他所有节点的直连节点先都标记为 1，根据直连关系记录路径的权重，不直连的权重记录为无穷大，此时 U 集合记录已经被计算过的顶点，即 $U=\{1\}$点，$U-V$ 集合记录剩下的没有被计算过的节点，即 $U-V=\{2,3,4,5,6\}$，如表 2.13 所示。根据结果，

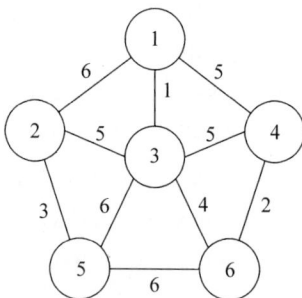

图 2.8　最小生成树实例

Figure 2.8　Minimum spanning tree instance

选取当前 $V-U$ 集合中,路径权重最小的节点进行下一步计算,即选择节点 3,并且此时顶点 1 到 3 的路径可以被确定下来。

Starting from the origin vertex 1, the directly connected vertexes of all other vertexes are initially marked as 1, and the weights of the paths are recorded based on their direct connection relationships. Path weights that are not directly connected are recorded as infinity. At this time, the set U records the vertexes that have been calculated, that is, $U=\{1\}$, and the $U-V$ set records the remaining vertexes that have not been calculated, that is, $U-V=\{2,3,4,5,6\}$. See Table 2.13. According to the result, the vertex with the smallest path weight in the current $V-U$ set is selected for the next step, that is, vertex 3 is selected, and the path from 1 to 3 can be determined at this time.

表 2.13　确定第一条路径

Table 2.13　Determine the first path

	vertex					graph
	2	3	4	5	6	
closest	1	1	1	1	1	
lowcost	6	1	5	∞	∞	
U	$\{1\}$					
$V-U$	$\{2,3,4,5,6\}$					

下一步选择顶点 3,将其最近顶点和权重记录清空,并且从 $V-U$ 集合转移到 U 集合。对比从顶点 3 出发能够直连到达的剩余顶点路径权重(不直连的为无穷大),如果比当前记录的权重小,则更新信息。

In the next step, select vertex 3, clear its recent vertex and weight records, and transfer it from the $V-U$ set to the U set. Compare the remaining vertex path weights that can be reached directly from vertex 3（those that are not directly connected are considered infinite）, and update the information if the weights are smaller than those currently recorded.

顶点 2 的权重为 5,比原来记录的权重小,因此更新顶点 2 的最近顶点信息为 3,权重为 5。

The weight of vertex 2 is 5, which is less than the weight of the original record. Therefore, the latest vertex information for node 2 is updated to 3, with a weight of 5.

顶点 3 与顶点 1 是直连的,但是 1 号顶点已经被计算过(在 U 集合中),因此不再计算。到顶点 4 的权重是 5,没有更小,因此不更新顶点 4 信息;到顶点 5 的权重为 6,比原来记录的小,更新顶点 5 的最近顶点为 3,权重为 6。

Vertex 3 is directly connected to vertex 1, but vertex 1 has already been computed（in the U set）, so it is not recomputed. The weight of vertex 4 is 5, which is not less, so the information for vertex 4 is not updated. The weight of vertex 5 is 6, which is smaller than the original record. The nearest vertex to update for node 5 is 3, with a weight of 6.

同样地,更新顶点 6 最近顶点为 3,权重为 4。根据更新结果,目前权重最小的顶点是节点 6,则下一步顶点 6 将被计算,并且顶点 3 到顶点 6 的路径已经确定,如表 2.14 所示。

Similarly, update the nearest vertex 6 to 3 and the weight to 4. According to the update result, the vertex with the least weight at present is vertex 6; thus, vertex 6 will be calculated in the next step, and the path from 3 to 6 has been determined. As shown in Table 2.14.

表 2.14 确定第二条路径

Table 2.14 Determine the second path

	vertex					graph
	2	3	4	5	6	
closest	3	–	1	3	3	
lowcost	5	–	5	6	4	
U	{1,3}					
$V-U$	{2,4,5,6}					

后续的计算过程与该步骤相同,逐个选取顶点、更新路径和存储信息,直到 $V-U$ 集合为空,即所有顶点都已经被计算过,停止计算,输出结果,如表 2.15～表 2.18 所示。

The subsequent calculation process is the same as the preceding step: select vertex one by one and update paths and storage information until the $V-U$ set is empty, that is, all vertexes have been calculated. The calculation stops, and the result is displayed, as shown in Table 2.15～Table 2.18.

表 2.15　确定第三条路径

Table 2.15　Identify the third path

	vertex					graph
	2	3	4	5	6	
closest	3	–	6	3	–	
lowcost	5	–	2	6	–	
U	{1,3,6}					
$V-U$	{2,4,5}					

表 2.16　确定第四条路径

Table 2.16　Determine the fourth path

	vertex					graph
	2	3	4	5	6	
closest	3	–	–	3	–	
lowcost	5	–	–	6	–	
U	{1,3,6,4}					
$V-U$	{2,5}					

表 2.17　确定第五条路径

Table 2.17　Determine the fifth path

	vertex					graph
	2	3	4	5	6	
closest	–	–	–	2	–	
lowcost	–	–	–	3	–	
U	{1,3,6,4,2}					
$V-U$	{5}					

表 2.18　确定第六条路径

Table 2.18　Determine the sixth path

	vertex					graph
	2	3	4	5	6	
closest	–	–	–	–	–	
lowcost	–	–	–	–	–	
U	{1,3,6,4,2,5}					
$V-U$	{}					

程序实现汇总采用 *list* 数据结构存储 *closest*、*lowcost*、集合 *U*、集合 *V–U*、最小生成树 *T*,选用邻接矩阵数据结构存储图。

The program uses *list* data structure to store *closest*, *lowcost*, set *U*, set *V–U* and the minimum spanning tree *T*, and uses adjacency matrix data structure to store graph.

算法需要导入 *sys* 依赖包。首先定义核心算法的实现函数 *prim_mst*(*graph*, *vertexes*),用于接收图的邻接矩阵和顶点集,并输出最小生成树和树的耗费。定义入口函数 *main*( ),其中用 *graph* 存储图的邻接矩阵,*vertexes* 存储顶点集,调用前面的定义函数用于求最小生成树和树的耗费,最后打印结果。

The algorithm needs to import the *sys* dependency package. First, define the core algorithm implementation function *prim_mst*(*graph*, *vertexes*), which is used to receive the adjacency matrix and the node set of the graph and output the minimum spanning tree and its cost. Define the entry *main*( ) function, where *graph* stores the adjacency matrix of the graph, and *vertexes* stores the vertex set. This function calls the previously defined prim_mst function to obtain the minimum spanning tree and its cost, and finally prints the result.

程序的输出结果如图 2.9 所示。代码如下。

The output of the program is shown in Figure 2.9. The code is as follows.

```python
import sys
def prim_mst(graph,vertexes):
    ulist = []
    ulist.append(vertexes[0])    #集合 U(Set U)
    tree_list = []               #最小生成树(Minimum spanning tree)
    closest = []                 #closest[i]表示生成树集合中与点 i 最近的点的编号(Represents the number of the
                                 #point closest to point i in the spanning tree collection)
    lowcost = []                 #lowcost[i]表示生成树集合中与点 i 最近的点构成的边最小权值 ,–1 表示 i 已经在
                                 #生成树集合中(Represents the minimum edge weight of the point closest to point
                                 #i in the spanning tree set. –1 indicates that i is already in the spanning tree set)
    lowcost.append(-1)
    closest.append(0)
    n = len(vertexes)
    for i in range(1,n):         #初始化 closest 数组和 lowcost 数组(Initialize the closest array and the lowcost array)
        lowcost.append(graph[0][i])
        closest.append(0)
    sum = 0
    for _ in range(1,n):         #n0-1 次贪心选择(n-1 greedy selection)
        minid = 0                #记录 V–U 中顶点最近的 U 中的顶点编号(Record the vertex number in the nearest U
                                 #of the vertexes in V–U)
        min = sys.maxsize
        for j in range(1,n):     #寻找每次插入生成树的权值最小 lowcost(Find the minimum lowcost of each weight
                                 #inserted into the spanning tree)
            if(lowcost[j] != -1 and lowcost[j]<min):
                minid = j
                min = lowcost[j]
        ulist.append(vertexes[minid])
        tree_list.append([vertexes[closest[minid]],vertexes[minid],lowcost[minid]])
        sum+= min
        lowcost[minid] = -1
```

续

```
        for j in range(1,n)：  #更新插入结点后 lowcost 数组和 closest 数组值(Update the values of the lowcost
                              #array and closest array after the node is inserted)
            if(lowcost[j]！ = -1 and lowcost[j]>graph[minid][j])：
                lowcost[j] = graph[minid][j]
                closest[j] = minid
    return sum,tree_list
if _ _name_ _== ' _ _main_ _'：
    MAX = sys.maxsize
    graph = [[0,6,1,5,99,99],[6,0,5,99,3,99],[1,5,0,5,6,4],[5,99,5,0,99,2],[99,3,6,99,
0,6],[99,99,4,2,6,0]]
    vertex = ['ONE','TWO','THREE','FOUR','FIVE','SIX']
    sum,tree_list = prim_mst(graph,vertex)
    for edge in tree_list：
        print(edge[0]+" --" +edge[1]+"   Weight:" +str(edge[2]))
    print("Total Cost/Weight: ",sum)
```

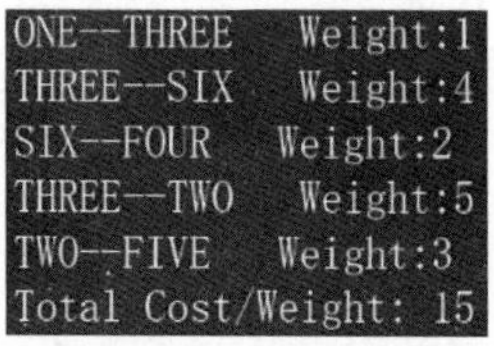

图 2.9　最小生成树的 Prim 算法运行结果

Figure 2.9　Result of Prim algorithm of minimum spanning tree

2.5.4　Kruskal 算法(Kruskal Algorithm)

Kruskal 算法是由 *J.B.Kruskal* 于 1956 年提出的,它从边的角度出发,每一次将图中权值最小的边提取出来,在不构成环的情况下,即判断该边的两个端点是否在一个连通分支,若在则舍弃边,若不在则将该边加入最小生成树,重复这个过程,直到图中所有的顶点都加入最小生成树中。

Kruskal's algorithm, proposed by *J.B. Kruskal* in 1956, extracts the edge with the smallest weight from the graph each time from the perspective of edges. It ensures there are no cycles by judging whether the two endpoints of the edge are in the same connected component. The process continues until all vertexes in the graph have been added to the minimum spanning tree.

Kruskal 算法的贪心策略是,权最小的边优先检查,若它的两个端点不在一个连通分支,则将改变加入最小生成树中,否则舍弃。

The greedy strategy of Kruskal's algorithm is to check the edges with the least weight first; if the two endpoints of an edge are not in the same connected component, the edge is added to the minimum spanning tree, otherwise, it is discarded.

1. 设计思想(Design Idea)

设 $G=(V,E)$ 是无向连通带权图,$V=\{1,2,\cdots,n\}$;设最小生成树 $T=(V,TE)$。

Let $G=(V,E)$ be an undirected connected weighted graph, where $V=\{1, 2,\cdots, n\}$; Let the minimum spanning tree $T=(V, TE)$.

初始时,最小生成树 T 的边集 $TE= \{\}$。不断做贪心选择:在边集 E 中选取权值最小的边 (i,j),判断端点 i、j 是否在同一个连通分支,若不在,则将边 (i,j) 加入边集 TE 中,即用边 (i,j) 将这两个连通分支合并连接成一个连通分支;否则,继续选择下一条最短边,直到 T 中所有顶点都在同一个连通分支为止。此时,选取到的 $n-1$ 条边恰好构成 G 的一棵最小生成树 T。

Initially, the edge set of the minimum spanning tree T is $TE= \{\}$. Keep making greedy choices: select the edge (i,j) with the smallest weight in edge set E, and judge whether the endpoints i and j are in the same connected component; if not, add the edge (i,j) to the edge set TE, that is, the edge (i,j) will merge the two connected components into a connected component. Otherwise, continue to select the next shortest edge until all vertexes in T are on the same connected component. In this case, the selected $n-1$ edges form exactly a minimum spanning tree T of G.

2. 算法伪代码(Algorithmic Pseudocode)

算法 2.5 贪心算法解决最小生成树 Kruskal 算法(Algorithm 2.5 Greedy algorithm of solves the minimum spanning tree Kruskal)

输入:图 G =(V,E)
Input:Graph G =(V,E)
输出:最小生成树 T
Output:Minimum spanning tree T

1.　sort(E)　//E 为边集(E is the edge set)
2.　T←Φ
3.　j←0//记录加入的边的条数(Records the number of edges added)
4.　for i←1 to |E| do//|E|是图中边的条数(|E| is the number of edges in the picture)
5.　　e←E[i]
6.　　if e 的两个端点不在同一个连通分支 then(if the two endpoints of e are not in the same connected branch then)
7.　　　T←{e}
8.　　　j←j+1
9.　　end if
10.　　if j=n-1//n 为图中顶点个数(n is the number of vertexes in the graph)
11.　　　break;
12.　　end if
13. end for
14. return T

3. 算法实例(Algorithm Instance)

用 *Kruskal* 算法对如图 2.9 所示的无向连通图构造一棵最小生成树。

Kruskal algorithm is used to construct a minimum spanning tree for the undirected, connected graph shown in Figure 2.9.

首先选择最短路径权重的两个顶点,即节点 1 和 3,确定其路径,并考虑两个顶点已经短接的关系情况,此时原来的 1 与 2、3 与 2 的路径已经视为一条,1 到 4、3 到 4 的路径也已经可以视为一条,但还是要绘制出来,并考虑下一步的选择,如图 2.10 所示。

First, the two vertexes with the shortest path weights, namely vertexes 1 and 3, are selected to determine their paths, and the relationship between these two vertexes is considered. At this point, the original paths from 1 to 2 and from 3 to 2 are already regarded as one, and the paths from 1 to 4 and from 3 to 4 are also considered as one. However, it is still necessary to draw these paths and consider the choice for the next step, as shown in Figure 2.10.

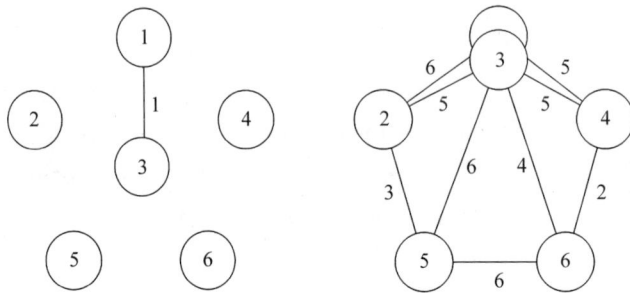

图 2.10　选择最短路径的两个顶点并合并

Figure 2.10　Select the two vertexes of the shortest path and merge them

在顶点 1 和顶点 3 短接后,选择下一条最短路径是顶点 4 与顶点 6,选择该路径并考虑将顶点 4 和顶点 6 短接,如图 2.11 所示。

After short-circuitingvertexes 1 and 3, the next shortest paths are between vertexes 4 and 6. Select the path and perform the short-circuiting on these vertexes, as shown in Figure 2.11.

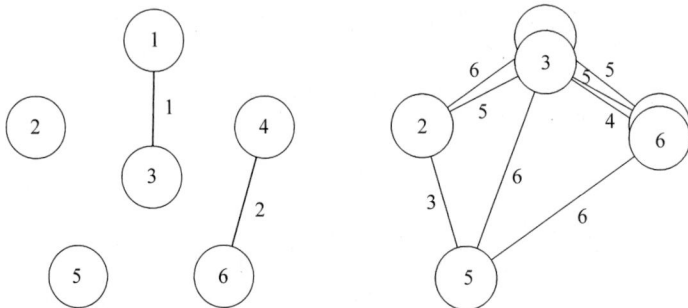

图 2.11　继续选择下一条最短路径并合并顶点

Figure 2.11　Proceed to select the next shortest path and merge the vertexes

下一条最短路径是 2 与 5,选择路径并短接两个顶点,如图 2.12 所示。

The next shortest path connects 2 and 5, select the path and short-circuit the two vertexes, as shown in Figure 2.12.

此时最短路径剩下 3 与 6 顶点之间的权重为 4 的路径,但是根据前面的短接结果,该路径与 1 到 4、3 到 4 的路径是同一条,则选择最短的 3 到 6 的路径,短接关系中将这三条“相同效果路径”都合并,如图 2.13 所示。

At this time, there is a path with a weight of 4 between vertexes 3 and 6 in the shortest path set. However, according to the previous short-circuiting results, this path

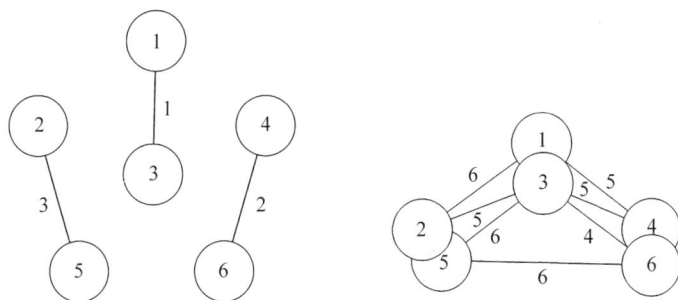

图 2.12 合并顶点 2 和 5

Figure 2.12 Merge vertexes 2 and 5

is identical to the paths from 1 to 4 and from 3 to 4. Therefore, the shortest path from 3 to 6 is selected, and the three ' same effect paths' are merged into a single short-circuiting relationship, as shown in Figure 2.13.

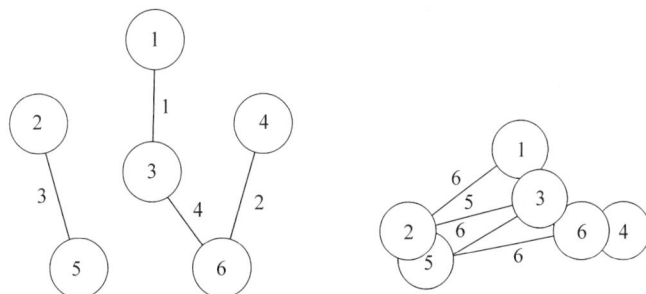

图 2.13 合并顶点 3 和 6

Figure 2.13 Merge vertexes 3 and 6

剩下的路径在短接效果中都是一样的,其中原来的顶点 2 到顶点 3 的路径最短,因此选择这一条,在短接图中去掉所有的相同路径,此时已经没有剩下的了,计算结束,得出结果,如图 2.14 所示。

The remaining paths all exhibit the same short-circuiting effect. Among them, the paths from node 2 to node 3 are the shortest; thus, we select this path and remove all identical paths from the short-circuiting diagram. At this point, no paths remain; the calculation is complete, and the result is obtained, as shown in Figure 2.14.

采用 *list* 存储图的边集,每个元素是一条边,用一个三元组(顶点 1,顶点 2,边权)表示。最小生成树 T 和标记顶点分支的数据结构 group 均用 list。

A *list* is used to store the edge set of the graph, where each element is an edge represented by a triple (vertex 1, vertex 2, edge weight). The list is used for both the minimum spanning tree T and the data structure that marks vertex branches.

定义函数 *kruskal*(*edge_list*, *vertexes*),用于接收图的顶点集和边集,返回最小生成树。然后,定义入口函数 *main*( ),初始化图的数据结构,调用函数生成最小生成树,最后打印结果。程序运行结果如图 2.15 所示。

Define the function *kruskal* (*edge_list*, *vertexes*) to receive the node set and edge

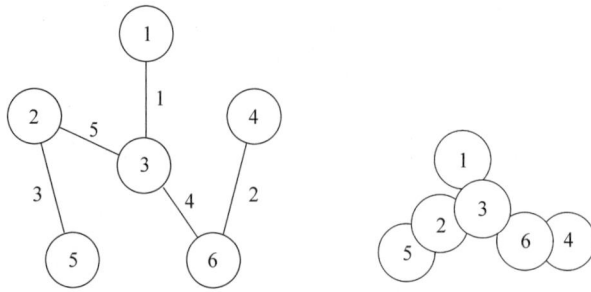

图 2.14　最终结果

Figrue2.14　Final result

set of the graph and return the minimum spanning tree. Then, define the entry main( ) function, initialize the graph's data structure, call the function to generate a minimum spanning tree, and finally print the result. The program's running result is shown in Figure 2.15.

代码如下(The code is as follows)。

```python
def kruskal(edge_list,vertexs):
    vertex_num = len(vertexs)
    edge_num = len(edge_list)
    tree_mst = []
    if vertex_num <= 0 or edge_num < vertex_num - 1:
        return tree_mst
    edge_list.sort(key= lambda a:a[2])      #按照边权由小到大排序(Sort by edge weight from smallest to largest)
    group = [[i] for i in range(1,vertex_num+1)]    #初始化,将图 G 的顶点看成 n 个孤立分支(Initialize,
                                            #treating the vertexes of the graph G as n isolated branches)
    for edge in edge_list:
        k = len(group)                          #获取分支数目(Get number of branches)
        if k == 1:  #如果只有一个分支,则跳出循环,算法结束(If there is only one branch, the loop exits and the
                    #algorithm ends)
            break
        for i in range(k):
            if edge[0] in group[i]:
                m = i
            if edge[1] in group[i]:
                n = i
        if m != n:
            tree_mst.append(edge)
            group[m] = group[m] + group[n]
            group.remove(group[n])
    return tree_mst
if __name__ == "__main__":
    edge_list = [[1,2,6],[1,3,1],[1,4,5],[2,3,5],[2,5,3],[3,4,5],[3,6,4],[3,5,6],[4,6,2],[5,6,6]]
    vertexs = [1,2,3,4,5,6]
    tree_mst =  kruskal(edge_list,vertexs)
    for edge in tree_mst:
        print(edge)
```

```
(1, 3,  1)
(4, 6,  2)
(2, 5,  3)
(3, 6,  4)
(2, 3,  5)
```

图 2.15　最小生成树 Kruskal 算法运行结果

Figure 2.15　Result of Kruskal algorithm in minimum spanning tree

2.6　背包问题(Knapsack Problem)

2.6.1　基本思想(Basic Idea)

1. 问题描述(Problem Description)

有 n 件物品和 1 个背包。对物品 $i(i=1,2,\cdots,n)$，其价值为 v_i，重量为 w_i，背包的容量为 W。如何选取物品装入背包，使背包中所装入的物品的总价值最大？其中，物品可以分割。

There are n items and 1 backpack. For item i ($i = 1$, 2, $\cdots$, n), its value is v_i, weight is w_i, and the capacity of the backpack is W. How to select items to pack in a backpack to maximize the total value of the items within it? Items can be divided.

2. 问题分析(Problem Analysis)

背包问题给定的已知条件包括背包的承载重量 W，n 件物品各自的重量 w_i 和价值 v_i，要求物品装载不能超过背包的承载重量，但是物品可以分割。

The given conditions for the knapsack problem include the backpack's carrying capacity W, the weight w_i and value v_i of n items. It is required that the total weight of the items cannot exceed the backpack's carrying capacity, but the items can be divided.

从已知条件分析，有以下三种策略。

From the analysis of the known conditions, there are three strategies.

(1) 根据价值最大的目标，选择价值最大的物品先装的策略。

According to the goal of the greatest value, choose the most valuable items first loading strategy.

(2) 根据背包承载限制考虑，重量小的物品先装。

According to the load limit of the backpack, the weight of the loaded items is small.

(3) 综合限制条件和价值目标，最好选择重量小且价值大的物品，即单位重量价值大的物品优先装入背包。

Comprehensive restrictions and value objectives, it is best to choose small weight and large value of items, that is, the value of large items per unit weight is limited into the backpack.

2.6.2　算法的设计与描述(Algorithm Design and Description)

1.设计思想(Design Idea)

输入：物品集合 $S=\{1,2,\cdots,n\}$，每件物品的重量 w_i 和价值 n_i，其中，$i=1,2,\cdots,n$，背

包能够承载的重量为 W。

Input. Set of items $S = \{1, 2, \cdots, n\}$, the weight w_i and value n_i of each item, where $i = 1, 2, \cdots, n$, the weight that the backpack can carry is W.

输出：解向量 $(x_1, x_2, \cdots, x_n)$。

Output. Solution vector $(x_1, x_2, \cdots, x_n)$.

目标函数（Goal function）：$\max \sum\limits_{i=i}^{n} v_i x_i$。

约束条件（Constraint condition）：$\begin{cases} \sum\limits_{i=1}^{n} w_i x_i \leqslant W \\ x_i \in [0, 1], i = 1, 2, \cdots, n \end{cases}$。

根据贪心策略，首先计算物品的单位重量的价值；然后按照单位重量的价值由小到大排序，只要没有达到背包承载的重量，就装入，直到物品 i 装不下时或物品已全部装入背包。如果物品全部装入背包，则算法结束；否则，从物品 i 分割出部分装入背包，把背包装满，算法结束。

According to the greedy strategy, first calculate the value of the item per unit weight; Then according to the value of the unit weight from small to large order, if it does not reach the weight of the backpack, it is loaded until the item i cannot fit or the item has been loaded into the backpack. If all items are loaded into the backpack, the algorithm ends; Otherwise, partition the part from item i into the backpack, fill the backpack, the algorithm ends.

2. 算法伪代码（Algorithmic Pseudocode）

算法 2.6　贪心算法解决背包问题（Algorithm 2.6　Greedy algorithm solves the knapsack problem）

输入：物品集合、物品重量、物品价值、背包承载的重量
Input：Item collection, item weight, item value, backpack weight
输出：物品装载情况和装入的价值
Output：Item load and load value

```
1.   i←1
2.   while i<=n do
3.      a[i]←v[i]/w[i]
4.      i←i+1
5.   end while
6.   sort a
7.   w←0
8.   p←0
9.   while i<=n and w+w[i]<=W do
10.     x[i]←1
11.     w←w+w[i]
12.     p←p+v[i]
13.     i←i+1
14.  end while
15.  if i<=n then
16.     x[i]←(W-w)/w[i]
17.     p←p+x[i]*v[i]
18.  end if
19.  return x,p
```

2.6.3　算法实现（Algorithm Implementation）

有5件物品,背包容量为10。物品编号、物品重量和物品价值如表2.19所示。计算单位质量物品价格并以此排序的结果如表2.20所示。

There are 5 items and the backpack capacity is 10. Item number, item weight and item value are shown in Table 2.19. The results of calculating the price per unit mass of items and ranking them are shown in Table 2.20.

表 2.19　背包问题初始数据
Table 2.19　Initial data of backpack problem

物品编号(Item number)	1	2	3	4	5
物品重量(Item weight)	2	2	6	5	4
物品价值(Item value)	6	3	6	4	6

表 2.20　计算物品的单位重量价格并排序
Table 2.20　Calculate the price per unit weight of the item and sort it

Number	1	2	5	3	4
Weight	2	2	4	6	5
Value	6	3	6	6	4
Value/Weight	3	1.5	1.5	1	0.8

选用三元组存储单个物品信息,并用三元组集合表示物品集,用 *list* 存储物品集和选中的物品。

Select triples to store individual item information, use triples to represent item sets, and use *list* to store item sets and selected items.

首先定义函数 *knapsack*(*capacity* = 0, *goods_set* = [ ]),参数为背包的承载重量和物品集,先执行排序,然后依次装入,返回排序结果。定义函数入口 *main*( )函数,初始化物品集并调用核心算法函数,得到装入的物品的总价值,最后打印结果。程序输出结果如图2.16所示。

First define the function knapsack (capacity = 0, goods_set = [ ]), the parameters of the backpack carrying weight and the set of items, first perform the sorting, then load in sequence, return the sorting result. Define the function entry main ( ) function, initialize the item set and call the core algorithm function, get the loaded item sum total value, and finally print the result. The program output is shown in Figure 2.16.

```
def knapsack(capacity = 0, goods_set = [ ]):
    #按单位价值量排序
    goods_set.sort(key = lambda goods:goods[1]/goods[2], reverse = True)
    result = [ ]
    sum_v = 0
```

续

```
    for goods in goods_set:
        if capacity < goods[2]:
            result.append([goods[0],capacity * goods[1]/goods[2],capacity])
            sum_v +=capacity * goods[1]/goods[2]
            break
        result.append(goods)
        sum_v +=goods[1]
        capacity -=goods[2]
    return result,sum_v
if _ _name_ _ =="_ _main_ _":
    some_goods =[(0, 4, 2), (1, 6, 8), (2, 3,5), (3, 8, 2), (4, 2, 1)]
    res,sum_v =knapsack(6, some_goods)
    for goods in res:
        print('物品编号 Number:' + str(goods[0]) +', 放入的重量 Weight:' + str(goods[2]) +', 放入的价值
Value:' + str(goods[1]))
    print("总价值 Total Value: " + str(sum_v))
```

```
物品编号Number:3 ,放入的重量Weight:2,放入的价值Value:8
物品编号Number:0 ,放入的重量Weight:2,放入的价值Value:4
物品编号Number:4 ,放入的重量Weight:1,放入的价值Value:2
物品编号Number:1 ,放入的重量Weight:1,放入的价值Value:0.75
总价值 Total Value: 14.75
```

图 2.16　贪心算法解决背包问题的运行结果

Figure 2.16　Result of greedy algorithm to solve knapsack problem

习题 2(Exercises Two)

1. 举例说明一种不能使用贪心策略的情况。

Give an example of a situation in which the greedy strategy cannot be used.

2. 将最优装载问题的贪心算法推广到两艘船的情形,贪心算法仍然能产生最优解吗? 若能,给出证明;若不能,请给出反例。

If the greedy algorithm of the optimal loading problem is extended to the case of two ships, can the greedy algorithm still produce the optimal solution? If so, prove it. If not, give a counterexample.

3. 已知 {A,B,C,D,E,F,G} 7 个字母各自对应的出现频数为 {22,15,17,6,5,8,12},请手动绘制与程序运行结果相同的哈夫曼编码树。

If the occurrence frequency of the seven letters {A,B,C,D,E,F,G} is known to be {22,15,17,6,5,8,12}, please manually draw the Huffman code tree as the program running result.

4. 几个人一起出去吃饭是常有的事,但在结账时常常会出现一些争执。现在有 n 个人出去吃饭,他们总共消费了 S 元。其中,第 i 个人带了 a_i 元。幸运的是,所有人带的钱的总数是足够付账的,但现在问题来了: 每个人分别要出多少钱呢?

请考虑贪心选择策略,并尝试设计算法解决问题。

It is common for several people to eat out together. There are often some disputes about payment when finished eating. Now n people are going out to dinner, and they're spending S dollars, and among them i carry a_i dollars. Fortunately, the total amount of money each person brought was enough to pay the bill, but now the question arose: how much would each person contribute? Please consider the greedy strategy, and try to design an algorithm to solve the problem.

5. 在一条数轴上有 N 家商店,它们的坐标分别为 $A_1 \sim A_N$。现在需要在数轴上建立一个仓库,每天清晨,从仓库到每家商店都要运送一车商品。为了提高效率,要求算出把仓库建在何处,可以使得仓库到每家商店的距离之和最小。

请设计相关的贪心算法,并尝试编程实现。

There are N stores on a line with coordinates A_1 to A_N. Now it is necessary to set up a warehouse on the line, and every morning, a load of goods needs to be delivered from the warehouse to each store. In order to improve efficiency, the location of the warehouse can minimize the sum of distances from the warehouse to each store. Design greedy algorithms and try programming them.

第3章

分治算法

Chapter 3   Divide- and- Conquer Algorithm

分治算法即分而治之的方法。分治法(Divide-and-Conquer Method)是最著名的算法设计技术,作为解决问题的一般性策略,分治法在政治和军事领域也是克敌制胜的法宝。

The divide-and-conquer algorithm is a method that follows the principle of divide-and-conquer. It is the most famous algorithm design technology, as a general strategy for solving problems, it is also a magic weapon for defeating enemies in political and military fields.

用计算机求解问题所需的时间一般都和问题规模有关,问题规模越小,求解问题所需的计算时间也越少,从而也较容易处理。分治法将一个难以直接解决的大问题划分成一些规模较小的子问题,分别求解各个子问题,再合并子问题的解得到原问题的解。

The time it takes to solve a problem with a computer is generally related to the size of the problem. The smaller the size of the problem, the less time it takes to solve the problem, and the easier it is to deal with. The divide-and-conquer method divides a large problem that is difficult to solve directly into some smaller subproblems, solves each subproblem separately, and then merges the solutions of the subproblems to obtain the solution of the original problem.

3.1　概述(Overview)

3.1.1　基本思想(Basic Idea)

分治法的设计思想是将一个难以直接解决的大问题,划分成一些规模较小的子问题,以便各个击破,分而治之。更一般地说,将要求解的原问题划分成 k 个较小规模的子问题,对这 k 个子问题分别求解。如果子问题的规模仍然不够小,则再将每个子问题划分为 k 个规模更小的子问题,如此分解下去,直到问题规模足够小,很容易求出其解为止,再将子问题的解合并为一个更大规模的问题的解,自底向上逐步求出原问题的解。

The design idea of divide-and-conquer method is to divide a large problem that is difficult to solve directly into some smaller subproblems, so that they can be divided and conquered one by one. More generally, the original problem to be solved is divided into k smaller scale subproblems, and these k subproblems are solved separately. If the subproblem is still too big to solve, and then divide each subproblem into k smaller subproblems, and so on, until the size of the problem is small enough to easily solve, and then combine the solution of the subproblem into the big problem, and the solution of the original problem is gradually solved from the bottom up.

3.1.2 算法的本质(Nature of Algorithm)

分治算法是将一个大问题分解成多个相同或相似的子问题,然后分别解决这些子问题,最后将它们的解合并起来,得到原始问题的解。这种方法通常通过递归的方式实现。

The divide-and-conquer algorithm is to decompose a large problem into multiple identical or similar subproblems, then solve these subproblems separately, and finally merge their solutions to obtain the solution of the original problem. This approach is usually implemented in a recursive way.

分治算法通常用于解决那些可以被分解为相互独立子问题的问题。经典的分治算法包括归并排序(Merge Sort)、快速排序(Quick Sort)、二叉树的遍历等。

Divide-and-conquer algorithm are often used to solve problems that can be broken down into mutually independent subproblems. Classic divide-and-conquer algorithm include Merge Sort, Quick Sort, binary tree traversal, and so on.

分治算法的优点在于它可以有效地利用多核处理器和并行计算,因为子问题是独立的,可以并行地求解。然而,它的缺点包括递归调用可能导致较大的开销和额外的存储空间需求。

The advantage of the divide-and-conquer algorithm is that it can make efficient use of multicore processors and parallel computing, since the subproblems are independent and can be solved in parallel. However, its disadvantages include that recursive calls may lead to large overhead and additional storage space requirements.

3.1.3 算法的解题步骤(Algorithm Solution Steps)

分治法的求解过程由以下三个阶段组成。

The solution process of the divide-and-conquer method consists of the following three step.

(1) 划分。需要把规模为 n 的原问题划分为 k 个规模较小的子问题,并尽量使这 k 个子问题的规模大致相等。

Divide. It is necessary to divide the original problem of size n into k smaller subproblems, and try to make the size of these k subproblems approximately equal.

(2) 求解子问题。各子问题的解法与原问题的解法通常是相同的,可以用递归的方法求解各个子问题,有时递归处理也可以用循环来实现。

Solve subproblems. The solution of each subproblem is usually the same as the solution of the original problem, and each subproblem can be solved by recursive methods, and sometimes recursive processing can also be implemented by loops.

（3）合并。把各个子问题的解合并起来,合并的代价因情况不同有很大差异,分治算法的有效性很大程度上依赖于合并的实现。

Merge. The cost of merging the solutions of the subproblems varies greatly from case to case, and the effectiveness of the divide-and-conquer algorithm depends heavily on the implementation of merging.

3.1.4　分治与递归（Divide-and-Conquer and Recursion）

由分治法产生的子问题往往是原问题的较小模式,反复应用分治手段,可以使子问题与原问题解法相同而其规模却不断缩小,最终使子问题缩小到很容易直接求解,这自然导致递归过程的产生。分治与递归就像一对孪生兄弟,经常同时应用在算法设计之中,并由此产生许多高效的算法。

The subproblems generated by the divide-and-conquer method are often smaller patterns of the original problem. Repeated application of the divide-and-conquer method can make the subproblems have the same solution as the original problem, but the scale of the subproblems is constantly reduced, and finally the subproblems are reduced to be easily solved directly, which naturally leads to the generation of a recursive process. Divide-and-conquer and recursion are often used together as twin brothers in algorithm design, resulting in many efficient algorithms.

递归（Recursion）是一种描述问题和解决问题的基本方法,递归程序直接调用自己或通过一系列调用语句间接调用自己,将待求解问题转换为解法相同的子问题,最终实现问题求解。

Recursion is a basic method to describe and solve problems. A recursive program directly calls itself or indirectly calls itself through a series of call statements, which transforms the problem to be solved into subproblems with the same solution method, and finally solves the problem.

因此,递归必须具备以下两个基本要素才能在有限次数的计算后得出结果。

Therefore, recursion must have the following two basic elements to arrive at a result after a finite number of evaluations.

（1）递归出口。确定递归到何时终止,即递归的结束条件。

Recursion exit. Determine when the recursion terminates, which is the end condition for the recursion.

（2）递归体。确定递归的方式,即原问题是如何分解为子问题的。

Recursion body. Determine the way of recursion, that is, how the original problem is broken into subproblems.

很多问题本身就是以递归形式给出的,可以用递归方法求解。例如,阶乘的递归定义。有些问题虽然定义本身不具有明显的递归特征,但其求解方法是递归的,如汉诺塔求解就是一个典型的代表。

Many problems are themselves formulated in recursive form and can be solved recursively. For example, the recursive definition of factorial. Although the definition of some problems does not have obvious recursive characteristics, their solving methods are recursive, such as the Tower of Hanoi solution is a typical representative.

3.2　排序问题算法(Sorting Problem Algorithm)

3.2.1　合并排序算法(Merge Sort Algorithm)

1. 算法思想(Algorithm Idea)

合并排序是采用分治策略实现对 n 个元素进行排序的算法,是分治算法的一个典型应用和完美体现。它是一种平衡、简单的二分分治策略,其计算过程分为三大步。

Merge sort is an algorithm that uses divide-and-conquer strategy to sort n elements. It is a typical application and perfect embodiment of algorithm. It is a balanced and simple bisection and conquer strategy, and its computation is carried out in three big steps.

(1) 分解。将待排序元素分成数目大致相同的两个子序列。

Decomposition. The elements to be sorted are divided into two sub-sequences with approximately the same number.

(2) 求解子问题。用合并排序法分别对两个子序列递归地进行排序。

Solve subproblem. The two subsequences are sorted recursively respectively by the merge sort method.

(3) 合并。将排好序的有序子序列进行合并,得到符合要求的有序序列。

Merge. The sorted subsequences are merged to obtain the ordered subsequences that meet the requirements.

合并问题给定的是合并两个已经排好序的序列。为此,可以把待排序元素分解成两个规模大致相等的子序列,如果不易解决,再将得到的子序列继续分解,直到子序列中包含的元素个数为1。众所周知,单个元素的序列本身是有序的,此时可进行合并,这就是分治策略的巧妙运用。

The merge problem is given by combining two already sorted sequences. To solve this problem, the element to be sorted can be decomposed into two approximately equal size subsequences. If it is difficult to solve the problem, the obtained subsequences are decomposed until the number of elements in the subsequences is 1. It's known that the sequence of individual elements is inherently ordered, so it can be merged. This is a clever use of the divide-and-conquer strategy.

2. 算法设计与描述(Algorithm Design and Description)

1) 合并过程(Merging Process)

合并排序的关键步骤在于如何合并两个已排好序的有序子序列。为了进行合并,引入一个辅助过程 $Merge(A, low, middle, high)$,该过程将排好序的两个子序列 $A[low:middle]$ 和 $A[middle+1:high]$ 进行合并。其中,low、$high$ 表示待排序范围在数组中的下界和上界,$middle$ 表示两个序列的分开位置,满足 $low \leqslant middle < high$。由于在合并过程中

可能会破坏原来的有序序列；因此，合并最好不要就地进行，本算法采用了辅助数组 $B[low:high]$ 来存放合并后的有序序列。

The key step of merge sort is how to merge two sorted subsequences. To merge it, an auxiliary procedure $Merge(A, low, middle, high)$ is introduced, which merges two sorted subsequences $A[low:middle]$ and $A[middle+1:high]$. Here, low and $high$ represent the lower and upper bounds of the range to be sorted in the array, and middle represents the separation position of the two sequences, satisfying $low \leqslant middle < high$. Since the original ordered sequence may be destroyed in the merging process, it is better not to merge in the original position. This algorithm uses the auxiliary array $B[low:high]$ to store the sorted sequence after merging.

合并方法如下。设置三个工作指针 i, j, k，其中，i 和 j 指示两个待排序序列中当前需比较的元素，k 指向辅助数组 B 中待放置元素的位置。比较 $A[i]$ 和 $A[j]$ 的大小关系，如果 $A[i]$ 小于或等于 $A[j]$，则 $B[k]=A[i]$，同时将指针 i 和 k 分别推进一步；反之，$B[k]=A[j]$，同时将指针 j 和 k 分别推进一步。如此反复，直到其中一个序列为空。最后，将非空序列中的剩余元素按原次序全部放到辅助数组 B 的尾部。

The merging method is as follows. Set three working pointers i, j, and k, where i and j indicate the current element to be compared in the two sequences to be sorted, and k refers to the location of the element to be placed in the auxiliary array B. Compare the size between $A[i]$ and $A[j]$, if $A[i]$ is less than or equal to $A[j]$, then $B[k] = A[i]$, and advance the pointers i and k one step respectively; Conversely, $B[k] = A[j]$, while advancing the pointers j and k one step respectively. This will repeat until one of the sequences is empty. Finally, the remaining elements of the non-empty sequence are placed in their original order at the end of the auxiliary array B.

最坏的情况是两个子序列 $A[low:middle]$ 和 $A[middle+1:high]$ 的数据是一个交替的序列，如 $(1,3,5,7)$ 和 $(2,4,6,8)$，则需要比较的次数为 $n-1=7$ 次。

In the worst case, if the data of the two subsequence $A[low:middle]$ and $A[middle+1:high]$ is an alternating sequence, such as $(1,3,5,7)$ and $(2,4,6,8)$, the number of comparisons required is $n-1=7$.

2）递归形式的合并排序算法（Recursive Form of Merge Sort Algorithm）

递归形式的合并排序算法就是把序列分为两个子序列，然后对子序列进行递归排序，最后把两个已排好序的子序列合并成一个有序的序列。

The recursive merge sort algorithm divides a sequence into two subsequences, recursively sorts the subsequences, and finally merges the two sorted subsequences into a sorted sequence.

可以将 $Merge$ 过程作为合并排序算法的一个子程序使用。设置 $MergeSort(A, low, high)$ 对子序列 $A[low:high]$ 进行排序，如果 $low \geqslant high$，则该子序列中至多只有一个元素；否则，依据分治算法的思想对问题进行分解，即计算出一个下标 $middle$，将 $A[low:high]$ 分解成 $A[low:middle]$ 和 $A[middle+1:high]$，分解原则是使二者的大小大致相等。

The $Merge$ procedure can be used as a subroutine of the merge sort algorithm.

Setting *MergeSort*(A, *low*, *high*) sorts the subsequence $A[low:high]$, if $low \geqslant high$, then there is at most one element in the subsequence. Otherwise, the problem is decomposed according to the idea of divide-and-conquer algorithm, that is, a subscript *middle* is calculated, and $A[low:high]$ is decomposed into $A[low:middle]$ and $A[middle+1:high]$, and the decomposition principle is to make the size of the two approximately equal.

3. 算法实现(Algorithm Implementation)

1) 实例构造(Instance Construction)

假设待排序序列 $A = <8,3,2,9,7,1,5,4>$,采用合并算法对序列 A 进行排序,具体排序过程如下。

Suppose that the sequence to be sorted $A = <8,3,2,9,7,1,5,4>$, and the merging algorithm is used to sort the sequence A, the specific sorting process is as follows.

首先将待排序列 $<8,3,2,9,7,1,5,4>$ 分解成两个规模大致相等的子序列 $<8,3,2,9>$ 和 $<7,1,5,4>$,由于子序列不够小,所以继续将子序列分解,直到分解成多个单个元素,即 $<8>$,$<3>$,$<2>$,…,以上为分治算法中的"分"。将待排序列分解完成后进行合并,现在待排序列为多个单个元素,首先将它们两两为一组,分为 4 组,每组中两个元素进行比较大小,元素小的在左边,元素大的在右边,由此得到 4 组元素。

Firstly, the sequence $<8,3,2,9,7,1,5,4>$ is decomposed into two approximately equal size sub-sequences $<8,3,2,9>$ and $<7,1,5,4>$. Since the sub-sequences are not small enough, the sub-sequences are decomposed until they are single elements, that is, $<8>$, $<3>$, $<2>$, …, the above is the "divide" in the divide-and-conquer algorithm. The sequence to be sorted is decomposed and merged. Now, the sequence to be sorted is multiple single elements, first, they are grouped in pairs and divided into four groups, and two elements in each group are compared in size, with the smaller element on the left and the larger element on the right. This results in four groups of elements.

每组元素内部都是排好顺序的,然后将这 4 组元素两两采用合并排序算法进行排序得到两组元素,最后将这两组元素再次采用合并排序算法进行排序,得到最终排序结果,以上为分治算法中的"治"。如图 3.1 所示,具体合并排序算法如图 3.2~图 3.6 所示。

Each group of elements is sorted internally, and then the four groups of elements are sorted by the pairs merge sort algorithm to get two groups of elements. Finally, the two groups of elements are sorted by the merge sort algorithm again to get the final sorting result. The above is the "conquer" in the divide-and-conquer algorithm. This is shown in Figure 3.1, and the specific merge sort algorithm is shown in Figure 3.2 to Figure 3.6.

具体合并排序算法步骤如图 3.2 所示。首先将需要合并的两组元素看作一个原始数组,分配 *left* 指针指向第一个元素,*right* 指针指向最后一个元素,*mid* 指针指向中间元素,再设置一个辅助数组用来存放排好序的元素。然后初始化数组,设置 i 指针代替 *left* 指针,j 指针代替 *mid*+1 指针,k 指针指向辅助数组第一个空位,初始值为 0。

The specific steps of merge sort algorithm are shown in Figure 3.2. Firstly, consider

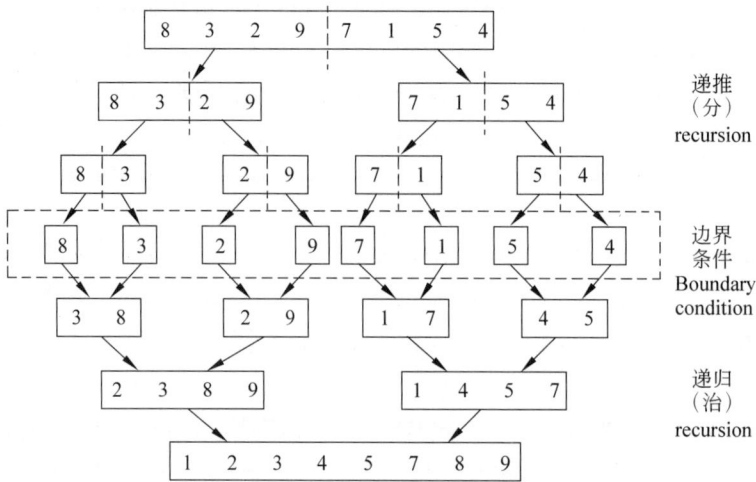

图 3.1　分治算法的"分"与"治"

Figure 3.1　"Divide" and "conquer" of divide-and-conquer algorithm

the two sets of elements to be merged as an original array, assign a *left* pointer to the first element, a *right* pointer to the last element, and a *mid* pointer to the middle element, and then set up an auxiliary array to hold the sorted elements. The array is then initialized by setting the i pointer instead of the *left* pointer, the j pointer instead of the $mid+1$ pointer, and the k pointer to the first empty in the auxiliary array, which is initially set to 0.

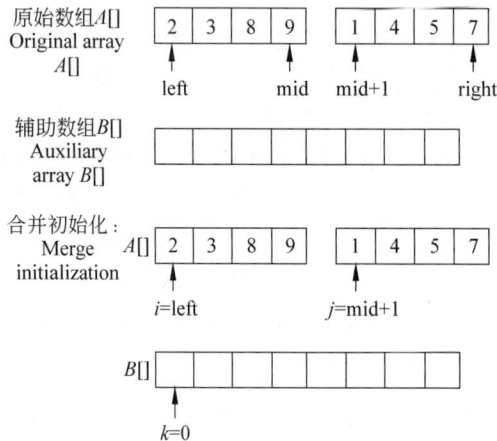

图 3.2　设置数组并初始化

Figure 3.2　Setting up the array and initializing

合并过程如图 3.3 所示。首先比较 i 指针和 j 指针元素,较小的元素放入辅助数组的空位,即 k 指针指向的位置,元素放入后,k 指针后移一位,指向辅助数组第二个位置,哪个数组元素被放进辅助数组,原本指向它的指针后移一位,指向下一个元素,另一个指针不动。

The merging process is shown in Figure 3.3. The smaller element is put into the empty of the auxiliary array, that is, the position pointed by the k pointer. After the element is placed, the k pointer moves back one to point to the second position of the auxiliary array. Which array element is put into the auxiliary array, the pointer originally pointing to it moves back one bit to point to the next element, and the other pointer does not move.

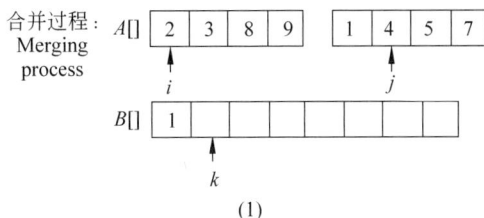

图 3.3　合并过程(1)

Figure 3.3　Merging process（1）

比如上例中 i 指针指向 2，j 指针指向 1，比较 1 和 2 的大小，1 较小，所以将 1 放入 B 数组，原本指向 1 的 j 指针后移指向 4。

Just as the example, the i pointer points to 2, and the j pointer points to 1. Comparing 1 and 2, the 1 is smaller, so the 1 is put into the B array, and the j pointer originally points to 1 is moved to 4.

指针移动后，i 指针指向 2，j 指针指向 4，比较 2 与 4 的大小，2 较小，所以将 2 放入 B 数组，即 k 指针指向的位置，然后将 k 指针后移一位，由于放入的元素是 2，所以 i 指针后移一位，j 指针不动，现在 i 指针指向 3，j 指针指向 4，k 指针指向 B 数组第三个位置，此位置为空，如图 3.4 所示。

After the pointer is moved, the i pointer points to 2, and the j pointer points to 4. Comparing 2 and 4, the 2 is smaller, so the 2 is put into the B array, which is the position pointed to by the k pointer, and then the k pointer is moved back one bit. Since the element put in is 2, the i pointer is moved back one bit, and the j pointer does not change, now the i pointer points to 3, the j pointer points to 4, and the k pointer points to the third position of the B array. This position is empty, as shown in Figure 3.4.

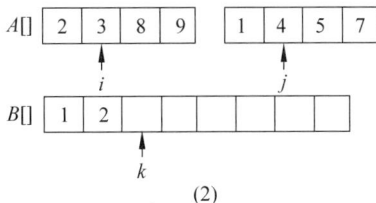

图 3.4　合并过程(2)

Figure 3.4　Merging process（2）

以此类推，继续将 i 指针和 j 指针元素进行比较，直到所有元素都比较结束，详细过程如图 3.5 所示。

And so on, continue to compare i pointer and j pointer elements until all elements are compared, the detailed process is shown in Figure 3.5.

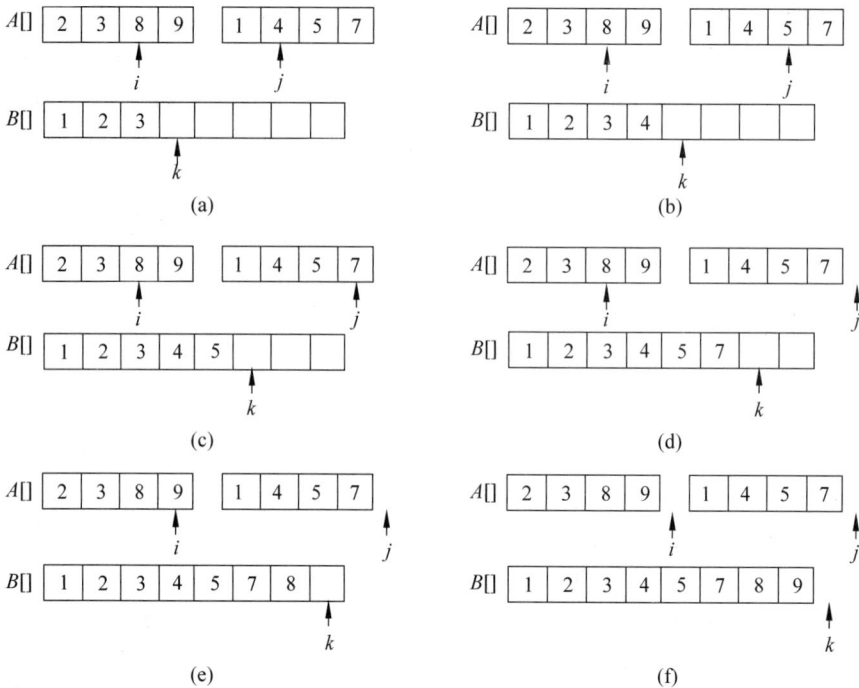

图 3.5　合并过程(3)

Figure 3.5　Merging process（3）

所有元素都比较完后,i、j 和 k 指针均指向空,B 数组即为待排序列合并排序完的结果,将 B 数组结果复制到原数组,合并排序完成,如图 3.6 所示。

After all elements are compared, pointers i, j and k point to empty, and the B array is the result of the merge sort of the sequence to be sorted. Copy the result of the B array to the original array, and the merge sort is completed, as shown in Figure 3.6.

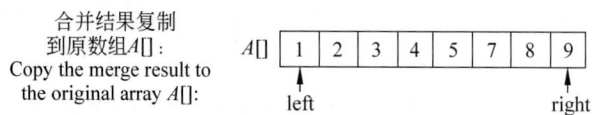

图 3.6　合并排序结果

Figure 3.6　Result of the merging sorted

2）实例实现(Instance Implementation)

设序列 $A=<8,3,2,9,7,1,5,4>$,采用合并算法对序列 A 进行排序,算法实现代码如下所示,实现结果如图 3.7 所示。

Let the sequence $A=<8,3,2,9,7,1,5,4>$, and use the merging algorithm to sort the sequence A. The algorithm implementation code is shown below, and the result is shown in Figure 3.7.

```
def merge( arr, left, mid, right) :
    n1 = mid - left + 1
    n2 = right - mid
```

续

```
#创建临时数组(Creating a temporary array)
L = arr[left:mid + 1]
R = arr[mid + 1:right + 1]
#归并临时数组到 arr[l..r](Merge temporary arrays into arr[l..r])
i = 0            #初始化第一个子数组的索引(Initialize the index of the first subarray)
j = 0            #初始化第二个子数组的索引(Initialize the index of the second subarray)
k = left         #初始归并子数组的索引(The index of the initial merged subarray)
while i < n1 and j < n2:
    if L[i] <= R[j]:
        arr[k] = L[i]
        i += 1
    else:
        arr[k] = R[j]
        j += 1
    k += 1
#复制 L[] 的保留元素(Copy the reserved elements of L[])
while i < n1:
    arr[k] = L[i]
    i += 1
    k += 1
#复制 R[] 的保留元素(Copy the reserved elements of R[])
while j < n2:
    arr[k] = R[j]
    j += 1
    k += 1
def mergeSort(arr, left, right):
    if left < right:
        mid = (left + right) // 2
        mergeSort(arr, left, mid)
        mergeSort(arr, mid + 1, right)
        merge(arr, left, mid, right)
if __name__ == "__main__":
    arr = [8,3,2,9,7,1,5,4]
    n = len(arr)
    mergeSort(arr, 0, n - 1)
    print("合并排序结果(Merge sort results):", arr)
```

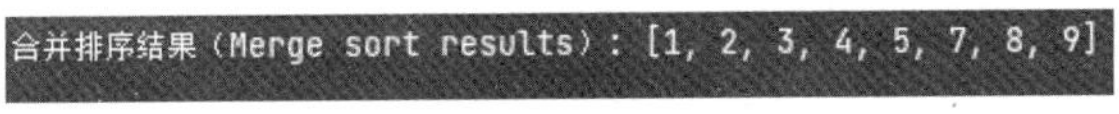

合并排序结果(Merge sort results): [1, 2, 3, 4, 5, 7, 8, 9]

图 3.7　程序运行结果

Figure 3.7　Result of the program of the algorithm

3.2.2　快速排序算法(Quick Sort Algorithm)

快速排序是由 *C.A.R. Hoare* 提出的一种划分交换排序,其基本思想是通过一趟扫描将待排序的元素分割成独立的三个序列:第一个序列中所有元素均不大于基准元素,第二个序列是基准元素,第三个序列中所有元素均大于基准元素。由于第二个序列已经处于正确位置,因此需要再按此方法对第一个序列和第三个序列分别进行排序,整个排序过程可以递归进行,最终可使整个序列变成有序序列。

Quick sort is a partition-exchange sort proposed by *C.A.R. Hoare*. The basic idea of Quick sort is to divide the elements to be sorted into three independent sequences by a single scan: the first sequence in which all elements are no greater than the base element, the second sequence in which all elements are the base element, and the third sequence in which all elements are greater than the base element. Since the second sequence is already in the correct position, the first and the third sequence need to be sorted in this way, and the whole sorting process can be carried out recursively, and finally the whole sequence can be turned into an ordered sequence.

1. 算法思想(Algorithm Idea)

快速排序的基本思想描述如下：假设当前待排序序列为 $R[low:high]$，其中，$low \leqslant high$，如果序列的规模足够小，则直接进行排序，否则分为三步处理。

The basic idea of quick sort is described as follows: suppose that the current sequence to be sorted is $R[low:high]$, where $low \leqslant high$. If the size of the sequence is small enough, it is sorted directly; otherwise, it is processed in three steps.

1) 分解(Decompose)

在 $R[low:high]$ 中选定一个元素作为基准元素($pivot$)，该基准元素的位置($pivotpos$) 在划分的过程中确定。以此基准元素为标准将待排序序列划分为两个子序列 $R[low: pivotpos-1]$ 和 $R[pivotpos+1:high]$，并使序列 $R[low:pivotpos-1]$ 中所有元素的值均小于或等于 $R[pivotpos]$，序列 $R[pivotpos+1:high]$ 中所有元素均大于或等于 $R[pivotpos]$。此时基准元素已位于正确的位置上，它无须参加后面的排序。划分序列的关键是要计算出所选定的基准元素所在的位置 $pivotpos$，其中，$low \leqslant pivotpos \leqslant high$。

An element is selected in $R[low:high]$ as the *pivot* element, and the position of the *pivot* element is determined during partitioning. The sequence to be sorted is divided into two subsequences $R[low:pivotpos-1]$ and $R[pivotpos+1:high]$, the values of all elements in the sequence $R[low:pivotpos-1]$ are less than or equal to $R[pivotpos]$, and all elements in the sequence $R[pivotpos+1:high]$ are greater than or equal to $R[pivotpos]$. Now the base element is in the correct position, it does not need to participate in subsequent sorting. The key to partitioning the sequence is to calculate the position of the selected base element $pivotpos$, where $low \leqslant pivotpos \leqslant high$.

2) 求解子问题(Solving Subproblem)

对两个子序列 $R[low:pivotpos-1]$ 和 $R[pivotpos+1:high]$，分别通过递归调用快速排序算法来进行排序。

Two subsequences $R[low:pivotpos-1]$ and $R[pivotpos+1:high]$ are sorted by recursively calling the quick sort algorithm, respectively.

3) 合并(Combine)

由于对 $R[low:pivotpos-1]$ 和 $R[pivotpos+1:high]$ 的排序是就地进行的，所以在 $R[low:pivotpos-1]$ 和 $R[pivotpos+1:high]$ 都已排好序后，合并步骤并不需要做什么，序列 $R[low:high]$ 就已排好序了。

Since $R[low:pivotpos-1]$ and $R[pivotpos+1:high]$ are sorted locally, there is nothing to do in the merge step after $R[low:pivotpos-1]$ and $R[pivotpos+1:high]$ have been sorted, $R[low:high]$ is already in order.

基准元素的选择方法有很多种,常用的有以下几种方法。

There are many ways to select the base element, and the following methods are commonly used.

(1) 第一个元素。即以待排序序列的首元素作为基准元素。

The first element. The first element of the sequence to be sorted as the base element.

(2) 最后一个元素。即以待排序的尾元素作为基准元素。

The last element. The tail element of the sequence to be sorted as the base element.

(3) 位于中间位置的元素。即以待排序序列的中间位置的元素作为基准元素。

The middle position element. The element in the middle position of the sequence to be sorted as the base element.

(4) "三者取中的规则"。即在待排序序列中,将该序列的第一个元素、最后一个元素和中间位置的元素进行比较,取三者中的中间值作为基准元素。

" Middle of the three". That is in the sequence to be sorted, the first element, the last element and the middle element of the sequence are compared, and the middle value of the three is taken as the base element.

(5) 选择 low 和 $high$ 之间的一个随机数 $k(low \leqslant k \leqslant high)$,用 $R[k]$ 作为基准元素。即采用随机函数产生一个位于 low 和 $high$ 之间的随机数 $k(low \leqslant k \leqslant high)$,用 $R[k]$ 作为基准,这相当于规定 $R[low:high]$ 中的元素是随机分布的。

Choose a random number k between low and $high$ $(low \leqslant k \leqslant high)$, and use $R[k]$ as the base element. That is, a random function is used to generate a random number k between low and $high$ $(low \leqslant k \leqslant high)$, and $R[k]$ is used as the base, which is equivalent to specifying that the elements in $R[low:high]$ are randomly distributed.

选择后三种选取方法时,需要先将选取的基准元素空位与初始 low 指针指向的元素进行位置互换,然后进行排序。

To select the latter three selection methods, you need to swap the position of the selected base element with the element pointed by the initial low pointer and then sort it.

2. 算法设计与描述(Algorithm Design and Description)

将待排序的 n 个元素放到列表 $list$ 中,用 $left$ 和 $right$ 记录子问题在 $list$ 中的位置及规模,基准元素记为 $pivot$。首先定义一个 $partition()$ 函数完成分解任务,然后采用递归方法完成子问题的排序。

Put the n items to be sorted into a $list$ table, and use $left$ and $right$ to record the position and size of the subproblem in the $list$, the base element is referred to as $pivot$. Firstly define a $partition()$ function to decompose the subproblems, and then use a recursive method to sort the subproblems.

在划分过程中,选取待排序序列的第一个元素作为基准元素,然后从左向右、从右向左两个方向轮流找出位置不正确的元素,将其交换位置。该过程一直持续到所有元素都比较结束,最后将基准元素放到正确的位置。

In the process of partitioning, the first element of the sequence to be sorted is selected as the base element, and then the incorrect elements are found from left to right and from right to left, and their positions are swapped. This process continues until all elements have been compared and the reference element is placed in the correct position.

3. 算法实现（Algorithm Implementation）

1）实例构造（Instance Construction）

假设待排序序列 $A = <54,74,97,24,87,78,68,35,16>$，采用快速排序算法对序列 A 进行排序，具体排序过程如图 3.8～图 3.18 所示。

Suppose to collate a sequence $A = <54,74,97,24,87,78,68,35,16>$, using the quick sort algorithm to sort sequence A, and a specific sorting process as shown in Figure 3.8 to Figure 3.18.

选取 54 作为基准元素，i 指针代表 low，j 指针代表 $high$。将基准元素移到外面的空位，将 i、j 指针中未指向空位的指针所指向的元素与基准元素比较大小，较小的元素放进 i 指针，较大的元素放进 j 指针。此时实例中 i 指针指向空，j 指针指向 16，将 16 与 54 比较，16 较小，将 16 放入 i 指针的空位，i 指针后移指向 74，j 指针原本指向 16，但因为 16 调走了，所以 j 指针目前指向空，如图 3.8 所示。

Select 54 as the base element, the i pointer represents low and the j pointer represents $high$. Move the base element to the vacancy outside, compare the size of the element pointed to by the i and j pointers that do not point to the vacancy with the base element, and put the smaller element into the i pointer and the big element into the j pointer. In this case, the i pointer points to empty, and the j pointer points to 16. Comparing 16 with 54, 16 is smaller, so 16 is put into the empty bit of the i pointer, and the i pointer moves back to 74. The j pointer originally pointed to 16, but because 16 was moved away, the j pointer now points to empty, as shown in Figure 3.8.

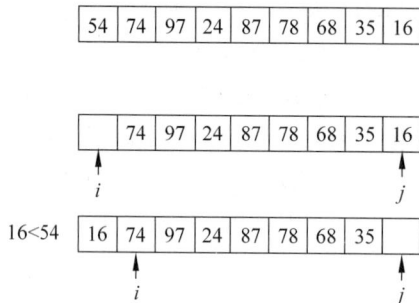

54	74	97	24	87	78	68	35	16

图 3.8　快速排序图(1)

Figure 3.8　Graph of quick sorting (1)

此时 i 指针指向 74，j 指针指向空，所以比较 74 与 54 的大小，74 较大，将 74 移进 j 指针指向的空位中，j 指针前移一位，i 指针指向空，此时 i 指针指向空，j 指针指向 35，如图 3.9 所示。

Now the i pointer points to 74, and the j pointer points to empty, so comparing the 74 and 54, 74 is larger, and 74 is moved into the empty pointed by the j pointer, the j pointer is moved forward one bit, and the i pointer points to empty, then the i pointer points to it, and the j pointer points to 35, as shown in Figure 3.9.

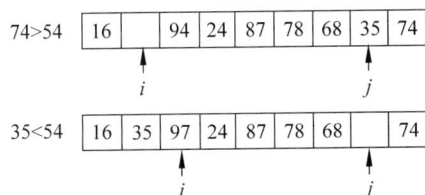

图 3.9 快速排序图（2）

Figure 3.9 Graph of quick sorting （2）

继续将指针指向的元素与基准元素做比较，调整各元素之间的顺序，直到 i j 指针指向同一个位置，如图 3.10 所示。

Continue to adjust the order of the elements by comparing the pointer element to the base elements until the i and j pointers point to the same position, as shown in Figure 3.10.

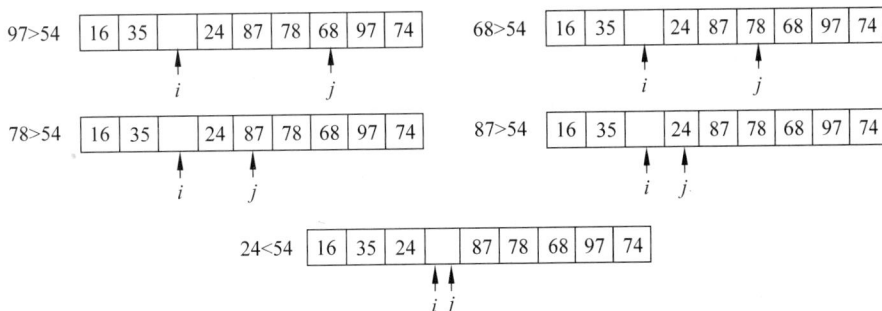

图 3.10 快速排序图（3）

Figure 3.10 Graph of quick sorting （3）

此时所有元素与基准元素做比较并排序完成，将基准元素放入 i j 指针所指向的空位，第一趟排序完成，得到一个初步排序结果，在基准元素 54 左边的元素都小于 54，在 54 右边的元素都大于 54，如图 3.11 所示。

Now all elements are compared with the base element and sorted, and the base element is put into the space pointed by the pointers i and j. The first sorting is completed, and a preliminary sorting result is obtained. The elements to the left of the base element 54 are all less than 54, and the elements to the right of the base element 54 are all more than 54, as shown in Figure 3.11.

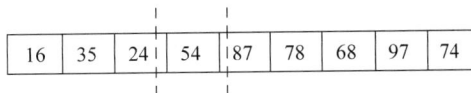

图 3.11 第一趟排序结果

Figure 3.11 Result of the first sorting

以基准元素为分界线，分别对两边继续进行第二趟排序。

Proceed with a second sort on both sides, using the base element as the dividing line.

先看左边子序列，如图 3.12 所示。选取第一个元素即 16 为基准元素，将基准元素移到外面空位，此时 i 指针指向空，j 指针指向 24,24 比 16 大，所以 24 应该放进 j 指针中，但由于 j 指针原本就指向 24，所以 24 位置不动，j 指针前移一位指向 35,35 比 16 大，所以 j 指针再次前移，与 i 指针一起指向空，将 16 放进空位。此时左边子序列第二趟排序完成，观察排序后的元素，发现以基准元素 16 为分界线，左边没有元素，因此对右边两个元素进行第三趟排序。

Now look at the left subsequence, as shown in Figure 3.12. The first element 16 is selected as the base element, and the base element is moved to the outside space. So the i pointer points to empty, the j pointer points to 24, and 24 is greater than 16, so 24 should be put into the j pointer, but since the j pointer originally points to 24, the position of 24 is not moved, and the j pointer is moved forward to 35, 35 is larger than 16, so, the j pointer is moved forward again, together with the i pointer to the empty, put 16 in the empty bit. And now, the second sorting of the left subsequence is completed. Observing the sorted elements, it is found that there is no element on the left side with the base element 16 as the dividing line, so the third sorting is performed on the right two elements.

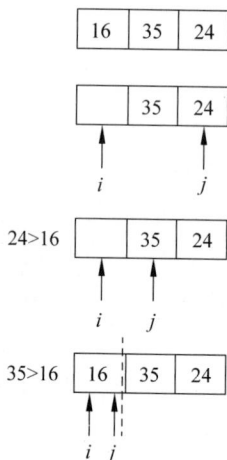

图 3.12　左边子序列第二趟排序结果
Figure 3.12　Result of the second sorting of the left subsequence

左边子序列第三趟排序如图 3.13 所示。选取 35 作为基准元素，此时 i 指针指向空，j 指针指向 24,24 小于 35，所以将 24 放进 i 指针指向的空位中，i 指针后移与 j 指针一起指向空位，将 35 放进空位，第三趟排序完成。

The third sorting of the left subsequence is shown in Figure 3.13. Pick 35 as the base element, so the i pointer points to empty and the j pointer points to 24, and 24 is less than 35, so put 24 into the empty pointed by the i pointer, the i pointer moves back together with the j pointer to the empty, and put 35 into the empty, and the third sorting is done.

左边子序列多次排序结果如图 3.14 所示。

The result of multiple sorting of the left subsequence is shown in Figure 3.14.

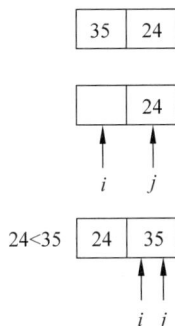

| 35 | 24 |

| | 24 |
　　i　j

24<35 | 24 | 35 |
　　　i j

| 16 | 24 | 35 |

图 3.13 左边子序列第三趟排序结果

Figure 3.13 Result of the third sorting
of the left subsequence

图 3.14 左边子序列多次排序结果

Figure 3.14 Result of the multiple sorting
of left subsequence

以此类推，右边子序列第二趟排序结果如图 3.15 所示。

Similarly, result of the second sorting of the right subsequences is shown in Figure 3.15.

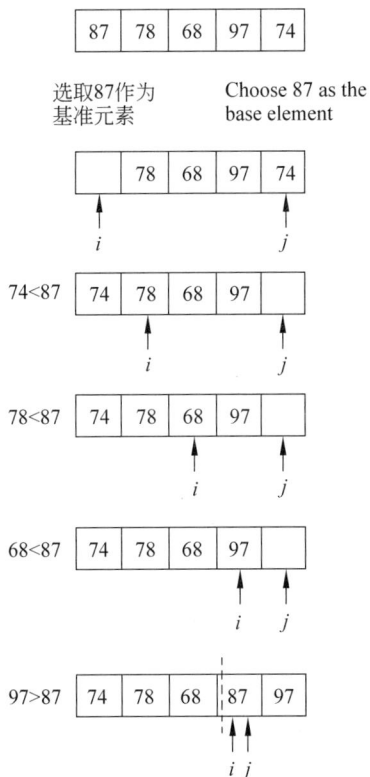

| 87 | 78 | 68 | 97 | 74 |

选取87作为
基准元素

Choose 87 as the
base element

| | 78 | 68 | 97 | 74 |
　i　　　　　　　j

74<87 | 74 | 78 | 68 | 97 | |
　　　　　i　　　　j

78<87 | 74 | 78 | 68 | 97 | |
　　　　　　　i　　　j

68<87 | 74 | 78 | 68 | 97 | |
　　　　　　　　　i　j

97>87 | 74 | 78 | 68 | 87 | 97 |
　　　　　　　　　i j

图 3.15 右边子序列第二趟排序结果

Figure 3.15 Result of the second sorting of the right subsequence

右边子序列第二趟排序完成，观察排序后的元素，发现以基准元素 87 为分界线，右边无须再排序，因此对左边三个元素进行第三趟排序，排序过程如图 3.16 所示。

After the second sorting of the right subsequence is completed, observing the sorted elements, it is found that the right element does not need to be sorted with the reference element 87 as the dividing line, so the third sorting of the three elements on the left is carried out. The sorting process is shown in Figure 3.16.

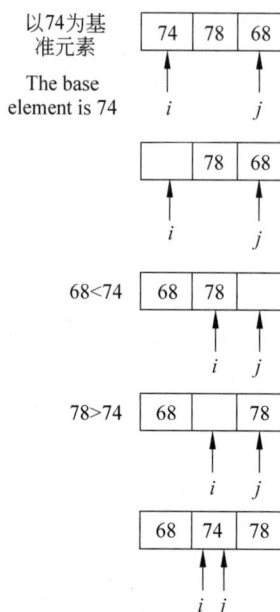

以74为基
准元素

The base
element is 74

| 74 | 78 | 68 |
| i | | j |

| | 78 | 68 |
| i | | j |

68<74
| 68 | 78 | |
| | i | j |

78>74
| 68 | | 78 |
| | i | j |

| 68 | 74 | 78 |
| i j | | |

图 3.16　右边子序列第三趟排序结果

Figure 3.16　Result of the third sorting run of the right subsequence

右边子序列多次排序结果如图 3.17 所示。

The result of multiple sorting of the right subsequence is shown in Figure 3.17.

| 68 | 74 | 78 | 87 | 97 |

图 3.17　右边子序列多次排序结果

Figure 3.17　Result of the multiple sorting of the right subsequence

合并左子序列和右子序列排序后的结果即为整个序列排序结果，如图 3.18 所示。

The result of merging the sorting of the left and right sub-sequences is the sorting result of the whole sequence, as shown in Figure 3.18.

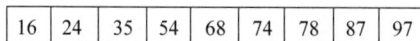

| 16 | 24 | 35 | 54 | 68 | 74 | 78 | 87 | 97 |

图 3.18　序列排序结果

Figure 3.18　Result of the sequence sorting

2）实例实现（Implementation）

上述实例实现代码如下，具体实现结果如图 3.19 所示。

The above example implementation code is shown below, and the result is shown in Figure 3.19.

```
def partition(list,left,right):          #待排序元素 left,起始索引 right,结束索引(Element to be sorted left, start index
                                         #right, end index)
    i = left
    j = right+1
    pivot = list[left]                   #用序列的第一个元素作为基准元素(Use the first element of the sequence as the
                                         #base element)
    while(True):
        i += 1                           #向右推动指针(Push the pointer to the right)
        while(list[i]<pivot):            #从首往后扫描,找到比基准元素大的立即停止,该元素位置不正确(Scan from head
                                         #to back and find a stop larger than the base element, which is incorrectly
                                         #positioned)
            i += 1                       #向右推动指针(Push the pointer to the right)
        j -= 1                           #向左推动指针(Push the pointer to the left)
        while(list[j]>pivot):            #从尾往向前扫描,找到比基准元素小的停止,该元素位置不正确(Scan from tail to
                                         #forward to find a stop smaller than the base element, which is incorrectly
                                         #positioned)
            j -= 1                       #向左推动指针(Push pointer left)
        if(i>= j):
            break
        list[i],list[j] = list[j],list[i]   #交换 list[i]和 list[j],交换后 i 执行加1操作(Exchange list[i] and
                                            #list[j]. After the exchange, i adds 1)
    list[j],list[left] = list[left],list[j]
    return j
def quick sort(list,left,right):
    if left < right:                     #边界条件判断(Boundary condition judgment)
        j = partition(list,left,right)   #分解(Decompose)
        quick sort(arr,left,j-1)         #递归左子序列(Recursive left subsequence)
        quick sort(arr,j+1,right)        #递归右子序列(Recursive right subsequence)
if _ _name_ _== " _ _main_ _":
    arr = [54,74,97,24,87,78,68,35,16]
    n = len(arr)
    quick sort(arr,0,n-1)
    print(" 排序后的数组(The sorted array):")
    print(arr)
```

```
排序后的数组(The sorted array):
[16, 24, 35, 54, 68, 74, 78, 87, 97]
```

图 3.19 快速排序问题的实验结果图

Figure 3.19 Result of the diagram of quick sorting problem

3.3 查找问题算法(Search Problem Algorithm)

3.3.1 顺序查找(Sequential Search)

1. 算法思想(Algorithm Idea)

顺序查找算法,也称为线性查找算法,是一种基本的搜索算法。它的基本思想是逐一地检查数据结构中的元素,直到找到目标元素或遍历完数据结构中的所有元素。

The sequential search algorithm, also known as the linear search algorithm, is a basic search algorithm. The basic idea is to check the elements in the data structure one by one until the target element is found or all elements in the data structure are traversed.

先从数据结构的第一个元素开始，逐一检查每个元素，对比当前元素与目标元素是否相等。如果相等，则找到目标元素，返回其位置(索引)；如果不相等，则继续向后检查下一个元素。重复以上步骤，直到找到目标元素或遍历完数据结构中的所有元素。

Start at the first element of the data structure and check each element one by one to see if the current element is equal to the target element. If equal, the target element is found and return its position (index), and if it is not equal, then go back to the next element. The above steps are repeated until the target element is found or all elements in the data structure are traversed.

2. 算法设计与描述（Algorithm Design and Description）

1) 算法设计(Algorithm Design)

(1) 输入参数(Input Parameter)。数据结构：待搜索的数据结构，可以是数组、列表等。目标元素：要查找的目标元素。

Data Structure：the data structure to be searched can be an array, list, and so on. Target Element：the target element to find.

(2) 遍历数据结构(Traverse the Data Structure)。使用循环遍历数据结构中的每个元素。

Use loops to iterate through each element in the data structure.

(3) 比较操作(Compare Operation)。对每个元素与目标元素进行比较。

Each element is compared to the target element.

(4) 找目标元素(Find the Target Element)。如果找到目标元素则算法终止，如果遍历完全部数据结构仍未找到目标元素，则返回一个特殊值表示未找到。

The algorithm terminates if the target element is found. If the element is not found after traversing the entire data structure, a special value is returned to indicate that it was not found.

2) 算法描述(Algorithm Description)

输入：非降顺序排列的数组 L，元素数 n，数 x。

Input：non-descending array L, number of elements n, number x.

输出：j，若 x 在 L 中，j 是 x 首次出现的下标；否则返回 Not Found。

Output：j, if x is in L, j is the subscript where x first appears; otherwise, it is Not Found.

基本运算：x 与 L 中元素的比较；$j=1$ 时，将 x 与 $L[j]$ 比较，如果 $x=L[j]$，则算法停止，输出 j；$j \neq 1$，则把 j 加 1，继续 x 与 $L[j]$ 的比较，如果 $j>n$，则停止并输出 Not Found。

Basic operation：comparison of x with the elements of L, when $j=1$, compare x with $L[j]$. If $x=L[j]$, the algorithm stops and outputs j. If $j \neq 1$, add 1 to j and continue comparing x with $L[j]$. If $j>n$, stop and print Not Found.

3. 算法实现(Algorithm Implementation)

1) 实例构造(Instance Construction)

假设有一个包含整数的列表$[3,1,4,1,5,9,2,6,5,3,5]$,现在要查找元素 6 是否在列表中,并返回它的索引,如图 3.20 所示。在列表中寻找 6,从列表第一个元素开始,对比元素是否等于要查找的元素,如果不等于则比较下一个元素,如果等于则返回此元素的位置索引。实例中 $3\neq6,1\neq6,4\neq6,\cdots,2\neq6,6=6$,找到目标元素,返回索引 index$=7$。

Suppose there is a list of integers $[3,1,4,1,5,9,2,6,5,3,5]$, and we need to find out if element 6 is in the list and return its index, as shown in Figure 3.20. Look for the 6 in the list, start from the first element in the list and check whether the element is equal to the element we're looking for, and if not, check the next element. If so, return the index of the element. In this example, $3\neq6,1\neq6,4\neq6,\cdots, 2\neq6,6=6$, find the target element, and return index$=7$.

3	1	4	1	5	9	2	6	5	3	5

index   0   1   2   3   4   5   6   7   8   9   10

图 3.20　顺序查找

Figure 3.20　Sequential searching

2) 实例实现(Implementation)

上述实例算法代码如下,实现结果如图 3.21 所示。

The above example algorithm code is as follows, and the implementation results are shown in Figure 3.21.

```
def sequential_search(data, target):
    """
    顺序查找算法实现(Sequential search algorithm implementation)
    Parameters:
    - data: 待搜索的数据结构(如数组或列表)(Data structures to be searched (such as arrays or lists))
    - target: 要查找的目标元素(The target element to find)
    Returns:
    -如果找到目标元素,返回其索引;如果未找到,返回特殊值 -1(If the target element is found, return its index; If not found, return the special value -1)
    """
    for index, element in enumerate(data):
        if element == target:
            return index
    return -1
#示例用法(Example usage)
data_list = [3, 1, 4, 1, 5, 9, 2, 6, 5, 3, 5]
target_element = 6
result = sequential_search(data_list, target_element)
if result != -1:
    print(f" 元素 {target_element} 在列表中的索引是 {result} \nElement {target_element} the index in the list is {result}")
    else:
    print(f" 元素 {target_element} 未在列表中找到\nThe element {target_element} was not found in the list")
```

```
元素 6 在列表中的索引是 7
Element 6 the index in the list is 7
```

图 3.21　顺序查找问题的程序运行结果

Figure 3.21　Result of the order searching problem

3.3.2　折半查找算法（Binary Search Algorithm）

1. 算法思想（Algorithm Idea）

二分查找又称为折半查找，它要求数据元素必须是按关键字大小有序排列的。假设有一个已排好序的 n 个元素 $s_1, \cdots, s_n$，要在这 n 个元素中查找一个特定元素 x 是否存在，若存在，则返回 x 在序列中的位置；否则，返回 -1。

Binary search, also known as split half search, requires that the data elements must be sorted by keyword size. Suppose there is a sorted n elements $s_1, \cdots, s_n$, which searches for the presence of a particular element x among these n elements and returns the position of x in the sequence if it exists. Otherwise, -1 is returned.

假定用 a_list 表示 n 个元素的有序序列，该序列由小到大排序。

Suppose an ordered sequence of n elements is represented by a_list, which is ordered from smallest to largest.

（1）分解。将有序序列分成规模大致相等的两部分，即每部分的规模大致为 $n/2$。

Decomposition. Divides an ordered sequence into two parts of roughly equal size, that is, each part has a size of roughly $n/2$.

（2）治理。用序列中间位置的元素与特定元素 x 比较，如果 x 等于中间元素，那么算法终止；如果 x 小于中间元素，那么在序列的左半部递归查找；否则，在序列的右半部递归查找。递归停止的条件是序列规模为 0 或 1。

Governance. Compare the element in the middle of the sequence with a specific element x. If x is equal to the middle element, the algorithm terminates; if x is less than the middle element, search recursively in the left half of the sequence; otherwise, search recursively in the right half of the sequence. The condition for the recursion to stop is that the sequence size is 0 or 1.

该算法的思想如下。假定元素序列已经由小到大排好序，将有序序列分成规模大致相等的两部分，然后取中间元素与特定查找元素 x 进行比较，如果 x 等于中间元素，则算法终止；如果 x 小于中间元素，则在序列的左半部继续查找，即在序列的左半部重复分解和治理操作；否则，在序列的右半部继续查找，即在序列的右半部重复分解和治理操作。可见，二分查找算法重复利用了元素间的次序关系。

The idea of the algorithm is as follows. Assuming that the elements of the sequence have been sorted from small to large, the ordered sequence is divided into two parts of approximately equal size, and the middle element is selected to compare with a specific search element x. If x is equal to the middle element, the algorithm terminates. If x is less than the middle element, the search continues in the left half of the sequence, that is, the decomposition and governance operations are repeated in the left half of the sequence; Otherwise, the search continues in the right half of the sequence, that is, the decomposition and governance operations are repeated in the right half of the series. The binary search algorithm reuses the order relationship between elements.

2. 算法设计与描述(Algorithm Design and Description)

通过问题分析可知,不同的子问题有不同的规模,并且在有序列 a_list 中的位置不同。采用辖定边界的方法统一表示不同问题的规模及在序列中的位置。

According to the problem analysis, different subproblems have different scales and different positions in the sequence a_list. It uses the method of governing the boundary to uniformly represent the scale and position of different problems in the sequence.

若令 $left$ 表示子问题的下边界,$right$ 表示子问题的上边界,则 $a\_list[left:right]$ 表示不同的子问题。其分解步骤表示为 $mid=(left+right)/2$(向下取整);递归的边界条件表示为 $left>right$。治理表示为:①判断 x 是否与 $a\_list[mid]$ 相等,若相等,则返回 mid;②判断 x 是否大于 $a\_list[mid]$,若大于,则在右边的子问题递归查找 x;否则,在左边的子问题递归查找 x。

If $left$ represents the lower bound of the subproblem and $right$ represents the upper bound of the subproblem, then $a\_list[left:right]$ represents different subproblems. The decomposition step is expressed as $mid=(left+right)/2$ (rounded down); the recursive boundary conditions are expressed as $left>right$. Check ①whether x is equal to $a\_list[mid]$, and if so, return mid; ② Determine whether x is greater than $a\_list[mid]$, if yes, then the right subproblem recursively finds x; Otherwise, the left subproblem recursively looks for x.

算法的求解步骤设计如下。

The algorithm's solution steps are designed as follows.

(1) 确定合适的数据结构。设置数组 $s[n]$ 来存放 n 个已排好序的元素;变量 low 和 $high$ 分别表示查找范围在数组中的下界和上界;$middle$ 表示查找范围的中间位置;x 为特定元素。

Determine the appropriate data structure. Setsan array $s[n]$ to hold n sorted elements; the low and $high$ variables represent the lower and upper bounds of the search range in the array, respectively; $middle$ indicates the middle of the search range, x is a specific element.

(2) 初始化。令 $low=0$,即指示 s 中第一个元素;$high=n-1$,即指示 s 中最后一个元素。

Initialization. Set $low=0$, indicating the first element in s; $high=n-1$, which indicates the last element in s.

(3) 查找范围的中间元素,$middle=(low+high)/2$。

Find the middle element of the range, $middle=(low+high)/2$.

(4) 判定 $low \leqslant high$ 是否成立,如果成立,转步骤(5);否则,算法结束。

Check whether $low \leqslant high$ is true. If it is true, go to step (5). Otherwise, the algorithm ends.

(5) 判断 x 与 $s[middle]$ 的关系。如果 x 等于 $s[middle]$，算法结束；如果 $x>s[middle]$，则令 $low=middle+1$；否则令 $high=middle-1$，转步骤(3)。

Determine the relationship between x and $s[middle]$. If x is equal to $s[middle]$, the algorithm ends; If $x>s[middle]$, let $low=middle+1$; If not, set $high=middle-1$. Go to step (3).

3. 算法实现(Algorithm Implementation)

1) 实例构造(Instance Construction)

假设待排序序列 A=<2,5,7,17,23,25,31,35,42,76,88>，采用折半查找算法查找元素 76 所在位置，如图 3.22 所示。

Suppose to collate sequence A=<2,5,7,17,23,25,31,35,42,76,88>, using binary search algorithm to find the position of the element 76, as shown in Figure 3.22.

	2	5	7	17	23	25	31	35	42	76	88
index	0	1	2	3	4	5	6	7	8	9	10
	left										right

图 3.22　折半查找

Figure 3.22　Binary search

具体查找过程如图 3.23~图 3.25 所示。

The lookup procedure is shown in Figure 3.23-Figure 3.25.

先将序列分为两个子序列，*left* 指针指向第一个元素，*right* 指针指向最后一个元素，*mid* 指针指向中间元素，$mid=(0+10)/2=5$，index=5，指向元素 25，以 25 为中间，分为左右两个子序列，如图 3.23 所示。

First, the sequence is divided into two subsequences, the *left* pointer points to the first element, the *right* pointer points to the last element, the *mid* pointer points to the middle element, $mid=(0+10)/2=5$, index=5, points to the element 25, with 25 as the middle, the left and right subsequence are divided into two subsequences, as shown in Figure 3.23.

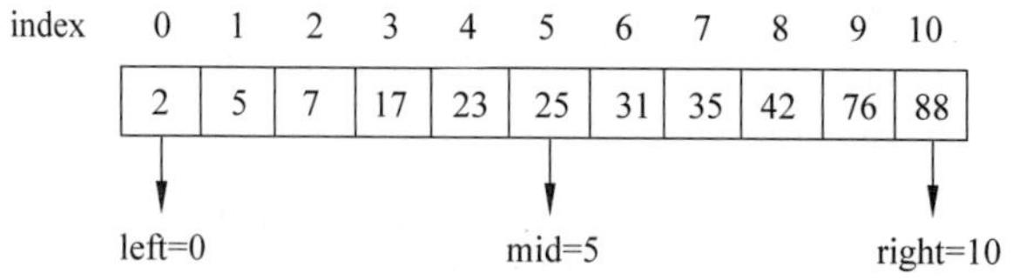

图 3.23　折半查找(1)

Figure 3.23　Binary search (1)

此时序列为<2,5,7,17,23,25,31,35,42,76,88>，由于要查找的元素为 76，76>25，所以在右边子序列<31,35,42,76,88>中查找目标元素，改变指针指向，左指针指向元素 31，右指针指向元素 88，$mid=(6+10)/2=8$，index=8，指向元素 42，如图 3.24 所示。

The sequence of <2,5,7,17,23,25,31,35,42,76,88>, because the element to be searched is 76, 76 > 25, search for the target element in the right sequence < 31,35,42, 76,88>, change the pointer direction so that the left pointer points to element 31, the right pointer points to element 88, $mid = (6+10)/2 = 8$, index = 8, points to element 42, as shown in Figure 3.24.

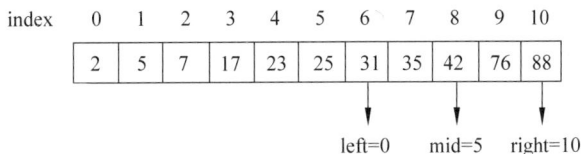

图 3.24　折半查找(2)

Figure 3.24　Binary search (2)

此时序列变为<31,35,42,76,88>,由于 76>42,所以在右边子序列<76,88>中查找元素,改变指针指向,左指针指向元素 76,右指针指向元素 88,$mid = (9+10)/2 = 9.5$(向下取整),index = 9,指向元素 76,如图 3.25 所示。此时 mid 指针指向的元素正是需要查找的元素,所以折半查找完成,返回找到的元素的索引值即 9。

At this point, the sequence becomes <31,35,42,76,88>. Since 76>42, look for the element in the right subsequence <76,88>, change the pointer direction, the left pointer points to element 76, the right pointer point to element 88, $mid = (9 + 10)/2 = 9.5$ (rounded down), index = 9, points to element 76, See Figure 3.25. At this point, the mid pointer is the element that is being searched for, so the binary search is completed and the index of the found element is returned 9.

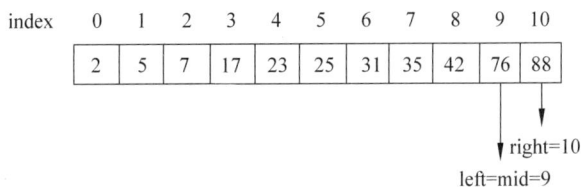

图 3.25　折半查找(3)

Figure 3.25　Binary search (3)

2) 实例实现(Implementation)

上述实例算法代码如下,具体实现结果如图 3.26 所示。

The above example algorithm code is as follows, and the specific implementation results are shown in Figure 3.26.

```
def bin_search(items, key):
    '''折半查找(Binary Search)'''
    start, end = 0, len(items) −1
    while start <= end:
        mid = (start +end) // 2
        if key > items[mid]:
```

续

```
        start = mid +1
    elif key < items[mid]:
        end = mid -1
    else:
        return mid
    return -1
print("元素所在的位置(从 0 开始)为: \nThe position of the element (starting from 0) is:",bin_search([2,5,7,17,
23,25,31,35,42,76,88], 76))
```

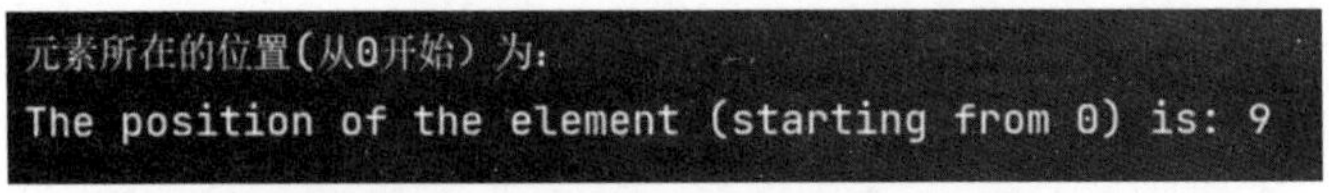

```
元素所在的位置(从0开始) 为:
The position of the element (starting from 0) is: 9
```

图 3.26　折半查找问题的实例结果

Figure 3.26　Result of the binary search algorithm

3.4　组合问题算法(Combinatorial Problem Algorithm)

3.4.1　最大子段和问题(Maximum Subsegment Sum Problem)

1. 算法思想(Algorithm Idea)

给定长度为 n 的整数序列 $(a_0,a_1,\cdots,a_{n-1})$，最大子段和问题(Maximum Subegment Sum Problem)要求该序列形如 $a_i+\cdots+a_j$ 的最大值 $(0 \leqslant i \leqslant j \leqslant n-1)$。例如，$(-2,11,-4,13,-5,2)$ 的最大子段和为 20，所求子区间为 $[1,3]$。

Given a sequence of integers of length n $(a_0,a_1,\cdots,a_{n-1})$, and the Sum of Largest Subsegment Problem requires the sequence to have the maximum value of $a_i+\cdots+a_j(0 \leqslant i \leqslant j \leqslant n-1)$. For example, the sum of the largest subsegments of $(-2,11,-4,13,-5,2)$ is 20 and the desired subinterval is $[1,3]$.

当序列只有一个元素时，最大的和只有一个可能，就是选取本身；当序列有两个元素时，只有三种可能，选取左边元素、选取右边元素、两个都选，这三个可能中选取一个最大的就是当前情况的最优解；对于多个元素时，最大的和也有三种情况，从左区间产生、从右区间产生、左右区间各选取一段。

When the sequence has only one element, there is only one possible maximum sum, which is the selection itself. When there are two elements in the sequence, there are only three possibilities - choose the left element, choose the right element, or choose both. The largest of the three possibilities is the optimal solution for the current situation. For multiple elements, the maximum sum also has three cases, generated from the left interval, generated from the right interval, and selected a segment of the left and right intervals.

因此不难看出，这个算法是基于分治思想的，每次二分序列，直到序列只有一个元素或者两个元素。当只有一个元素时就返回自身的值，有两个时返回三个中最大的，有多个元素时返回左、右、中间的最大值。分治思想如图 3.27 所示。

Therefore, it is not difficult to see that the algorithm is based on the idea of divide-and-conquer, each time the sequence is divided until the sequence has only one element or two elements. It returns the value of itself if there is only one element, the maximum of three if there are two, and the maximum of the left, right, and middle if there are multiple elements. The divide-and-conquer idea is illustrated in Figure 3.27.

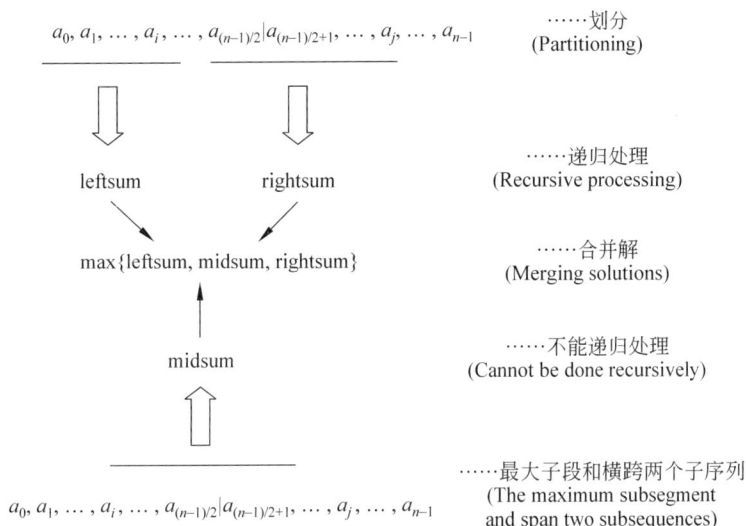

$$a_0, a_1, \ldots, a_i, \ldots, a_{(n-1)/2} | a_{(n-1)/2+1}, \ldots, a_j, \ldots, a_{n-1}$$

……划分
(Partitioning)

leftsum　　rightsum

……递归处理
(Recursive processing)

max{leftsum, midsum, rightsum}

……合并解
(Merging solutions)

midsum

……不能递归处理
(Cannot be done recursively)

$$a_0, a_1, \ldots, a_i, \ldots, a_{(n-1)/2} | a_{(n-1)/2+1}, \ldots, a_j, \ldots, a_{n-1}$$

……最大子段和横跨两个子序列
(The maximum subsegment and span two subsequences)

图 3.27　最大子段和问题的分治思想

Figure 3.27　Idea of the sum of divide-and-conquer largest subsegment problem

2. 算法设计与描述(Algorithm Design and Description)

最大子段和问题的算法设计有三大部分。

The algorithmic design of the maximum subsegment sum problem has three major parts.

1) 划分(Divide)

按照平衡子问题的原则,将序列$(a_0, a_1, \cdots, a_{n-1})$划分成长度相同的两个子序列$(a_0, a_1, \cdots, a_{(n-1)/2})$和$(a_{(n-1)/2+1}, \cdots, a_{n-1})$,则会出现以下三种情况。

Following the principle of balanced subproblems, the sequence $(a_0, a_1, \cdots, a_{n-1})$ divides two subsequences of the same size $(a_0, a_1, \cdots, a_{(n-1)/2})$ and $(a_{(n-1)/2+1}, \cdots, a_{n-1})$, then the following three cases will occur.

(1) 序列$(a_0, a_1, \cdots, a_{n-1})$的最大子段和等于$(a_0, a_1, \cdots, a_{(n-1)/2})$的最大子段和。

The sum of the largest subsegments of sequence $(a_0, a_1, \cdots, a_{n-1})$ is equal to the sum of the largest subsegments of $(a_0, a_1, \cdots, a_{(n-1)/2})$.

(2) 序列$(a_0, a_1, \cdots, a_{n-1})$的最大子段和等于$(a_{(n-1)/2+1}, \cdots, a_{n-1})$的最大子段和。

The sum of the largest subsegments of sequence $(a_0, a_1, \cdots, a_{n-1})$ is equal to the sum of the largest subsegments of $(a_{(n-1)/2+1}, \cdots, a_{n-1})$.

(3) 序列$(a_0, a_1, \cdots, a_{n-1})$的最大子段和等于$a_i + a_{i+1} + \cdots + a_j$的最大子段和,且$0 \leqslant i \leqslant$

$(n-1)/2, (n-1)/2+1 \leqslant j \leqslant n-1$。

The sum of the largest subsegments of sequence $(a_0, a_1, \cdots, a_{n-1})$ is equal to the sum of the largest subsegments of $a_i + a_{i+1} + \cdots + a_j$, and $0 \leqslant i \leqslant (n-1)/2, (n-1)/2+1 \leqslant j \leqslant n-1$.

2) 求解子问题(Solving Subproblem)

对于划分阶段的情况(1)和(2)可递归求解,情况(3)需要分别计算。

For the partition phase cases(1) and (2) can be solved recursively, while case (3) needs to be calculated separately.

$$s_1 = \max\left\{\sum_{k=i}^{\frac{(n-1)}{2}} a_k\right\}\left(0 \leqslant i \leqslant \frac{(n-1)}{2}\right),$$

$$s_2 = \max\left\{\sum_{k=\frac{(n-1)}{2}+1}^{j} a_k\right\}\left(\frac{(n-1)}{2}+1 \leqslant i \leqslant n-1\right)$$

则 $s_1 + s_2$ 为情况(3)的最大子段和。

Then $s_1 + s_2$ is the largest subsegment sum of case (3).

3) 合并(Combine)

比较在划分阶段三种情况下的最大子段和,取三者之中的较大者为原问题的解。前两种情形符合子问题递归特性,所以递归可以求出。对于第三种情形,则需要单独处理。

Compare the sum of the largest subsegments in the three cases of divided stages, and take the largest of the three as the solution of the original problem. The first two cases conform to the recursion property of the subproblem, so recursion can be solved. For the third case, it needs to be handled separately.

第三种情形必然包括$(n-1)/2$ 和$(n-1)/2+1$ 两个位置,这样就可以利用穷举的思路求出：以$(n-1)/2$ 为终点,往左移动扩张,求出子段和最大的一个 *leftmax*1；以$(n-1)/2+1$ 为起点,往右移动扩张,求出子段和最大的一个 *rightmax*1；则 *leftmax*1+*rightmax*1 是第三种情况可能的最大值。

The third case should include two positions $(n-1)/2$ and $(n-1)/2+1$, so it can be used the idea of exhaustion to find out: take $(n-1)/2$ as the end point, move to the left to expand, find the subsegment and the largest *leftmax*1. Take $(n-1)/2+1$ as the starting point, move to the right and expand to find the subsegment and the largest *rightmax*1; *leftmax*1+*rightmax*1 is the maximum value possible in case 3.

分治法的难点在于第三种情形的理解,这里应该抓住第三种情形的特点,也就是中间有两个定点,然后分别往两个方向扩张,以遍历所有属于第三种情形的子区间,求得最大的$(n-1)/2$ 个。

The difficulty of the dive-and-conquer method lies in the understanding of the third case, which should grasp the characteristics of the third case, that is, there are two fixed points in the middle, and then expand in two directions respectively to traverse all the subintervals belonging to the third case, and obtain the maximum $(n-1)/2$.

最大子段和 *Maxsum* 算法设计描述如下。

The sum of the Largest Subsegment of *Maxsum* algorithm is described as follows.

输入: 序列 $a[0:n-1]$。

Input: Sequence $a[0:n-1]$.

输出: 最大子段和 *max*。

Output: The sum of the Largest Subsegment of *max*.

(1) 设置初始值: $max=0$; $left=0$, $right=n-1$。

Set initial value: $max=0$; $left=0$, $right=n-1$.

(2) 如果 $left=right$, 则 *max* 取 $(0, a[left])$ 的较大者。

If $left=right$, then *max* is the greater of $(0, a[left])$.

(3) 取中间点 $mid=(left+right)/2$, 有以下三种情况: 递归计算左区间 $[left, mid]$ 的最大子段和 *leftsum*(第 1 种); 递归计算右区间 $[mid+1, right]$ 的最大子段和 *rightsum*(第 2 种); 分别计算区间 $[left, mid]$ 的最大子段和 *leftsum*1, 区间 $[mid+1, right]$ 的最大子段和 *rightsum*1、*leftsum*1+*rightsum*1 赋值给 *midsum*(第 3 种)。

Take the middle point $mid = (left+right)/2$, there are three cases: recursively compute the Sum of Largest Subsegment of the left interval $[left, mid]$ (the first kind); recursively compute the Sum of Largest Subsegment of the right interval $[mid+1, right]$ (the second kind); calculated respectively the Sum of Largest Subsegment of *leftsum*1 and the interval $[left, mid]$, and the largest subsegment and *rightsum*1 and *leftsum*1+*rightsum*1 of the interval $[mid+1, right]$ are assigned to *midsum* (the third kind).

(4) 计算 *leftsum*、*rightsum* 和 *midsum* 的最大者, 即为 *max*, 将 *max* 返回。

Calculated the largest of *leftsum*, *rightsum*, and *midsum*, which is *max*, and *max* is returned.

3. 算法实现(Algorithm Implementation)

1) 实例构造(Instance Construction)

假定有数组 $[-2,1,-3,4,-1,2,1,-5,4]$, 求此数组的最大子段和, 如图 3.28 所示。将数组分为规模大致相等的两部分 $[-2,1,-3,4,-1]$ 和 $[2,1,-5,4]$, 再将这两个数组分解, 左边数组分为 $[-2,1]$、$[-3]$ 和 $[4,-1]$, 右边数组分为 $[2,1]$ 和 $[-5,4]$, 对于子数组 $[-2,1]$, 它有两个元素, 它的最大子段和为选取元素 1, 将数组 $[-2,1]$ 和 $[-3]$ 合并, 它的最大子段和仍为选取元素 1, 对于子数组 $[4,-1]$, 它的最大子段和为选取元素 4, 所以对于左边子序列 $[-2,1,-3,4,-1]$, 它的最大子段和为选取元素 4; 对于子数组 $[2,1]$, 它的最大子段和为选取元素 2 和 1, 为 3, 对于子数组 $[-5,4]$, 它的最大子段和为选取元素 4, 所以对于右边子序列, 最大子段和为选取元素 4; 此时需要计算横跨两个子序列的最大子段和。

Suppose there is an array $[-2, 1, -3, 4, -1, 2, 1, -5, 4]$, find the maximum subsegment sum of this array, as shown in Figure 3.28. Array can be divided into two roughly equal scale parts $[-2,1,-3,4,-1]$ and $[2,1,-5,4]$, then decompose the two arrays, the array on the left is divided into $[-2,1]$, $[-3]$ and $[4,-1]$, the array on the right side is divided into $[2,1]$ and $[-5,4]$, for subarray $[2,1]$, it has two elements, its maximum subsegment sum is selected element 1. If array $[-2,1]$ and $[-3]$ are merged, its maximum subsegment sum is still selected element 1. For the subarray $[4,-1]$, its maximum subsegment sum is selected element 4, so for the left subsequence $[-2,1,-3,4,-1]$, its maximum subsegment sum is selected element 4. For the subarray $[2,1]$, its maximum subsegment sum is to select elements 2 and 1, which is 3, for the subarray $[-5,4]$, its maximum subsegment sum is to select element 4, so for the right subsequence, the maximum subsegment sum is to select element 4. At this point, the maximum subsegment sum across the two subsequences needs to be computed.

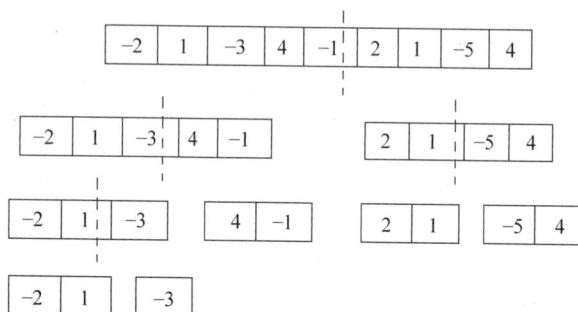

图 3.28　实例求解过程

Figure 3.28　Processing of the example solving

计算横跨两个子序列的最大子段和如图 3.29 所示。从中间元素出发，先向左边建立子数组并求和，第一个子数组为 $[-1]$，它的和为 -1，第二个子数组为 $[4,-1]$，它的和为 3，第三个子数组为 $[-3,4,-1]$，它的和为 0，第四个子数组为 $[1,-3,4,-1]$，它的和为 1，最后一个子数组为 $[-2,1,-3,4,-1]$，它的和为 -1，比较每个子数组的和的大小，得出左边子序列最大的子数组是 $[4,-1]$，它的和为 3；同样地，得出右边子序列最大的子数组是 $[2,1]$，它的和为 3；因此，横跨两个子序列的最大子段和为 3+3=6，上面已经计算出左边子序列最大子段和为 4，右边子序列最大子段和为 4，横跨两个子序列的最大子段和为 6，比较三个数值的大小，选取最大值即 6 作为整个序列最大子段和，包含的元素为 $[4,-1,2,1]$。

Calculate the largest subsegment sum across the two subsequences is computed as shown in Figure 3.29. Starting from the middle element, it can be first built subarrays to the left and sum them up, the first subarray is created to the left and summed, the first subarray is $[-1]$, its sum is -1, the second subarray is $[4,-1]$, its sum is 3, the third subarray is $[-3,4,-1]$, its sum is 0, the fourth subarray is $[1,-3,4,-1]$, its sum is 1, the last subarray is $[-2,1,-3,4,-1]$, and its sum is -1. Comparing the size of the sum of each subarray, the largest subarray on the left is $[4,-1]$, and its sum is 3. Similarly, the largest subarray on the right is $[2,1]$, which sums to 3; and the maximum sum of subsegments across the two subsequences is 3+3=6, which has been calculated above that the maximum sum of subsegments in the left subsequence is 4, the maximum sum of subsegments in the right subsequence is 4, and the maximum sum of subsegments across the two subsequences is 6. Comparing the size of the three values, the maximum value of 6 is selected as the maximum sum of subsegments in the whole sequence, and the elements are $[4,-1,2,1]$.

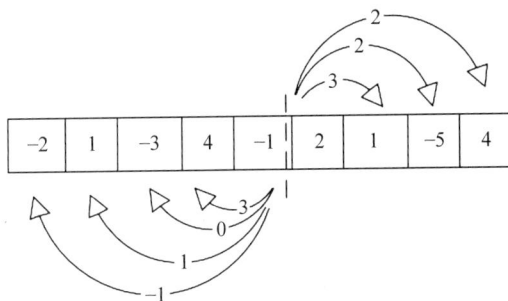

图 3.29　横跨两个子序列最大子段和

Figure 3.29　Maximum subsegment sum of across the two subsequences

2）实例实现（Instance Implementation）

上述实例实现代码如下所示，具体实现结果如图 3.30 所示。输入为 $[-2,1,-3,4,-1,2,1,-5,4]$，输出为 6，程序结果表示连续子数组 $[4,-1,2,1]$ 的和是最大的，和为 6。

The above example implementation code is shown below, and the specific implementation result is shown in Figure 3.30. Input is $[-2,1,-3,4,-1,2,1,-5,4]$, the output is 6, application results indicate that the sum of the continuous subarray $[4,-1,2,1]$ is the largest, and the sum is 6.

```python
def max_crossing_sum(arr, low, mid, high):
    #计算包含中间元素的左侧最大子段和(Calculates the sum of the largest subsegment on the left that contains the
    #middle element)
    left_sum = float('-inf')
    sum_temp = 0
    for i in range(mid, low - 1, -1):
        sum_temp += arr[i]
        left_sum = max(left_sum, sum_temp)
    #计算包含中间元素的右侧最大子段和(Calculates the sum of the largest subsegment on the right that contains
    #the middle element)
    right_sum = float('-inf')
    sum_temp = 0
    for i in range(mid + 1, high + 1):
        sum_temp += arr[i]
        right_sum = max(right_sum, sum_temp)
    #返回左右两侧的最大子段和之和(Returns the sum of the largest subsegments on the left and right sides)
    return left_sum + right_sum
def max_subarray_sum(arr, low, high):
    if low == high:
        # 基本情况：只有一个元素(Basic situation：only one element)
        return arr[low]
    #计算数组的中间位置(Calculates the middle of the array)
    mid = (low + high) // 2
    #递归计算左右两侧的最大子段和(Calculate the sum of the largest subsegments of the left and right)
```

续

```
left_sum = max_subarray_sum(arr, low, mid)
right_sum = max_subarray_sum(arr, mid +1, high)
#计算跨越中间位置的最大子段和(Computes the sum of the largest subsegments spanning the middle position)
cross_sum = max_crossing_sum(arr, low, mid, high)
#返回三者中的最大值(Returns the maximum of the three)
return max(left_sum, right_sum, cross_sum)
#示例用法(Example usage)
arr = [-2, 1, -3, 4, -1, 2, 1, -5, 4]
result = max_subarray_sum(arr, 0, len(arr) -1)
print("最大子段和为(The maximum sum of subsegments is):", result)
```

最大子段和为(The maximum sum of subsegments is): 6

图 3.30　最大子段和问题的实验结果图

Figure 3.30　Result of the maximum sum of subsegments problem

3.4.2　棋盘覆盖问题(Chessboard Covering Problem)

1. 算法思想(Algorithm Idea)

在一个由 $2^k \times 2^k (k \geq 0)$ 个方格组成的棋盘中，恰有一个方格与其他方格不同，则称该方格为特殊方格，且称该棋盘为一个特殊棋盘。显然，特殊方格在棋盘上出现的位置有 4^k 种情形，因而有 4^k 种不同的特殊棋盘。如图 3.31(a)所示的特殊棋盘是当 $k=2$ 时 16 个特殊棋盘中的一个。

In a chessboard consisting of $2^k \times 2^k (k \geq 0)$ squares, if exactly one square is different from the others, the square is called a special square, and the chessboard is called a special board. Obviously, there are 4^k cases in which special squares can appear on the board, so there are 4^k different special chessboards, and the special chessboard shown in Figure 3.31 (a) is one of the 16 special chessboards when $k=2$.

棋盘覆盖问题(Chessboard Covering Problem)要求用如图 3.31(b)所示的 4 种不同形状的 L 形骨牌覆盖给定棋盘上除特殊方格以外的所有方格，且任何两个 L 形骨牌不得重叠覆盖。容易知道，在任何一个 $2^k \times 2^k$ 的棋盘覆盖中，用到的 L 形骨牌个数恰为 $(4^k-1)/3$。

The Chessboard Covering Problem requires to cover all but special squares on a given chessboard with four L-shaped dominoes of different shapes as shown in Figure 3.31 (b), and no two L-shaped dominoes may overlap in coverage. It's easy to know that in any $2^k \times 2^k$ chessboard overlay, the number of L-dominos used is exactly $(4^k-1)/3$.

整个算法的过程就是不断地"分"，然后分别地"治"，将棋盘"分"到足够小的时候，"治"起来就变得简单轻松了。分治的技巧在于如何划分棋盘，使划分后的棋盘的大小相同，并且每个子棋盘均包含一个特殊方格，从而将原问题分解为规模较小的棋盘覆盖问题。对于 $k>0$ 时，可将 $2^k \times 2^k$ 的棋盘划分为 4 个 $2^{k-1} \times 2^{k-1}$ 的子棋盘，如图 3.32(a)所示。

The whole process of the algorithm is to constantly " divide" and then " conquer" respectively, " divide" the chessboard until it is small enough, " conquer" becomes easy and easy. The trick of divide-and-conquer is how to divide the chessboard so that the

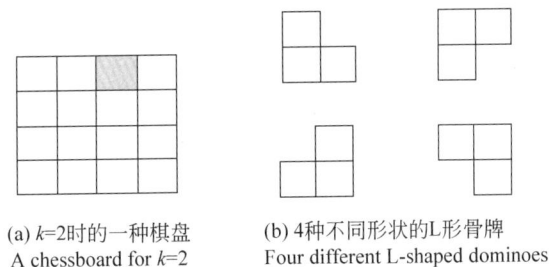

(a) k=2时的一种棋盘
A chessboard for k=2

(b) 4种不同形状的L形骨牌
Four different L-shaped dominoes

图 3.31　棋盘覆盖问题结构图

Figure 3.31　Diagram of chessboard covering problem structure

size of the divided chessboard is the same and each sub-chessboard contains a special square, so that the original problem is decomposed into a smaller chessboard covering problem. For $k>0$, the $2^k \times 2^k$ board can be divided into four $2^{k-1} \times 2^{k-1}$ sub-boards, as shown in Figure 3.32 (a).

这样划分后，由于原棋盘只有一个特殊方格，所以，这 4 个子棋盘中只有一个子棋盘包含该特殊方格，其余三个子棋盘中没有特殊方格。为了将这三个没有特殊方格的子棋盘转换为特殊棋盘，以便采用递归方法求解，可以用一个 L 形骨牌覆盖这三个较小棋盘的会合处，如图 3.32(b)所示，从而将原问题转换为 4 个较小规模的棋盘覆盖问题。递归地使用这种划分策略，直至将棋盘分割为 1×1 的子棋盘。

After this division, only one of the four sub-boards contains a special square, and the other three have no special squares, because the original board has only one special square. To transform these three sub-boards without special squares into special boards that can be solved recursively, an L-shaped domino can be used to cover the meeting of the three smaller chessboards, as shown in Figure 3.32 (b), thereby transforming the original problem into four smaller chessboard covering problems. Use this partitioning strategy recursively until the board is divided into 1×1 sub-boards.

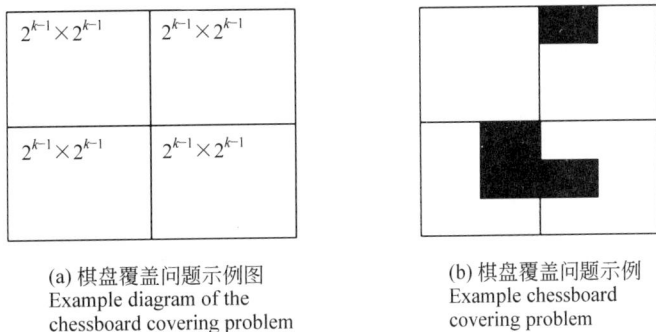

(a) 棋盘覆盖问题示例图
Example diagram of the chessboard covering problem

(b) 棋盘覆盖问题示例
Example chessboard covering problem

图 3.32　棋盘的分割与构造图

Figure 3.32　Diagram of division and chessboard construction

2. 算法设计与描述(Algorithm Design and Description)

采用上述分治算法可以编写成一个递归函数，从而解决了棋盘覆盖问题。棋盘及特殊方格的表示如图 3.33 所示。数据结构的设计如下。

Using the above divide-and-conquer algorithm can be written as a recursive function, thus solving the chessboard cover problem. The chessboard and special square representation are shown in Figure 3.33. The data structure is designed as follows.

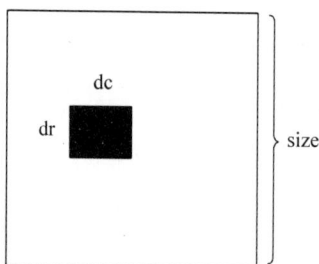

图 3.33 棋盘及特殊方格的表示

Figure 3.33   Representation of chessboard and special squares

（1）棋盘。可以用一个二维数组 $board[size][size]$ 表示一个棋盘，其中，$size = 2^k$，$size$ 表示棋盘的行数或列数。为了在递归处理的过程中使用同一个棋盘，将数组 $board$ 设为全局变量。

Chessboard. A chessboard can be represented by a two-dimensional array $board$ $[size][size]$, where $size = 2^k$ and $size$ is the number of rows or columns of the chessboard. To use the same board recursively, you need to set the array $board$ to a global variable.

（2）子棋盘。整个棋盘用二维数组 $board[size][size]$ 表示，其中的子棋盘由棋盘左上角的下标 tr、tc 和棋盘大小 s 表示。

Child chessboard. The entire board is represented by the two-dimensional array $board[size][size]$, where the sub-boards are represented by the subscript tr, tc, and board size s in the upper left corner of the board.

（3）特殊方格。用 $board[dr][dc]$ 表示特殊方格，dr 和 dc 是该特殊方格在二维数组 $board$ 中的下标。

Special squares. A special square is represented by $board[dr][dc]$, where dr and dc are the subscripts of the special square in the two-dimensional array $board$.

（4）L形骨牌。一个 $2^k \times 2^k$ 的棋盘中只有一个特殊方格，所以，用到 L 形骨牌的个数为 $(4^k-1)/3$，将所有 L 形骨牌从 1 开始连续编号，用一个全局变量 $tile$ 表示，该全局变量初值为 0。

L-shaped dominoes. A $2^k \times 2^k$ chessboard has only one special square, so the number of L-shaped dominoes used is $(4^k - 1)/3$, and all L-shaped dominoes are numbered consecutive-starting from 1, denoted by a global variable $tile$, which has an initial value of 0.

3. 算法实现(Algorithm Implementation)

1）实例构造(Instance Construction)

假设有一个 $2^2 \times 2^2$ 的棋盘，其中右下角的方格缺失，如图 3.34 所示。每个数字代表一个骨牌，而 X 表示缺失的方格，目标是用 L 形的骨牌覆盖整个棋盘。

Suppose there is a $2^2 \times 2^2$ chessboard with the bottom right square missing, as shown in Figure 3.34. Each number represents a domino, while X represents the missing squares, and the goal is to cover the entire board with L-shaped dominoes.

```
1 1 | 2 2

1 3 | 2 4

------ | ------

5 5 | 6 6

7 7 | 8 X
```

图 3.34　$2^2 \times 2^2$ 棋盘

Figure 3.34　Chessboard of $2^2 \times 2^2$

首先观察 4 个 L 形骨牌,发现如果给每个骨牌都增加一个小方格,那么它们都可以变成一个完整的正方形,如图 3.35 所示。所以对于 2×2 的棋盘,可以假定缺失的方格是需要增加的方格,那么无论是哪个位置缺失方格,其余的三个位置都可以用一个 L 形骨牌覆盖,这是解决复杂棋盘覆盖问题的基础。

First, to look at the four L-shaped dominoes and find that if it could add a small square to each domino, then it will become a complete square, as shown in Figure 3.35. So, for a 2×2 chessboard, it can be assumed that the missing square is the square that needs to be added, so no matter which position is missing, the remaining three positions can be covered by an L-shaped domino. This is the basis for solving complex chessboard covering problems.

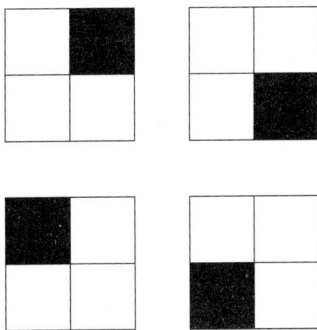

图 3.35　补充 L 形骨牌

Figure 3.35　Supplementary of L-shaped dominoes

根据 2×2 棋盘覆盖的解法,对于上述实例,可以将其简化成 4×4 的棋盘并将其分为 4 个 2×2 的棋盘,如图 3.36 所示。那么 4×4 的棋盘整体看来就是一个大型的 2×2 的棋盘,由于缺失的方格在右下角,所以可以用 4 种 L 形骨牌中的其中一种来覆盖其余三个子棋盘的会合处。为了区分每个 L 形骨牌的覆盖范围,将此时覆盖的方格记为 1,然后分别观察每一个子棋盘,对于左上角的子棋盘,由于刚才在会合处已经覆盖了一个小方格,可以将被覆盖的方格看作另一种特殊方格,那么左上角子棋盘变为有一个特殊方格的 2×2 棋盘,继续利用图 3.35 的思想，可以用一个 L 形骨牌覆盖没有被覆盖的方格，记为 2，以此类推，整个棋盘除了缺失的方格，其余部分可以正好被 5 个 L 形骨牌覆盖完成。

According to the solution of 2×2 chessboard covering, for the above example, it can be simplified into a 4×4 chessboard and divided into four 2×2 chessboards, as shown in Figure 3.36. The 4×4 chessboard looks like a large 2×2 chessboard. Since the missing square is in the bottom right corner, one of the four L-shaped dominoes can be used to cover the meeting point of the other three sub-boards. To distinguish the coverage of each L-shaped domino, the covered square is denoted as 1, and then each sub-board is observed separately. For the upper left sub-board, since a small square has just been covered at the junction, the covered square can be regarded as another special square, then the upper left sub-board becomes a 2×2 board with a special square, continuing to use the idea of Figure 3.35, an L-shaped domino can be used to cover the square which is not covered, denoted as 2, and so on. Except for the missing squares, the rest of the board can be covered by exactly five L-shaped dominoes.

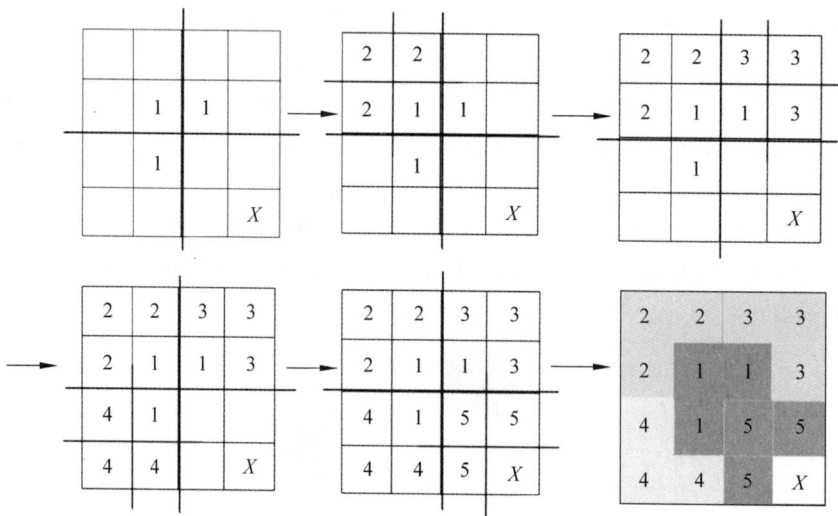

图 3.36 棋盘覆盖分析

Figure 3.36 Analysis of chessboard covering

2）实例实现（Implementation）

上述实例的算法代码如下所示，具体实现结果如图 3.37 所示。

The algorithm code of the above example is shown below, and the results are shown in Figure 3.37.

```
def chessboard_cover(board, tr, tc, dr, dc, size, tile):
    """
    在棋盘上覆盖 L 形骨牌(Cover the chessboard with L-shaped dominoes)
    :param board:棋盘(Chessboard)
    :param tr:棋盘的起始行(Starting row of the board)
```

续

```
    :param tc:棋盘的起始列(Starting column of the board)
    :param dr:缺失方格的行(Row of the missing square)
    :param dc:缺失方格的列(Column of the missing square)
    :param size:棋盘的大小(Size of the chessboard)
    :param tile:当前使用的骨牌编号(Number of the domino currently in use)
    """
    global tile_count
    if size == 1:
        return
    t = tile_count
    tile_count += 1
    s = size // 2
    #覆盖左上子棋盘(Cover the upper left sub-board)
    if dr < tr + s and dc < tc + s:
        chessboard_cover(board, tr, tc, dr, dc, s, t)
    else:
        board[tr + s -1][tc + s -1] = t
        chessboard_cover(board, tr, tc, tr + s -1, tc + s -1, s, t)
    #覆盖右上子棋盘(Cover the upper right sub-board)
    if dr < tr + s and dc >= tc + s:
        chessboard_cover(board, tr, tc + s, dr, dc, s, t)
    else:
        board[tr + s -1][tc + s] = t
        chessboard_cover(board, tr, tc + s, tr + s -1, tc + s, s, t)
    #覆盖左下子棋盘(Cover the lower left sub-board)
    if dr >= tr + s and dc < tc + s:
        chessboard_cover(board, tr + s, tc, dr, dc, s, t)
    else:
        board[tr + s][tc + s -1] = t
        chessboard_cover(board, tr + s, tc, tr + s, tc + s -1, s, t)
    #覆盖右下子棋盘(Cover the lower right sub-board)
    if dr >= tr + s and dc >= tc + s:
        chessboard_cover(board, tr + s, tc + s, dr, dc, s, t)
    else:
        board[tr + s][tc + s] = t
        chessboard_cover(board, tr + s, tc + s, tr + s, tc + s, s, t)
#示例用法(Example usage)
size = 2  #棋盘大小为 2^size × 2^size(The size of the board is 2^size × 2^size)
board = [[0] * (2 **size) for _ in range(2 **size)]
tile_count = 1  #起始的骨牌编号(The starting domino number)
chessboard_cover(board, 0, 0, 3, 3, 2 **size, 0)
#打印覆盖后的棋盘(Print the covered board)
for row in board:
    print(row)
```

```
[2, 2, 3, 3]
[2, 1, 1, 3]
[4, 1, 5, 5]
[4, 4, 5, 0]
```

图 3.37 棋盘覆盖问题程序运行结果图

Figure 3.37 Result of the chessboard covering problem

3.5 循环赛日程表（Round Robin Schedule）

3.5.1 算法思想（Algorithm Idea）

设有 $n=2^k$ 个运动员要进行羽毛球循环赛，现要设计一个满足以下要求的比赛日程表：每个选手必须与其他 $n-1$ 个选手各赛一次；每个选手一天只能比赛一次；循环赛一共需要进行 $n-1$ 天。由于 $n=2^k$，显然 n 为偶数。

With $n=2^k$ players to play a round-robin badminton tournament, it needs to design a tournament schedule that meets the following requirements: Each player must play once against other $n-1$ players; each player can compete only once a day; the round robin takes a total of $n-1$ days. Since $n=2^k$, obviously n is an even number.

1. 分解（Decompose）

根据分治算法的思想，将选手一分为二，n 个选手的比赛日程表可以通过对 $n/2=2^{k-1}$ 个选手设计的比赛日程表来实现，而 2^{k-1} 个选手的比赛日程表可通过对 $2^{k-1}/2=2^{k-2}$ 个选手设计的比赛日程表来实现，以此类推，2^2 个选手的比赛日程表可通过对两个选手设计的比赛日程表来实现。此时，问题的求解将变得非常简单。

According to the idea of divide-and-conquer algorithm, the players are divided into two parts. The schedule of n players can be realized by designing the schedule of $n/2=2^{k-1}$ players, and the schedule of 2^{k-1} players can be realized by designing the schedule of $2^{k-1}/2=2^{k-2}$ players, and so on. A competition schedule of 2^2 players can be realized by a competition schedule designed for two players. At this point, the solution of the problem will become very simple.

2. 治理（Conquer）

递归解决两个规模为 2^{k-1} 个选手的子问题，然后让两组选手对打，就可以排出整个循环赛日程表。递归停止的条件为问题的规模为 1 时，1 个选手不用安排比赛日程。

Recursively solving two subproblems of 2^{k-1} player size and then having the two pairs of players play each other can complete the round-robin schedule. The condition of recursive stop is that when the size of the problem is 1, and 1 player does not have to schedule a match.

如 $n=2$，分为两组，每组 1 个选手，边界条件，不用安排。然后两组对打，安排如图 3.38 所示。

For example, if $n=2$, divided into two groups, 1 player in each group, boundary conditions, no arrangement. Then the two groups play against each other, and the arrangement is shown in Figure 3.38.

如 $n=4$,分为两组(1,2)和(3,4),每组两个选手,递归安排两组,然后两组对打,安排如图 3.39 所示。

If $n=4$, divide into two groups (1,2) and (3,4), each group has two players, arrange two groups recursively, and then the two groups play against each other, as shown in Figure 3.39.

天数(days) 编号(Number)	1
1	2
2	1

图 3.38 $n=2$ 日程表

Figure 3.38 Schedule of $n=2$

天数(days) 编号(Number)	1	2	3
1	2	3	4
2	1	4	3
3	4	1	2
4	3	2	1

图 3.39 $n=4$ 日程表

Figure 3.39 Schedule of $n=4$

3.5.2 算法设计与描述(Algorithm Design and Description)

这种解法是把求解 2^k 个选手比赛日程问题划分成依次求解 2^1、2^2、…、2^k 个选手的比赛日程问题,换言之,2^k 个选手的比赛日程是在 2^{k-1}个选手的比赛日程的基础上通过迭代的方法求得的。在每次迭代中,将问题划分为 4 部分。

This method divides the problem of solving 2^k players' competition schedule into solving 2^1, 2^2, …, 2^k, in other words, the competition schedule of 2^k players is calculated by an iterative method based on the competition schedule of 2^{k-1} players, in each iteration, the problem is divided into four parts.

(1) 左上角。左上角为 2^{k-1}个选手在前半程的比赛日程。

Top left corner. Top left corner is the competitions schedule of 2^{k-1} players in the first half.

(2) 左下角。左下角为另 2^{k-1}个选手在前半程的比赛日程,由左上角加 2^{k-1}得到,例如,2^2 个选手比赛,左下角由左上角直接加 2 得到;2^3 个选手比赛,左下角由左上角直接加 4 得到。

Lower left corner. The lower left corner is the competition schedule of another 2^{k-1} player in the first half, which is obtained by adding 2^{k-1} in the upper left corner, for example, when 2^2 players play, the lower left corner is obtained by adding 2 directly to the top left corner; when 2^3 players play, the lower left corner is obtained by adding 4 directly from the top left corner.

（3）右上角。将左下角直接抄到右上角得到另外 2^{k-1} 个选手在后半程的比赛日程。

Top right corner. Copy the lower left corner directly to the top right corner to get the other 2^{k-1} players in the second half of the competition schedule.

（4）右下角。将左上角直接抄到右下角得到 2^{k-1} 个选手在后半程的比赛日程。

Lower right corner. Copy the top left corner directly to the lower right corner to get the competition schedule of 2^{k-1} players in the second half of the course.

算法设计的关键在于寻找这 4 部分元素之间的对应关系。

The key of algorithm design is to find the corresponding relationship between these four elements.

3.5.3　算法实现（Algorithm Implementation）

1. 实例构造（Instance Construction）

假设有 8 个运动员要进行羽毛球循环赛，要求每个选手必须与其他 7 个选手各赛一次，每个选手一天只能比赛一次，循环赛一共需要进行 7 天。即安排 $n=2^3$ 个选手 $n-1$ 天的比赛日程，安排过程如下。

Suppose there are eight players to play a round-robin badminton match, and each player must play once with the other seven players. Each player can only play once a day, and the round-robin match takes seven days in total. That is, arrange a competition schedule for $n=2^3$ players for $n-1$ days, as follows.

将 8 个选手（1,2,3,4,5,6,7,8）分为两组（1,2,3,4）和（5,6,7,8），每组 4 个选手。

Divide the eight players (1,2,3,4,5,6,7,8) into two groups (1,2,3,4) and (5,6,7,8) with four players in each group.

将 4 个选手分为两组，每组 2 个选手，分别为（1,2）、（3,4）、（5,6）、（7,8）。

The four players were divided into two groups with two players in each group, whichwere (1,2), (3,4), (5,6), and (7,8) respectively.

将两个选手分为两组，每组一个选手，到了递归的边界条件，开始回归。回归时，组内比赛日程已经安排好，剩下只需两组对打。第一天的比赛安排如图 3.40 所示。

Divide the two players into two groups with one player in each group. When the recursion boundary condition is reached, the regression is started. When it comes back, the schedule is already set, and only two sets of matches are left. The tournament schedule for the first day is shown in Figure 3.40.

继续回归，每组两个选手的已经安排好，将两组对打，便可以得到前三天的比赛日程表，如图 3.41 所示。

Continue to return, 2 players in each group have been arranged, and the competition schedule of the first 3 days can be obtained by beating the two groups, as shown in Figure 3.41.

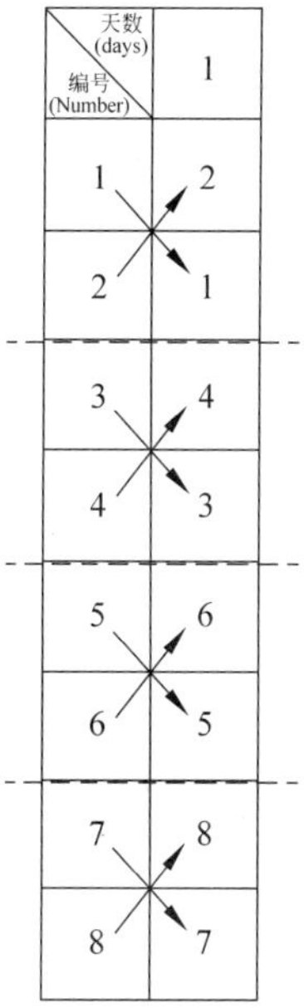

图 3.40 $n=8$ 的第一天竞赛安排示意图

Figure 3.40 Schematic diagram of the competition schedule for the first day with $n=8$

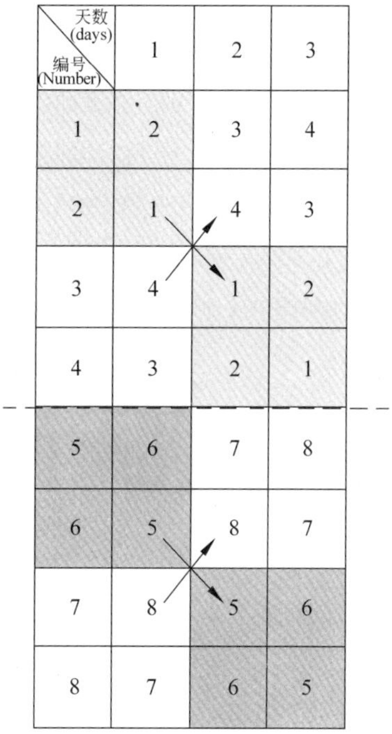

图 3.41 $n=8$ 的前三天竞赛安排示意图

Figure 3.41 Schematic diagram of the competition schedule for the first three days with $n=8$

继续回归,每组 4 个选手已经安排好,将两组对打,便可以得到 7 天的比赛日程表,如图 3.42 所示。

Continue to return, each group of 4 players has been arranged, and the two groups will play against each other to get a 7-day competition schedule, as shown in Figure 3.42.

天数(days) 编号(Number)	1	2	3	4	5	6	7
1	2	3	4	5	6	7	8
2	1	4	3	6	5	8	7
3	4	1	2	7	8	5	6
4	3	2	1	8	7	6	5
5	6	7	8	1	2	3	4
6	5	8	7	2	1	4	3
7	8	5	6	3	4	1	2
8	7	6	5	4	3	2	1

图 3.42 $n=8$ 时 7 天的竞赛安排示意图

Figure 3.42   Schematic diagram of the seve n-day competition arrangement when $n=8$

2. 实例实现(Instance Implementation)

上述实例实现代码如下,具体实现结果如图 3.43 所示。

The above example implementation code is as follows, and the result is shown in Figure 3.43.

```
def arrange(p, q, n, arr):
    if(n>1):                              #规模大于1时,分解(If the scale is greater than 1, decompose)
        arrange(p, q, n//2, arr)          #递归解决子问题(Solve subproblems recursively)
        arrange(p, q+n//2, n//2, arr)     #递归解决子问题(Solve subproblems recursively)
        #两组对打 (Double sparring)
        #填左下角 (Fill in the bottom left corner)
        for i in range(p+n//2, p+n):
            for j in range(q, q+n//2):
                arr[i][j] = arr[i-n//2][j+n//2]
        # 填右下角 (Fill in the bottom right corner)
        for i in range(p+n//2, p+n):
            for j in range(q+n//2, q+n):
                arr[i][j] = arr[i-n//2][j-n//2]
```

续

```
        return arr
if _ _name_ _ = = '_ _main_ _':
    import numpy as np
    k = 3
    n = 2 **k
    arr = np.zeros((n, n), dtype = int)
    for i in range(n):
        arr[0][i] = i+1
    arrange(0, 0, n, arr)
    print(arr)
```

```
[[1 2 3 4 5 6 7 8]
 [2 1 4 3 6 5 8 7]
 [3 4 1 2 7 8 5 6]
 [4 3 2 1 8 7 6 5]
 [5 6 7 8 1 2 3 4]
 [6 5 8 7 2 1 4 3]
 [7 8 5 6 3 4 1 2]
 [8 7 6 5 4 3 2 1]]
```

图 3.43 循环赛日程表运行结果

Figure 3.43 Result of the program for round-robin schedule problems

运行结果中,第一行为 1 2 3 4 5 6 7 8,1 表示本行都是 1 号选手和其他选手的比赛,如果第 2 个数为 2(其下标可以看成 1)表示第一天 1 号和 2 号比赛,第 4 个数为 4(其下标可以看成 3),表示第三天 1 号和 4 号比赛,以此类推。

In the running results, the first row 1 2 3 4 5 6 7 8, 1 means that the row is all a match between player 1 and other players. If the second number is 2 (its subscript can be viewed as 1), it means game 1 and 2 on the first day, and the fourth number is 4 (its subscript can be viewed as 3), it means game 1 and 4 on the third day, and so on.

习题 3(Exercises Three)

1. 简单叙述分治法具有哪些特征。

Please briefly describe the characteristics of divide-and-conquer algorithm.

2. 简述合并算法的思想。

Please briefly describe the idea of the merging algorithm.

3. 采用快速排序的思想将给定序列 $T[1:9] = \{45,23,65,57,18,2,90,12,84\}$ 由小到大排序。要求:先描述快速排序的思想,然后写出首次分解得到两个子问题的过程、递归的结果、子问题的解合并成原问题的解的过程。

Using the idea of quick sorting, sort the given sequence $T[1:9] = \{45,23,65,57,18,$

2,90,12,84} from small to large. Requirements: First describe the idea of quick sorting, and then write the process of first decomposition to get two subproblems, recursive results, and the process of merging the solution of the subproblem into the solution of the original problem.

4. 用分治法设计的合并排序算法描述如下,其中,$A[low:high]$存放待排序序列,请填写空缺部分的语句。

The divide-and-conquer merge sort algorithm is described as follows, where $A[low:high]$ stores the sequence to be sorted, please fill in the blanks.

```
void MergeSort (int A[ ],int low,int high)
{    int middle;
        if (low<high)
          {
                _____;
                MergeSort(A,low,middle);
                _____;
                Merge(A,low,middle,high);
          }
}
```

5. 给定一个整数数组[3,9,6,2,8,5,1,7],设计一个分治算法来找到数组中的最大元素,请用 *Python* 代码实现。

Given an array of integers [3, 9, 6, 2, 8, 5, 1, 7], design a divide-and-conquer algorithm to find the largest element in the array, do this in *Python* code.

第 4 章

动态规划算法

Chapter 4 Dynamic Programming Algorithm

动态规划(Dynamic Programming)是在 20 世纪 50 年代由美国数学家贝尔曼(Richard Bellman)为研究最优控制问题而提出的,"programming"的含义是计划和规划的意思。在计算机科学界,动态规划算法成为一种通用的算法设计技术用来求解多阶段决策最优化问题。

Dynamic programming algorithm was proposed by Richard Bellman, an American mathematician, in the 1950s for the study of optimal control problems. The meaning of "programming" is to plan. In computer science, dynamic programming has become a general algorithm design technique for solving multi-stage decision optimization problems.

4.1 概述(Overview)

动态规划算法通常用于求解具有某种最优性质的问题,在这类问题中,可能会有许多可行解,每一个可行解都对应一个值,而最终的目的是找到具有最优价值的解。动态规划算法是求解最优化问题的一种途径、一种方法,而不是一种特殊算法。针对最优化问题,由于各个问题的性质不同,确定最优解的条件也互不相同,因而动态规划算法的设计也各具特色,而不存在一种万能的动态规划算法可以解决各类最优化问题。

Dynamic programming algorithms are often used to solve problems with certain optimal properties, where there may be many feasible solutions, each of which corresponds to a value, and the ultimate goal is to find the solution with the optimal value. Dynamic programming method is a way and a method to solve optimization problems, not a special algorithm. For optimization problems, due to the different nature of each problem, the conditions for determining the optimal solution are different, so the design methods of dynamic programming have their own characteristics, and there is no universal dynamic programming algorithm that can solve various optimization problems.

4.1.1 基本思想（Basic Idea）

动态规划算法的思想比较简单,其实质是分治思想和解决冗余,因此它与分治法和贪心法类似,都是将待求解问题分解为更小的、相同的子问题,然后对子问题进行求解,最终产生一个整体最优解。

The idea of dynamic programming algorithm is relatively simple. Its essence is the divide-and-conquer idea and the solution of redundancy, so it is like the divide-and-conquer method and the greedy method, which decompose the problem into smaller, identical subproblems, and then solve the subproblems, eventually producing an overall optimal solution.

适合采用动态规划算法求解的问题,经分解得到的各个子问题往往不是相互独立的。在求解过程中,将已解决的子问题的解进行保存,在需要时可以轻松找出。这样就避免了大量无意义的重复计算,从而降低算法的时间复杂性。如何对已解决的子问题的解进行保存呢? 通常采用表的形式,即在实际求解过程中,一旦某个子问题被计算过,不管该问题以后是否用得到,都将其计算结果填入该表,需要时就从表中找出该子问题的解。具体的动态规划算法多种多样,但它们具有相同的填表格式。

When the problem is suitable to be solved by dynamic programming method, the subproblems obtained by decomposition are often not independent of each other. During the solving process, the solutions of the solved subproblems are saved and can be easily found out when needed. In this way, many meaningless repeated calculations are avoided, and the time complexity of the algorithm is reduced. How to save the solution of a solved subproblem? It is usually in the form of a table, that is, in the actual solution process, once a subproblem has been calculated, regardless of whether the problem will be used in the future, its calculation results will be filled in the table, and the solution to the subproblem will be found from the table when needed. The specific dynamic programming algorithms are varied, but they all have the same form filling format.

4.1.2 算法的解题步骤（Algorithm Solution Steps）

动态规划算法适合用来求解最优化问题,通常可按下列步骤对算法的解题过程进行设计。

Dynamic programming algorithm is suitable for solving optimization problems. Usually, the process of solving the algorithm can be designed according to the following steps.

（1）分析最优解的性质,刻画最优解的结构特征,考查是否适合采用动态规划算法。

The properties of the optimal solution are analyzed, the structural characteristics of the optimal solution are depicted, and whether the dynamic programming method is suitable is examined.

（2）递归定义最优价值（即建立递归式或动态规划方程）。

Define the optimal value recursively （that is, create a recursion or dynamic programming equation）。

（3）以自底向上的方式计算出最优价值,并记录相关信息。

The optimal value is calculated in a bottom-up manner and the relevant information is recorded.

（4）根据计算最优价值时得到的信息,构造出最优解。

Based on the information obtained when calculating the optimal value, the optimal solution is constructed.

另外,在进一步探讨动态规划的设计方法及应用之前,有两点需要注意：首先是问题能否用动态规划进行求解,不恰当的方式将使问题的描述不具有最优子结构性质,从而无法建立最优价值的递归关系,动态规划的应用也就无从谈起。

In addition, before further discussing the design method and application of dynamic programming, there are two points that need to be noted：First, whether the problem can be solved by dynamic programming is very important, and inappropriate way will make the description of the problem without the property of optimal substructure, so that the recursive relationship of optimal values cannot be established, and the application of dynamic programming will not be possible.

因此,步骤（1）是最关键的一步。另外在算法的实现过程中,应充分利用子问题的重叠性质来提高解题效率。具体地说,应采用递推（迭代）的方法来编程计算由递归式定义的最优价值,而不是采用直接递归的方法。

Therefore, step（1）is the most critical step. Second, in the implementation of the algorithm, we should make full use of the overlapping property of the subproblems to improve the efficiency of solving the problem. Specifically, a recursive （iterative） method should be used to program the calculation of the optimal value defined by the recursion, rather than direct recursive methods.

4.1.3　动态规划的基本要素（Essential Elements of Dynamic Programming）

任何一种算法的思想方法都有其局限性,超出特定条件,它就失去了作用。同样,动态规划算法并非适合于求解所有的最优化问题,采用该算法求解的问题应具备三个基本要素：最优子结构性质、子问题重叠性质和自底向上的求解方法。在这三大要素的指导下,可以对某问题是否适合采用动态规划算法求解进行预判。

Any kind of algorithm thinking method has its limitations, beyond certain conditions, it will lose its role. Similarly, the dynamic programming algorithm is not suitable for solving all optimization problems. The problems solved by the dynamic programming algorithm should have three basic elements：the optimal substructure property, the subproblem overlap property, and the bottom-up solution method. Under the guidance of these three elements, we can predict whether a problem is suitable to be solved by a dynamic programming algorithm.

1. 最优子结构性质（Optimal Substructural Properties）

最优子结构性质，通俗地讲就是问题的最优解包含其子问题的最优解。最优子结构性质是动态规划的基础，任何问题，如果不具备该性质，就不可能用动态规划方法来解决。总之，根据最优子结构性质导出的动态规划基本方程是解决一切动态规划问题的基本方法。

The property of optimal substructure is that the optimal solution of a problem contains the optimal solution of its subproblem. The optimal substructure property is the basis of dynamic programming. Any problem without this property cannot be solved by dynamic programming method. In a word, the basic equation of dynamic programming derived from the properties of optimal substructure is the basic method to solve all dynamic programming problems.

2. 子问题重叠性质（Overlapping Properties of Subproblems）

递归算法求解问题时，每次产生的子问题并不总是新问题，有些子问题出现多次，这种性质称为子问题的重叠性质。

When recursive algorithms solve problems, the subproblems generated each time are not always new problems, and some subproblems appear multiple times, which is called the overlapping property of subproblems.

在应用动态规划时，对于重复出现的子问题，只需在第一次遇到时就加以解决，并把已解决的各个子问题的解存储在表中，便于以后遇到时直接引用，从而不必重新求解，可大大提高解题的效率。子问题重叠性质并不是动态规划适用的必要条件，但是如果该性质无法满足，动态规划算法同其他算法相比就不具备优势。

In the application of dynamic programming, for recurring subproblems, only need to be solved at the first time, and the solution of each subproblem that has been solved is stored in the table, which is convenient for direct reference in the future, so that it does not have to solve again, which can greatly improve the efficiency of solving problems. The property of overlapping subproblems is not a necessary condition for dynamic programming, but if the property is not satisfied, the dynamic programming algorithm has no advantage over other algorithms.

3. 自底向上的求解方法（Bottom-up Solution Method）

由于动态规划解决的问题具有子问题重叠性质，求解时需要采用自底向上的方法。首先选择合适的表格，将递归的停止条件填入表格的相应位置；然后将问题的规模一级一级放大，求出每一级子问题的最优价值，并将其填入表格的相应位置，直到问题所要求的规模，此时便可以求出原问题的最优价值。

Because the problems solved by dynamic programming have overlapping subproblems, the bottom-up method is needed to solve them. Firstly, the appropriate table is selected and the recursive stopping condition is filled in the corresponding position of the table. Then, the scale of the problem is enlarged step by step, and the optimal value of each subproblem is obtained, and it is filled in the corresponding position of the table until the required scale of the problem, at which time, the optimal value of the original problem can be obtained.

4.2　矩阵连乘问题（Matrix Multiplication Problem）

4.2.1　问题分析（Problem Analysis）

1. 问题描述（Problem Description）

给定 n 个矩阵 $\{A_1, A_2, A_3, \cdots, A_n\}$，其中，$A_i$ 与 $A_{i+1}(i=1,2,\cdots,n-1)$ 是可乘的。用加括号的方法表示矩阵连乘的次序，不同加括号的方法所对应的计算次序是不同的。其所对应的计算量也是不同的，甚至差别很大。由于在矩阵相乘的过程中，仅涉及加法和乘法两种基本运算，乘法耗时远远大于加法耗时，故采用矩阵连乘所需乘法的次数来对不同计算次序的计算量进行衡量。

Given n matrices $\{A_1, A_2, A_3, \cdots, A_n\}$, where A_i and $A_{i+1}(i=1,2,\cdots,n-1)$ are multiplicative. The order of matrix multiplication is denoted by the method of adding parentheses. The order of calculation is different for different methods of adding parentheses. The amount of computation is also different, even very different. In the process of matrix multiplication, only two basic operations are involved, addition and multiplication, and the time of multiplication is much larger than that of addition. Therefore, the number of multiplications required by matrix multiplication is used to measure the calculation amount of different calculation orders.

即矩阵连乘问题就是对于给定 n 个连乘的矩阵，找出一种加括号的方法，使得矩阵连乘的计算量最小。

That is, the problem of matrix multiplication is to find a way to add parentheses to minimize the amount of calculation of matrix multiplication for a given n consecutive matrices.

2. 建立最优价值的递归关系式（Establish the Recursive Relation of the Optimal Value）

$A_iA_{i+1}\cdots A_j$ 矩阵连乘，其中，矩阵 A_m 的行数为 p_m，列数为 $q_m(m=i,i+1,\cdots,j)$ 且相邻矩阵是可乘的（即 $q_m=p_{m+1}$）。设它们的最佳计算次序所对应的乘法次数为 $m[i][j]$，则 $A_iA_{i+1}\cdots A_k$ 的最佳计算次序对应的乘法次数为 $m[i][k]$，$A_{k+1}A_{k+2}\cdots A_j$ 的最佳计算次序对应的乘法次数为 $m[k+1][j]$。

$A_iA_{i+1}\cdots A_j$ matrix multiplication, where the number of rows of the matrix A_m is p_m and the number of columns is q_m ($m=i, i+1, \cdots, j$) and the adjacent matrices are multiplicative (i.e. $q_m=p_{m+1}$). Suppose that the number of multiplications corresponding to their optimal computation order is $m[i][j]$, then the number of multiplications corresponding to the optimal computation order of $A_iA_{i+1}\cdots A_k$ is $m[i][k]$, The number of multiplications corresponding to the optimal computation order of $A_{k+1}A_{k+2}\cdots A_j$ is $m[k+1][j]$.

当 $i=j$ 时，只有一个矩阵，不用相乘，故 $m[i][i]=0$；

When $i=j$, there is only one matrix and no multiplication, so $m[i][i]=0$.

当 $i<j$ 时, 有

$$m[i][j] = \min_{i \leqslant k < j}\{m[i][k] + m[k+1][j] + p_i q_k q_j\} \qquad (4\text{-}1)$$

When $i<j$, so

$$m[i][j] = \min_{i \leqslant k < j}\{m[i][k] + m[k+1][j] + p_i q_k q_j\}$$

将 n 个矩阵的行数和列数存储在一维数组 $P[0:n]$ 中 (由于 $q_m = p_{m+1}$, 因此 P 中只需要存储 $n+1$ 个元素), 则第 i 个矩阵的行数存储在 P 的第 $i-1$ 个位置, 列数存储在 P 的第 i 个位置, 则上述递归式可改写为

If the rows and columns of the n matrices are stored in a one-dimensional array $P[0:n]$ (since $q_m = p_{m+1}$, only $n+1$ elements need to be stored in P), then the number of rows of the i th matrix is stored in the $i-1$ position of P, and the number of columns is stored in the i position of P, then the above recursion can be rewritten as follows.

$$m[i][j] = \begin{cases} 0, & i = j \\ \min\limits_{i \leqslant k < j}\{m[i][k] + m[k+1][j] + P_{i-1}P_k P_j\}, & i < j \end{cases} \qquad (4\text{-}2)$$

4.2.2 算法设计与描述（Algorithm Design and Description）

采用自底向上的方法求最优价值。

A bottom-up method is used to find the optimal value.

（1）确定合适的数据结构（Determine the Appropriate Data Structure）。

采用二维数组 m 来存放各个子问题的最优价值, 二维数组 s 来存放各个子问题的最优决策 (如果 $s[i][j]=k$, 则最优加括号方法为 $(A_i, \cdots, A_k)(A_{k+1}, \cdots, A_j)$), 一维数组 P。

A two-dimensional array m is used to store the optimal value of each subproblem, and two-dimensional array s is used to store the optimal decision of each subproblem (if $s[i][j] = k$, the optimal parenthesis method is $(A_i, \cdots, A_k)(A_{k+1}, \cdots, A_j)$), one-dimensional array P.

（2）初始化（Initialization）。

令 $m[i][i]=0, k[i][i]=0$, 其中, $i=1,2,\cdots,n$。

Let $m[i][i]=0$, $k[i][i]=0$, where $i=1,2,\cdots,n$.

（3）循环阶段（Cycle Stage）。

① 按照递归关系式计算两个矩阵 $A_i A_{i+1}$ 相乘时的最优价值并将其存入 $m[i][i+1]$, 同时将最优决策记入 $s[i][i+1]$, $i=1,2,\cdots,n-1$。

Calculate the optimal value of the multiplication of two matrices $A_i A_{i+1}$ according to the recursive relationship and store it in $m[i][i+1]$, and record the optimal decision in $s[i][i+1]$, $i=1,2, \cdots, n-1$.

② 按照递归关系式计算三个矩阵 $A_i A_{i+1} A_{i+2}$ 相乘时的最优价值并将其存入 $m[i][i+2]$, 同时将最优决策记入 $s[i][i+2]$, $i=1,2,\cdots,n-2$。

Calculate the optimal value of the multiplication of three matrices $A_i A_{i+1} A_{i+2}$ according to the recursive relationship and store it in $m[i][i+2]$, and record the optimal decision in $s[i][i+2]$, $i=1, 2, \cdots, n-2$.

③ 以此类推,直到按照递归关系式计算 n 个矩阵 $A_1A_2\cdots A_n$ 相乘时的最优价值并将其存入 $m[1][n]$,同时将最优决策记入 $s[1][n]$。

And so on, until the optimal value of multiplying n matrices $A_1A_2\cdots A_n$ is calculated according to the recursive relation and stored in $m[1][n]$, while the optimal decision is recorded in $s[1][n]$.

至此,$m[1][n]$ 即为原问题的最优价值。

So far, $m[1][n]$ is the optimal value of the original problem.

(4) 根据二维数组 s 记录的最优决策信息来构造最优解。

The optimal solution is constructed according to the optimal decision information recorded in two-dimensional array s.

① 递归构造 $A_1\cdots A_{s[1][n]}$ 的最优解,直到包含一个矩阵结束。

The optimal solution for $A_1\cdots A_{s[1][n]}$ is recursively constructed until a matrix is included.

② 递归构造 $A_{s[1][n]+1}\cdots A_n$ 的最优解,直到包含一个矩阵结束。

The optimal solution for $A_{s[1][n]+1}\cdots A_n$ is recursively constructed until a matrix is included.

③ 将①和②递归的结果加括号。

Parentheses the results of the ① and ② recursion.

4.2.3　算法实现(Algorithm Implementation)

1. 实例构造(Instance Construction)

求矩阵 $A_1(3\times2)$、$A_2(2\times5)$、$A_3(5\times10)$、$A_4(10\times2)$ 和 $A_5(2\times3)$ 连乘的最佳计算次序。计算过程如下。

Find the optimal calculation order of the matrix $A_1(3\times2)$, $A_2(2\times5)$, $A_3(5\times10)$, $A_4(10\times2)$ and $A_5(2\times3)$. The calculation process is as follows.

(1) 初始化(Initialization)。

令 $m[i][i]=0$,$s[i][i]=0$,其中,$i=1,2,\cdots,5$。

Let $m[i][i]=0$, $s[i][i]=0$, where $i=1, 2, \cdots, 5$.

(2) 计算两个矩阵最优价值(Calculate the optimal values for two matrices)。

按照递归关系式计算两个矩阵 A_iA_{i+1} 相乘时的最优价值,其中 $i=1,2,3,4$。

Calculate the optimal value of the multiplication of two matrices A_iA_{i+1} according to the recursive relationship, where $i=1, 2, 3, 4$.

当 $i=1$ 时,

when $i=1$

$m[1][2]=\min\{m[1][1]+m[2][2]+P_0P_1P_2\}=0+0+3\times2\times5=30\}$;$s[1][2]=1$。

当 i=2 时,

when $i=2$

$m[2][3] = \min\{m[2][2] + m[3][3] + P_1P_2P_3 = 0 + 0 + 2 \times 5 \times 10 = 100\}; s[2][3] = 2$。

以此类推，求得。

And so on.

$m[3][4] = 100, s[3][4] = 3; m[4][5] = 60, s[4][5] = 4$。

（3）计算三个矩阵最优价值（Calculate the optimal values for three matrices）。

按照递归关系式计算三个矩阵 $A_iA_{i+1}A_{i+2}$ 相乘时的最优价值，其中，$i = 1, 2, 3$。

Calculate the optimal value of the multiplication of three matrices $A_iA_{i+1}A_{i+2}$ according to the recursive relationship, where $i = 1, 2, 3$.

当 $i = 1$ 时，

when $i = 1$

$$m[1][3] = \min\begin{cases} m[1][1] + m[2][3] + P_0P_1P_3 = 0 + 100 + 3 \times 2 \times 10 = 160 \\ m[1][2] + m[3][3] + P_0P_2P_3 = 30 + 0 + 3 \times 5 \times 10 = 180 \end{cases}; s[1][3] = 1$$。

当 $i = 2$ 时，

when $i = 2$

$$m[2][4] = \min\begin{cases} m[2][2] + m[3][4] + P_1P_2P_4 = 0 + 100 + 2 \times 5 \times 2 = 120 \\ m[2][3] + m[4][4] + P_1P_3P_4 = 100 + 0 + 2 \times 10 \times 2 = 140 \end{cases}; s[2][4] = 2$$。

当 $i = 3$ 时，

when $i = 3$

$$m[3][5] = \min\begin{cases} m[3][3] + m[4][5] + P_2P_3P_5 = 0 + 60 + 5 \times 10 \times 3 = 210 \\ m[3][4] + m[5][5] + P_2P_4P_5 = 100 + 0 + 5 \times 2 \times 3 = 130 \end{cases}; s[3][5] = 4$$。

（4）计算四个矩阵最优价值（Calculate the optimal values for four matrix）。

按照递归关系式计算四个矩阵 $A_iA_{i+1}A_{i+2}A_{i+3}$ 相乘时的最优价值，其中，$i = 1, 2$。

Calculate the optimal value of the multiplication of four matrices $A_iA_{i+1}A_{i+2}A_{i+3}$ according to the recursive relationship, where $i = 1, 2$.

当 $i = 1$ 时，

when $i = 1$

$$m[1][4] = \begin{cases} m[1][1] + m[2][4] + P_0P_1P_4 = 0 + 120 + 3 \times 2 \times 2 = 132 \\ m[1][2] + m[3][4] + P_0P_2P_4 = 30 + 100 + 3 \times 5 \times 2 = 160; s[1][4] = 1 \\ m[1][3] + m[4][4] + P_0P_3P_4 = 160 + 0 + 3 \times 10 \times 2 = 220 \end{cases}$$。

当 $i = 2$ 时，

when $i = 2$

$$m[2][5] = \begin{cases} m[2][2] + m[3][5] + P_1P_2P_5 = 0 + 130 + 2 \times 5 \times 3 = 160 \\ m[2][3] + m[4][5] + P_1P_3P_5 = 100 + 60 + 2 \times 10 \times 3 = 220; s[2][5] = 4 \\ m[2][4] + m[5][5] + P_1P_4P_5 = 120 + 0 + 2 \times 2 \times 3 = 132 \end{cases}$$

（5）计算五个矩阵最优价值（Calculate the optimal values for five matrixs）。

按照递归关系式计算五个矩阵 $A_1A_2A_3A_4A_5$ 相乘时的最优价值。

Calculate the optimal value of the multiplication of five matrices $A_1A_2A_3A_4A_5$ according to the recursive relationship.

$$m[1][5]=\min\begin{cases} m[1][1]+m[2][5]+P_0P_1P_5=0+132+3\times2\times3=150 \\ m[1][2]+m[3][5]+P_0P_2P_5=30+130+3\times5\times3=205 \\ m[1][3]+m[4][5]+P_0P_3P_5=160+60+3\times10\times3=310 \\ m[1][4]+m[5][5]+P_0P_4P_5=132+0+3\times2\times3=150 \end{cases};s[1][5]=1。$$

具体结果如表 4.1 和表 4.2 所示。

The specific results are shown in Table 4.1 and Table 4.2.

表 4.1 实例最优价值 $m[i][j]$
Table 4.1 Optimal value of example $m[i][j]$

$m[i][j]$	A_1	A_2	A_3	A_4	A_5
A_1	0	30	160	132	150
A_2		0	100	120	132
A_3			0	100	130
A_4				0	60
A_5					0

表 4.2 实例最优决策 $s[i][j]$
Table 4.2 Example optimal decision $s[i][j]$

$s[i][j]$	A_1	A_2	A_3	A_4	A_5
A_1	0	1	1	1	1
A_2		0	2	2	4
A_3			0	3	4
A_4				0	4
A_5					0

（6）构造最优解（Construct the Optimal Solution）。

根据表 4.2 中记录的最优决策信息来构造最优解，递归构造 $A_1\cdots A_{s[1][5]}$ 和 $A_{s[1][5]+1}\cdots A_5$ 等子问题的最优解，直到包含一个矩阵结束，最后将递归的结果加括号，得到最终结果。具体过程如图 4.1 所示。

According to the optimal decision information recorded in Table 4.2 to construct the optimal solution, recursively construct the optimal solution of subproblems such as $A_1\cdots A_{s[1][5]}$ and $A_{s[1][5]+1}\cdots A_5$ until a matrix is included. Finally, the result of the recursion is parenthesis to obtain. The specific process is shown in Figure 4.1.

2. 代码实现（Code Implementation）

在 *Python* 中，选用二维数组 m 和 s 分别存储各个子问题的最优价值和最优决策，用一维数组 p 存储矩阵行列值，*res* 存储计算次序。

In *Python* programming, two-dimensional arrays m and s are used to store the optimal values and optimal decisions of each subproblem, one-dimensional arrays p are used to store matrix column values, and *res* is used to store the calculation order.

定义一个 *MatrixChain*( ) 函数，接收矩阵行列数据 p 和问题规模 n，输出最优价值二维表 m 和最优决策二维表 s。

Define a *MatrixChain*( ) function that receives matrix row and column data p and problem size n, and outputs a two-dimensional table of optimal values m and a two-dimensional table of optimal decisions s.

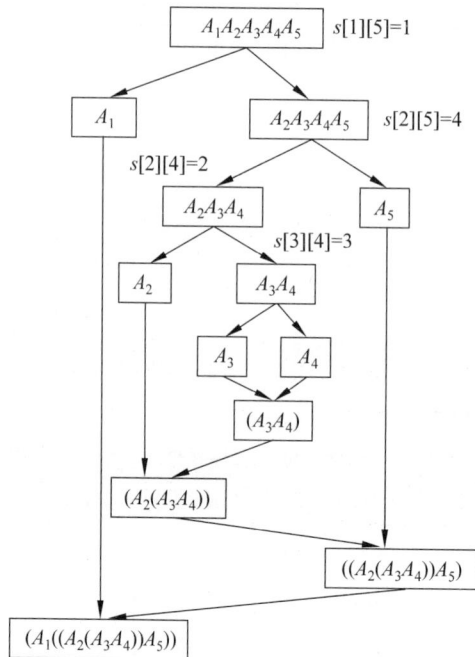

图 4.1　递归构造 $A_1A_2A_3A_4A_5$ 的最优解

Figure 4.1　Optimal solution of recursively constructed $A_1A_2A_3A_4A_5$

定义一个 *Traceback*( ) 函数构造问题的最优解，用 i,j 代表矩阵连乘子问题的范围（即 $A_i\cdots A_j$）、决策矩阵 s，输出最优计算次序 *res*。在 *main*( ) 函数中，给定一个实例的所有矩阵行列 *arr*，然后将其压缩存储在列表 p 中，调用 *MatrixChain*( ) 函数计算最优价值，调用 *Traceback*( ) 函数构造最优解，最后将计算次序输出到显示器上。程序运行结果如图 4.2 所示。代码如下。

Define a *Traceback*( ) function to construct the optimal solution of the problem, use i and j to represent the scope of the matrix multiplication subproblem(i.e. $A_i\cdots A_j$), decision matrix s, output the optimal calculation order *res*. In the *main*( ) function, given an instance of all matrix rows and columns *arr*, and then compressed store them in the list p, call the *MatrixChain*( ) function to calculate the optimal value, call the *Traceback*( ) function to construct the optimal solution, and finally the order of calculation is output to the display. The result of running the program is shown in Figure 4.2. The code is as follows.

```
import numpy as np
def MatrixChain(p, n):
    m = np.zeros((n + 1, n + 1))
    s = np.zeros((n + 1, n + 1))
    for i in range(n + 1):
        m[i][i] = 0
        s[i][i] = 0
```

续

```
    for r in range(2, n + 1):
        for i in range(1, n -r + 2):
            j = i + r -1
            m[i][j] = m[i + 1][j] + p[i -1] * p[i] * p[j]
            s[i][j] = i
            for k in range(i + 1, j):
                t = m[i][k] + m[k + 1][j] + p[i -1] * p[k] * p[j]
                if t < m[i][j]:
                    m[i][j] = t
                    s[i][j] = k
    return m, s


def Traceback(i, j, s):
    global res
    if i == j:
        res.append('A' + str(i))
    else:
        res.append('(')
        Traceback(i, int(s[i][j]), s)
        Traceback(int(s[i][j]) + 1, j, s)
        res.append(')')


if _ _name_ _ == "_ _main_ _":
    arr = [[3, 2], [2, 5], [5, 10], [10, 2], [2, 3]]
    n = len(arr)
    res = []
    p = []
    for i in range(n):
        if i == 0:
            p.append(arr[0][0])
            p.append(arr[0][1])
        else:
            p.append(arr[i][1])
    m, s = MatrixChain(p, n)
    Traceback(1, n, s)
    print('矩阵连乘的最优次序为(The optimal order of matrix multiplication is as follows):')
    print(''.join(res))
```

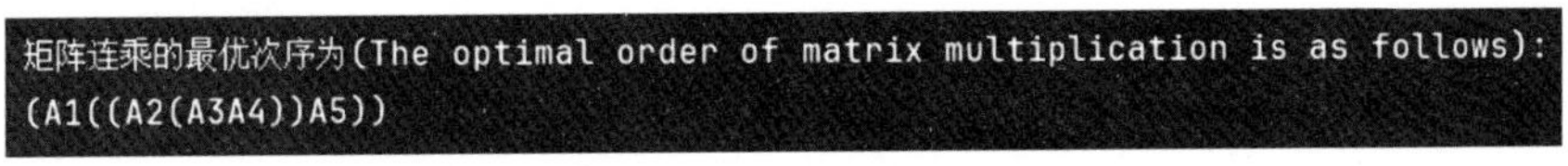

矩阵连乘的最优次序为(The optimal order of matrix multiplication is as follows):
(A1((A2(A3A4))A5))

图 4.2 矩阵连乘程序运行结果

Figure 4.2   Result of the matrix multiplication program

4.3 最长公共子序列问题（Longest Common Subsequence Problem）

4.3.1 问题分析（Problem Analysis）

1. 基本概念（Basic Conception）

1) 子序列（Subsequence）

给定序列 $X = \{x_1, x_2, x_3, \cdots, x_n\}$、$Z = \{z_1, z_2, z_3, \cdots, z_k\}$，若 Z 是 X 的子序列，当且仅当存在一个严格递增的下标序列 $\{i_1, i_2, i_3, \cdots, i_k\}$，对 $\forall j \in \{1, 2, 3, \cdots, k\}$ 有 $z_j = x_{i_j}$。如序列 $X = \{A, B, C, B, D, A, B\}$ 的子序列有 $\{A, B\}$、$\{B, C, A\}$、$\{A, B, C, D, A\}$ 等。

Given the sequence $X = \{x_1,\ x_2,\ x_3,\ \cdots,\ x_n\}$, $Z = \{z_1,\ z_2,\ z_3,\ \cdots,\ z_k\}$, if Z is a subsequence of X, when and only when there exists a strictly increasing sequence of subscripts $\{i_1,\ i_2,\ i_3,\ \cdots,\ i_k\}$, for $\forall j \in \{1,\ 2,\ 3,\ \cdots,\ k\}$ have $z_j = x_{i_j}$. Such as sequence $X = \{A,\ B,\ C,\ B,\ D,\ A,\ B\}$ subsequences are $\{A,\ B\}$, $\{B,\ C,\ A\}$, $\{A,\ B,\ C,\ D,\ A\}$, etc.

2) 公共子序列（Common Subsequence）

给定序列 X 和 Y，若序列 Z 是 X 的子序列，也是 Y 的子序列，则称 Z 是 X 和 Y 的公共子序列。如序列 $X = \{A, B, C, B, D, A, B\}$ 和序列 $Y = \{A, C, B, E, D, B\}$ 的公共子序列有 $\{A, B\}$、$\{C, B, D\}$、$\{A, C, B, D, B\}$ 等。

Given a sequence X and Y, if the sequence Z is a subsequence of X and is also a subsequence of Y, then Z is said to be a common subsequence of X and Y. For example, the common subsequences of sequences $X = \{A,\ B,\ C,\ B,\ D,\ A,\ B\}$ and $Y = \{A,\ C,\ B,\ E,\ D,\ B\}$ are $\{A,\ B\}$, $\{C,\ B,\ D\}$, $\{A,\ C,\ B,\ D,\ B\}$, etc.

3) 最长公共子序列（Longest Common Subsequence）

包含元素最多的公共子序列即为最长公共子序列。如上述 X 序列和 Y 序列的最长公共子序列为 $\{A, C, B, D, B\}$。

The longest common subsequence is that contains the most elements. For example, the longest common subsequence of the X and Y sequences above is $\{A,\ C,\ B,\ D,\ B\}$.

即最长公共子序列问题就是给定两个序列 $X = \{x_1, x_2, \cdots, x_m\}$ 和 $Y = \{y_1, y_2, \cdots, y_n\}$，找出 X 和 Y 的最长的公共子序列。

The longest common subsequence problem is that given two sequences $X = \{x_1, x_2, \cdots, x_m\}$ and $Y = \{y_1,\ y_2,\ \cdots,\ y_n\}$, to find the longest common subsequence of X and Y.

2. 建立最优价值的递归关系式（Establish the Recursive Relation of the Optimal Value）

假如 $L[i][j]$ 表示序列 X_i 和 Y_j 的最长公共子序列的长度，则最优价值的递推关系式如下。

Suppose $L[i][j]$ represents the length of the longest common subsequence of sequences X_i and Y_j. Then the optimal value recurrence relation is as follows.

$$L[i][j] = \begin{cases} 0, & i=0 \text{ or } j=0 \\ L[i-1][j-1]+1, & i,j>0 \text{ and } x_i = y_j \\ \max\{L[i][j-1], L[i-1][j]\}, & i,j>0 \text{ and } x_i \neq y_j \end{cases} \tag{4-3}$$

4.3.2 算法设计与描述(Algorithm Design and Description)

1. 确定合适的数据结构(Determine the Appropriate Data Structure)

采用二维数组 L 来存放各个子问题的最优价值,二维数组 b 来存放各个子问题最优价值的来源。$b[i][j]=1$ 表示 $L[i][j]$ 由 $L[i-1][j-1]+1$ 得到,$b[i][j]=2$ 表示 $L[i][j]$ 由 $L[i][j-1]$ 得到,$b[i][j]=3$ 表示 $L[i][j]$ 由 $L[i-1][j]$ 得到。数组 $x[1:m]$ 和 $[1:n]$ 分别存放 X 序列和 Y 序列。

Two-dimensional array c is used to store the optimal value of each subproblem, and two-dimensional array b is used to store the source of the optimal value of each subproblem. $b[i][j]=1$ is represent that $L[i][j]$ get from $L[i-1][j-1]+1$, $b[i][j]=2$ is represent that $L[i][j]$ get from $L[i][j-1]$, and $b[i][j]=3$ is $L[i][j]$ get from $L[i-1][j]$. The arrays $x[1:m]$ and $[1:n]$ hold the X and Y sequences, respectively.

2. 初始化(Initialization)

令 $L[i][0]=0, L[0][j]$,其中,$0 \leqslant i \leqslant m, 0 \leqslant j \leqslant n$。

Let $L[i][0]=0, L[0][j]$, where $0 \leqslant i \leqslant m, 0 \leqslant j \leqslant n$.

3. 循环阶段(Cycle Stage)

根据递归关系式,确定序列 X_i 和 Y_j 的最长公共子序列长度,$1 \leqslant i \leqslant m$。

According to the recursive relation, determine the length of the longest common subsequence of the sequences X_i and Y_j is determined, $1 \leqslant i \leqslant m$.

(1) 当 $i=1$ 时,求出 $L[1][j]$,同时记录 $b[1][j]$,$1 \leqslant j \leqslant n$。

When $i=1$, find $L[1][j]$ and record $b[1][j]$, $1 \leqslant j \leqslant n$.

(2) 当 $i=2$ 时,求出 $L[2][j]$,同时记录 $b[2][j]$,$1 \leqslant j \leqslant n$。

When $i=2$, find $L[2][j]$ and record $b[2][j]$, $1 \leqslant j \leqslant n$.

(3) 以此类推,直到 $i=m$ 时,求出 $L[m][j]$,同时记录 $b[m][j]$,$1 \leqslant j \leqslant n$。此时,$L[m][n]$ 便是序列 X 和 Y 的最长公共子序列长度。

And so on, until $i=m$, find $L[m][j]$, and record $b[m][j]$, $1 \leqslant j \leqslant n$. In this case, $L[m][n]$ is the longest common subsequence length of the sequences X and Y.

4. 根据二维数组 b 记录的相关信息以自底向上的方式来构造最优解

The optimal solution is constructed in a bottom-up manner according to the relevant information recorded in the two-dimensional array b.

(1) 初始时,$i=m, j=n$。

Initially, $i=m, j=n$.

（2）若 $b[i][j]=1$，则输出 $x[i]$，同时递推到 $b[i-1][j-1]$；若 $b[i][j]=2$，则递推到 $b[i][j-1]$，若 $b[i][j]=3$，则递推到 $b[i-1][j]$。

If $b[i][j]=1$, then output $x[i]$ and recursively to $b[i-1][j-1]$; If $b[i][j]=2$, the recursion to $b[i][j-1]$, if $b[i][j]=3$, the recursion to $b[i-1][j]$.

（3）重复执行步骤4中的（2），直到 $i=0$ 或 $j=0$，此时就可以得到序列 X 和 Y 的最长公共子序列。

Repeat（2）in step 4 until $i=0$ 或 $j=0$, at which point the longest common subsequence of the sequences X and Y is obtained.

4.3.3　算法实现（Algorithm Implementation）

1. 实例构造（Instance Construction）

给定序列 $X=\{A,B,C,B,D,A,B\}$ 和 $Y=\{B,D,C,A,B,A\}$，求它们的最长公共子序列。

A given sequences $X=\{A, B, C, B, D, A, B\}$ and $Y=\{B, D, C, A, B, A\}$, find their longest common subsequence.

（1）由题意得 $m=7,n=6$，将停止条件填入数组 L 和数组 b 中，即初始化，$L[i][0]=0,L[0][j]=0,b[i][0]=0,b[0][j]=0$，其中，$0\leqslant i\leqslant m,0\leqslant j\leqslant n$，如表4.3和表4.4所示。

According to the meaning of the question, $m=7$, $n=6$, fill the stopping condition into the array L and the array b, that is, initialization, $L[i][0]=0$, $L[0][j]=0$, $b[i][0]=0$, $b[0][j]=0$, where $0\leqslant i\leqslant m$, $0\leqslant j\leqslant n$, as shown in Table 4.3 and Table 4.4.

表 4.3　$i=0$ 或 $j=0$ 时的最优价值数组
Table 4.3　Optimal value array for $i=0$ or $j=0$

L		B	D	C	A	B	A
	0	0	0	0	0	0	0
A	0						
B	0						
C	0						
B	0						
D	0						
A	0						
B	0						

（2）当 $i=1$ 时，$X_1=\{A\}$，最后一个字符为 A；Y_j 的规模从1逐步放大到6，其最后一个字符分别为 B、D、C、A、B、A，根据递归关系式，当 $j=1$ 时，$B\neq A$，$L[1][1]=L[0][1]$，$b[1][1]=3$；当 $j=2$ 时，$D\neq A$，$L[1][2]=L[0][2]$，$b[1][2]=3$；当 $j=3$ 时，$C\neq A$，$L[1][3]=L[0][3]$，$b[1][3]=3$；当 $j=4$ 时，$A=A$，$L[1][4]=L[0][3]+1$，$b[1][4]=1$；当 $j=5$ 时，$B\neq A$，$L[1][5]=L[1][4]$，$b[1][5]=2$；当 $j=6$ 时，$A=A$，$L[1][6]=L[0][5]+1$，$b[1][6]=1$。结果如表4.5和表4.6所示。

表 4.4 $i=0$ 或 $j=0$ 时的状态数组

Table 4.4 State array for $i=0$ or $j=0$

			j				
b		B	D	C	A	B	A
	0	0	0	0	0	0	0
A	0						
B	0						
C	0						
B	0						
D	0						
A	0						
B	0						

(左侧列标注 i)

表 4.5 $i=1$ 时的最优价值数组

Table 4.5 Optimal value array for $i=1$

			j				
L		B	D	C	A	B	A
	0	0	0	0	0	0	0
A	0	0	0	0	1	1	1
B	0						
C	0						
B	0						
D	0						
A	0						
B	0						

(左侧列标注 i)

When $i=1$, $X_1 = \{A\}$, the last character is A; The scale of Y_j is gradually enlarged from 1 to 6, and its last character is B, D, C, A, B, A respectively. According to the recursive relationship, when $j=1$, $B \neq A$, $L[1][1] = L[0][1]$, $b[1][1] = 3$; When $j=2$, $D \neq A$, $L[1][2] = L[0][2]$, $b[1][2] = 3$; When $j=3$, $C \neq A$, $L[1][3] = L[0][3]$, $b[1][3] = 3$; When $j=4$, $A=A$, $L[1][4] = L[0][3]+1$, $b[1][4] = 1$; When $j=5$, $B \neq A$, $L[1][5] = L[1][4]$, $b[1][5] = 2$; When $j=6$, $A=A$, $L[1][6] = L[0][5]+1$, $b[1][6] = 1$. The results are shown in Tables 4.5 and 4.6.

（3）当 $i=2$ 时，$X_2 = \{A, B\}$，最后一个字符为 B；Y_j 的规模从 1 逐步放大到 6，其最后一个字符分别为 B、D、C、A、B、A，根据递归关系式，当 $j=1$ 时，$B=B$，$L[2][1] = L[1][0]+1$，$b[2][1] = 1$；当 $j=2$ 时，$D \neq B$，$L[2][2] = L[2][1]$，$b[2][2] = 2$；当 $j=3$ 时，$C \neq B$，$L[2][3] = L[2][2]$，$b[2][3] = 2$；当 $j=4$ 时，$A \neq B$，$L[2][4] = L[1][4]$，$b[2][4] = 3$；当 $j=5$ 时，$B=B$，$L[2][5] = L[1][4]+1$，$b[2][5] = 1$；当 $j=6$ 时，$A \neq B$，$L[2][6] = L[2][5]$，$b[2][6] = 2$。结果如表 4.7 和表 4.8 所示。

表 4.6 $i=1$ 时的状态数组

Table 4.6   State array for $i=1$

			j				
b		B	D	C	A	B	A
		0	0	0	0	0	0
i	A	0	3	3	3	1	2	1
	B	0						
	C	0						
	B	0						
	D	0						
	A	0						
	B	0						

When $i=2$, $X_2=\{A, B\}$, and the last character is B; The scale of Y_j is gradually enlarged from 1 to 6, and its last character is B, D, C, A, B, A respectively. According to the recursive relation, when $j=1$, $B=B$, $L[2][1]=L[1][0]+1$, $b[2][1]=1$; When $j=2$, $D\neq B$, $L[2][2]=L[2][1]$, $b[2][2]=2$; When $j=3$, $C\neq B$, $L[2][3]=L[2][2]$, $b[2][3]=2$; When $j=4$, $A\neq B$, $L[2][4]=L[1][4]$, $b[2][4]=3$; When $j=5$, $B=B$, $L[2][5]=L[1][4]+1$, $b[2][5]=1$; When $j=6$, $A\neq B$, $L[2][6]=L[2][5]$, $b[2][6]=2$. The results are shown in Tables 4.7 and 4.8.

表 4.7 $i=2$ 时的最优价值数组

Table 4.7   Optimal value array for $i=2$

			j				
L		B	D	C	A	B	A
		0	0	0	0	0	0
i	A	0	0	0	0	1	1	1
	B	0	1	1	1	1	2	2
	C	0						
	B	0						
	D	0						
	A	0						
	B	0						

表 4.8 *i* = 2 时的状态数组

Table 4.8 State array for *i* = 2

j

b		B	D	C	A	B	A
	0	0	0	0	0	0	0
A	0	3	3	3	1	2	1
B	0	1	2	2	3	1	2
C	0						
B	0						
D	0						
A	0						
B	0						

i (leftmost label for rows)

（4）以此类推，直到 $i=7$。$X_7 = \{A, B, C, B, D, A, B\}$，$X$ 的最后一个字符为 B；Y_j 的规模从 1 逐步放大到 6，其最后一个字符分别为 B、D、C、A、B、A，根据递归关系式，计算结果如表 4.9 和表 4.10 所示。

And so on until $i=7$. $X_7 = \{A, B, C, B, D, A, B\}$, the last character of X is B; The scale of Y_j is gradually enlarged from 1 to 6, and its last character is B, D, C, A, B, and A respectively. According to the recursive relationship, the results of calculation are shown in Table 4.9 and Table 4.10.

表 4.9 *i* = 7 时的最优价值数组

Table 4.9 Optimal value array for *i* = 7

j

L		B	D	C	A	B	A
	0	0	0	0	0	0	0
A	0	0	0	0	1	1	1
B	0	1	1	1	1	2	2
C	0	1	1	2	2	2	2
B	0	1	1	2	2	3	3
D	0	1	2	2	2	3	3
A	0	1	2	2	3	3	4
B	0	1	2	2	3	4	4

i (leftmost label for rows)

表 **4.10**　$i=7$ 时的状态数组

Table 4.10　State array for $i=7$

		j					
b		B	D	C	A	B	A
		0	0	0	0	0	0	0
	A	0	3	3	3	1	2	1
	B	0	1	2	2	3	1	2
i	C	0	3	3	1	2	3	3
	B	0	1	3	3	3	1	2
	D	0	3	1	3	3	3	3
	A	0	3	3	3	1	3	1
	B	0	1	3	3	3	1	3

（5）从 $i=7$，$j=6$ 处向前递推，由于 $b[7][6]=3$，递推到 $b[6][6]$；$b[6][6]=1$，输出 $X[6]$，即字符 A，递推到 $b[5][5]$；$b[5][5]=3$，递推到 $b[4][5]$；$b[4][5]=1$，输出 $X[4]$，即字符 B，递推到 $b[3][4]$；$b[3][4]=2$，递推到 $b[3][3]$；$b[3][3]=1$，输出 $X[3]$，即字符 C，递推到 $b[2][2]$；$b[2][2]=2$，递推到 $b[2][1]$；$b[2][1]=1$，输出 $X[2]$，即字符 B，递推到 $b[1][0]$；此时，$j=0$，算法结束，由于以上过程是逆推，所以 X 和 Y 的最长公共子序列应为 $\{B, C, B, A\}$。

From $i=7$, $j=6$ forward recursion, since $b[7][6]=3$, recursion to $b[6][6]$; $b[6][6]=1$, output $X[6]$, that is, character A, recursively to $b[5][5]$; $b[5][5]=3$, recursively to $b[4][5]$; $b[4][5]=1$, output $X[4]$, that is, character B, recursively to $b[3][4]$; $b[3][4]=2$, recursively to $b[3][3]$; $b[3][3]=1$, output $X[3]$, that is, character C, recursively to $b[2][2]$; $b[2][2]=2$, recursively to $b[2][1]$; $b[2][1]=1$, output $X[2]$, that is, character B, recursively to $b[1][0]$; At this point, $j=0$, the algorithm ends. Since the above process is a backward push, the longest common subsequence of X and Y should be $\{B, C, B, A\}$.

2. 代码实现（Code Implementation）

在 *Python* 语言中，选用二维列表 L 和 b 分别存储最优价值和相关信息。

In *Python* language, the two-dimensional list L and b are selected to store the optimal value and related information respectively.

首先定义一个函数 *lcs*( )求解最优价值，同时记录相关信息。*lcs* 接收字符串 X 和 Y 输出最优价值 L 和相关信息 b。定义 *printLcs*( )函数构造最优解，接收输入的相关信息 b、字符串 X、子问题 A_i，B_j，输出最长公共子序列。

First define a function *lcs* ( ) to solve for the optimal value, while recording the relevant information, *lcs* receives strings X and Y and outputs the optimal value L and the related information b. The *printLcs* ( ) function is defined to construct the optimal solution, receiving the relevant information of input b, string X, subproblem A_i, B_j, and output the longest common subsequence.

在 *main*( )函数中,给定字符串 X 和 Y,调用 *lcs*( )函数和 *printLcs*( )函数求字符串 X 和 Y 的最长公共子序列,最后将最长公共子序列输出到显示器上。程序运行结果如图 4.3 所示,代码如下。

In the *main*( ) function, given strings X and Y, call the *lcs*( ) function and the *printLcs*( ) function to find the longest common subsequence of strings X and Y, and finally output the longest common sequence to the display. The program running result is shown in Figure 4.3. The code is as follows.

```python
def lcs(X, Y):
    lenX = len(X)
    lenY = len(Y)
    #二维表 L 存放公共子序列的长度(Table L holds the length of a common subsequence)
    L = [[0 for i in range(lenY + 1)] for j in range(lenX + 1)]
    #二维表 flag 存放公共子序列的长度步进(Table flag holds the step length of the common subsequence)
    b = [[0 for i in range(lenY + 1)] for j in range(lenX + 1)]
    for i in range(lenX):
        for j in range(lenY):
            if X[i] == Y[j]:
                L[i + 1][j + 1] = L[i][j] + 1
                b[i + 1][j + 1] = 'ok'
            elif L[i + 1][j] > L[i][j + 1]:
                L[i + 1][j + 1] = L[i + 1][j]
                b[i + 1][j + 1] = 'left'
            else:
                L[i + 1][j + 1] = L[i][j + 1]
                b[i + 1][j + 1] = 'up'
    return L, b

def printLcs(b, X, i, j):
    if i == 0 or j == 0:
        return
    if b[i][j] == 'ok':
        printLcs(b, X, i - 1, j - 1)
        print(X[i - 1], end='')
    elif b[i][j] == 'left':
        printLcs(b, X, i, j - 1)
    else:
        printLcs(b, X, i - 1, j)

if __name__ == "__main__":
    X = ['A', 'B', 'C', 'B', 'D', 'A', 'B']
    Y = ['B', 'D', 'C', 'A', 'B', 'A']
    L, b = lcs(X, Y)
    print('最长公共子序列为(The longest common subsequence is):')
    printLcs(b, X, len(X), len(Y))
```

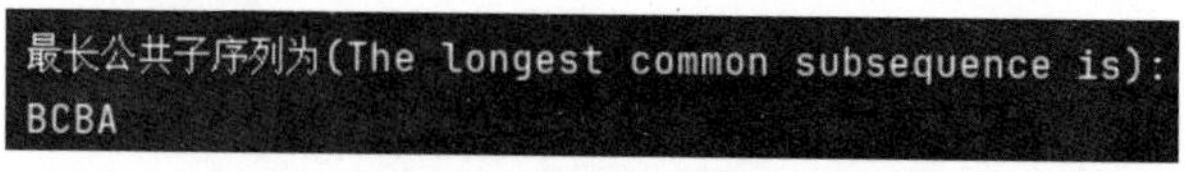

```
最长公共子序列为(The longest common subsequence is):
BCBA
```

图 4.3　最长公共子序列程序运行结果

Figure 4.3　Result of the longest common subsequence program

4.4　0/1 背包问题（0/1 Knapsack Problem）

4.4.1　问题分析（Problem Analysis）

1. 问题提出（Description of Problem）

0/1 背包问题可描述为 n 件物品和 1 个背包。对物品 i，其价值为 v_i，重量为 w_i，背包的容量为 W。如何选取物品装入背包，使背包中所装入的物品的总价值最大？物品不可分割。

The 0/1 backpack problem can be described as n items and 1 backpack. For item i, its value is v_i, its weight is w_i, and the capacity of the backpack is W. How to select items to put into the knapsack so that the total value of the items in the knapsack is maximized? Items are indivisible.

在选择装入背包的物品时，对于物品 i 只有两种选择，即装入背包或不装入背包。不能将物品 i 装入背包多次，也不能只装入物品 i 的一部分。假设 x_i 表示物品 i 被装入背包的状态，当 $x_i = 0$ 时，表示物品没有被装入背包；当 $x_i = 1$ 时，表示物品被装入背包。

When selecting items to pack in the backpack, there are only two options for item i, that is to pack in or not to pack. Item i cannot be loaded into the backpack more than once, nor can it be loaded into only part of item i. Suppose x_i represents the state of item i being loaded into the backpack, when $x_i = 0$, it means that the item is not loaded into the backpack; When $x_i = 1$, the item is loaded into the backpack.

根据问题描述，设计出如下的约束条件和目标函数。

According to the description of the problem, the following constraints and objective functions are designed.

约束条件（Constraint condition）：

$$\begin{cases} \sum_{i=1}^{n} w_i x_i \leqslant W \\ x_i \in \{0, 1\}, \quad 1 \leqslant i \leqslant n \end{cases} \tag{4-4}$$

目标函数（Objective function）：

$$\max \sum_{i=1}^{n} v_i x_i$$

于是，问题归结为寻找一个满足以上约束条件，并使目标函数达到最大的解向量 $X = (x_1, x_2, \cdots, x_n)$。

Thus, the problem boils down to finding a solution vector $X = (x_1, x_2, \cdots, x_n)$ that satisfies the above constraints and maximizes the objective function.

2. 建立最优价值的递归关系式（Establish the Recursive Relation of the Optimal Value）

设 $C[i][j]$ 表示子问题的最优价值，则最优价值的递归定义式如下。

Let $C[i][j]$ represent the optimal value of the subproblem, then the recursive definition of the optimal value is as follows.

$$C[0][j] = C[i][0] = 0 \tag{4-5}$$

$$C[i][j] = \begin{cases} C[i-1][j], & j < w_i \\ \max\{C[i-1][j], C[i-1][j-w_i] + v_i\} & j \geq w_i \end{cases} \tag{4-6}$$

4.4.2 算法设计与描述（Algorithm Design and Description）

1. 设计算法所需的数据结构（Design the Data Structure Required by the Algorithm）

采用数组 $w[n]$ 来存放 n 件物品的重量，数组 $v[n]$ 用来存放 n 件物品的价值，背包容量为 W，数组 $C[n+1][W+1]$ 用来存放每一次迭代的执行结果；数组 $x[n]$ 用来存储所装入背包的物品状态。

The array $w[n]$ is used to store the weight of n items, array $v[n]$ is used to store the value of n items, backpack capacity is W, array $C[n+1][W+1]$ is used to store the execution result of each iteration; The array $x[n]$ is used to store the status of the items loaded into the backpack.

2. 初始化（Initialization）

按式（4-1）初始化数组 C。

Initialize array C with formula (4-1).

3. 循环阶段（Cycle Stage）

按式（4-2）确定前 i 件物品能够装入背包的情况下得到的最优价值。

Determine the optimal value when the first i items can be loaded into the backpack according to formula (4-2).

（1）$i=1$ 时，求出 $C[1][j]$，$1 \leq j \leq W$。

When $i=1$, to find $C[1][j]$, $1 \leq j \leq W$.

（2）$i=2$ 时，求出 $C[2][j]$，$1 \leq j \leq W$。

When $i=2$, to find $C[2][j]$, $1 \leq j \leq W$.

以此类推，直到 $i=n$ 时，求出 $C[n][W]$。此时，$C[n][W]$ 便是最优价值。

And so on, until $i=n$, we find $C[n][W]$. In this case $C[n][W]$ is the optimal value.

4. 确定装入背包的具体物品（Determine the Specific Items to Pack in the Backpack）

从 $C[n][W]$ 的值向前推，如果 $C[n][W] > C[n-1][W]$，表明第 n 件物品被装入背包，则 $x_n = 1$，前 $n-1$ 件物品被装入容量为 $W-w_n$ 的背包中；否则，第 n 件物品没有被装入背包，则 $x_n = 0$，前 $n-1$ 件物品被装入容量为 W 的背包中。以此类推，直到确定第 1 件物品是否被装入背包中为止。由此，得到以下关系式。

Push forward from the value of $C[n][W]$, if $C[n][W] > C[n-1][W]$, indicating that the n item is loaded into the backpack, then $x_n = 1$, and the first $n-1$ items are loaded into the backpack with capacity $W-w_n$; Otherwise, the n item is not loaded into the backpack, then $x_n = 0$, and the first $n-1$ items are loaded into the backpack of capacity W, and so on, until it is determined whether the first item is packed into the backpack. From this, the following relation is obtained.

$$\begin{cases} x_i = 0, j = j, & C[i][j] = C[i-1][j] \\ x_i = 1, j = j - w_i & C[i][j] > C[i-1][j] \end{cases} \tag{4-7}$$

按照式(4-7)，从 $C[n][W]$ 的值向前倒推，即 j 初始为 W，i 初始为 n，即可确定装入背包的具体物品。

According to formula (4-7), from the value of $C[n][W]$ backwards, that is, j is initially W, i is initially n, that can determine the specific items loaded into the backpack.

4.4.3　算法实现(Algorithm Implement)

1. 实例构造(Instance Construction)

有 5 件物品，其重量分别为<2, 2, 6, 5, 4>，价值分别为<6, 3, 5, 4, 6>。背包容量为 10，物品不可分割，求装入背包的物品和获得的最优价值。

There are five items whose weights are <2, 2, 6, 5, 4> and values are <6, 3, 5, 4, 6>. The backpack capacity is 10, and the items are inseparable, find the items to be loaded into the knapsack and the optimal value obtained.

（1）初始化(Initialization)。

采用二维数组 $C[6][11]$ 来存放各个子问题的最优价值，行 i 表示物品，列 j 表示背包容量，$C[i][j]$ 表示当背包容量为 j，可选择装入的物品为前 i 件物品时的最优价值。初始化第 0 行和第 0 列，结果如表 4.11 所示。

A two-dimensional array $C[6][11]$ is used to store the optimal value of each subproblem, row i represents the item, column j represents the backpack capacity, $C[i][j]$ represents the optimal value when the backpack capacity is j and the optional items are the first i items. Initialize row 0 and column 0, and the results are shown in Table 4.11.

表 4.11　初始化第 0 行和第 0 列

Table 4.11　Initializes row 0 and column 0

	0	1	2	3	4	5	6	7	8	9	10
0	0	0	0	0	0	0	0	0	0	0	0
1	0										
2	0										
3	0										
4	0										
5	0										

（2）当 $i=1$ 时的情况（When $i=1$）。

当 $i=1$ 时，求出 $C[1][j]$，$1 \leqslant j \leqslant W$。由于物品 1 的重量 $w_1=2$，价值 $v_1=6$，故分为以下两种情况讨论。

When $i=1$, find $C[1][j]$, $1 \leqslant j \leqslant W$. Since the weight of item 1 is $w_1=2$ and the value $v_1=6$, it is discussed in the following two cases.

① 如果 $j<w_1$，即 $j<2$ 时，$C[1][j]=C[0][j]$。

If $j<w_1$, that is, $j<2$, $C[1][j]=C[0][j]$.

② 如果 $j \geqslant w_1$，即 $j \geqslant 2$ 时，$C[1][j]=\max\{C[0][j], C[0][j-w_1]+v_1\}=\max\{C[0][j], C[0][j-2]+6\}$。

If $j \geqslant w_1$, that is, $j \geqslant 2$, $C[1][j]=\max\{C[0][j], C[0][j-w_1]+v_1\}=\max\{C[0][j], C[0][j-2]+6\}$.

故 $i=1$ 时的最优价值如表 4.12 所示。

Therefore, the optimal value when $i=1$ is shown in Table 4.12.

表 4.12 $i=1$ 时获得的最优价值

Table 4.12 Optimal values obtained for $i=1$

	0	1	2	3	4	5	6	7	8	9	10
0	0	0	0	0	0	0	0	0	0	0	0
1	0	0	6	6	6	6	6	6	6	6	6
2	0										
3	0										
4	0										
5	0										

（3）当 $i=2$ 时的情况（When $i=2$）。

当 $i=2$ 时，求出 $C[2][j]$，$1 \leqslant j \leqslant W$。由于物品 2 的重量 $w_2=2$，价值 $v_2=3$，故分为以下两种情况讨论。

When $i=2$, find $C[2][j]$, $1 \leqslant j \leqslant W$. Since the weight of item 2 is $w_2=2$ and the value $v_2=3$, it is discussed in the following two cases.

① 如果 $j<w_2$，即 $j<2$ 时，$C[2][j]=C[1][j]$。

If $j<w_2$, that is, $j<2$, $C[2][j]=C[1][j]$.

② 如果 $j \geqslant w_2$，即 $j \geqslant 2$ 时，$C[2][j]=\max\{C[1][j], C[1][j-w_2]+v_2\}=\max\{C[1][j], C[1][j-2]+3\}$。

If $j \geqslant w_2$, that is, $j \geqslant 2$, $C[2][j]=\max\{C[1][j], C[1][j-w_2]+v_2\}=\max\{C[1][j], C[1][j-2]+3\}$.

故 $i=2$ 时的最优价值如表 4.13 所示。

Therefore, the optimal value when $i=2$ is shown in Table 4.13.

表 4.13　$i=2$ 时获得的最优价值

Table 4.13　Optimal values obtained for $i=2$

	0	1	2	3	4	5	6	7	8	9	10
0	0	0	0	0	0	0	0	0	0	0	0
1	0	0	6	6	6	6	6	6	6	6	6
2	0	0	6	6	9	9	9	9	9	9	9
3	0										
4	0										
5	0										

（4）当 $i=5$ 时的情况（When $i=5$）。

以此类推，直到 $i=5$ 时，由于物品 5 的重量 $w_5=4$，价值 $v_5=6$，故分为以下两种情况讨论。

And so on, until $i=5$, since the weight of item 5 is $w_5=4$ and the value $v_5=6$, it is discussed in the following two cases.

① 如果 $j<w_5$，即 $j<4$ 时，$C[5][j]=C[4][j]$。

If $j<w_5$, that is, $j<4$, $C[5][j]=C[4][j]$.

② 如果 $j \geqslant w_5$，即 $j \geqslant 4$ 时，$C[5][j]=\max\{C[4][j], C[4][j-w_5]+v_5\}=\max\{C[4][j], C[4][j-4]+6\}$。

If $j \geqslant w_5$, that is, $j \geqslant 4$, $C[5][j]=\max\{C[4][j], C[4][j-w_5]+v_5\}=\max\{C[4][j], C[4][j-4]+6\}$.

故 $i=5$ 时的最优价值如表 4.14 所示。

Therefore, the optimal value when $i=5$ is shown in Table 4.14.

表 4.14　$i=5$ 时获得的最优价值

Table 4.14　Optimal values obtained for $i=5$

	0	1	2	3	4	5	6	7	8	9	10
0	0	0	0	0	0	0	0	0	0	0	0
1	0	0	6	6	6	6	6	6	6	6	6
2	0	0	6	6	9	9	9	9	9	9	9
3	0	0	6	6	9	9	9	9	11	11	14
4	0	0	6	6	9	9	9	10	11	13	14
5	0	0	6	6	9	9	12	12	15	15	15

此时，$C[5][10]=15$，即装入背包的物品的最优价值为 15。

In this case, $C[5][10]=15$, that is, the optimal value of the item loaded into the backpack is 15.

（5）最优解（Optimal Solution）。

从 $i=5, j=10$ 向前递推,如果 $C[i][j]=C[i-1][j]$,说明第 i 件物品没有装入背包,那么 $x_i=0$。如果 $C[i][j]>C[i-1][j]$,说明第 i 件物品被装入背包,那么 $x_i=1$,更新 $j=j-w_i$。由于 $C[5][10]=15>C[4][10]=14$,说明物品5被装入了背包,因此 $x_5=1$,更新 $j=j-w_5=10-4=6$;由于 $C[4][j]=C[4][6]=9=C[3][6]$,说明物品4没有被装入背包,因此 $x_4=0$;由于 $C[3][j]=C[3][6]=9=C[2][6]$,说明物品3没有被装入背包,因此 $x_3=0$;由于 $C[2][j]=C[2][6]=9>C[1][6]=6$,说明物品2被装入背包,因此 $x_2=1$,更新 $j=j-w_2=6-2=4$;由于 $C[1][j]=C[1][4]=6>C[0][4]=0$,说明物品1被装入背包,因此 $x_1=1$,更新 $j=j-w_1=4-2=2$。最终装入背包物品的最优解 $X=(1, 1, 0, 0, 1)$。

From $i=5$, $j=10$ recursively, if $C[i][j]=C[i-1][j]$, the i item was not packed into the backpack, then $x_i=0$. If $C[i][j]>C[i-1][j]$, the i th item is loaded into the backpack, then $x_i=1$, updating $j=j-w_i$. Since $C[5][10]=15>C[4][10]=14$, item 5 is loaded into the backpack, so $x_5=1$, update $j=j-w_5=10-4=6$; Since $C[4][j]=C[4][6]=9=C[3][6]$, item 4 is not packed into the backpack, so $x_4=0$; Since $C[3][j]=C[3][6]=9=C[2][6]$, item 3 is not packed into the backpack, so $x_3=0$; Since $C[2][j]=C[2][6]=9>C[1][6]=6$, item 2 is loaded into the backpack, so $x_2=1$, update $j=j-w_2=6-2=4$; Since $C[1][j]=C[1][4]=6>C[0][4]=0$, that item 1 is loaded into the backpack, so the $x_1=1$, update $j=j-w_1=4-2=2$. The optimal solution $X=(1, 1, 0, 0, 1)$ for the final backpack item.

2. 代码实现(Code Implementation)

首先定义一个 $knapsack()$ 函数接收物品的重量、价值和背包的容量,返回装入的最优价值和最优解。在 $main()$ 函数中,给定5件物品的实例,调用 $knapsack()$ 函数得到该实例的最优价值和最优解,并将结果打印输出到显示器上。程序运行结果如图4.4所示,代码如下。

First, a $knapsack()$ function is defined to accept the weight of the item, the value, and the capacity of the backpack, and return the optimal value and the optimal solution of the load. In the $main()$ function, given 5 instances of items, call the $knapsack()$ function to get the optimal value and the optimal solution of the instance, and display the result. The program running result is shown in Figure 4.4, and the code is as follows.

```
import numpy as np
def knapsack(w, v, W):
    n = len(w)
    c = np.zeros((n + 1, W + 1), dtype = np.int32)

    for i in range(1, n + 1):
        for j in range(1, W + 1):
            if w[i - 1] <= j:
                c[i][j] = max(c[i - 1][j - w[i - 1]] + v[i - 1], c[i - 1][j])
            else:
                c[i][j] = c[i - 1][j]
    x = [0] * n
    j = W
```

续

```
        for i in range(n, 0, -1):
            if c[i][j] > c[i - 1][j]:
                x[i - 1] = 1
                j -= w[i - 1]
        return c[n][W], x
if _ _name_ _ = = " _ _main_ _":
    w = [2, 2, 6, 5, 4]
    v = [6, 3, 5, 4, 6]
    w_most = 10
    bestp, x = knapsack(w, v, w_most)
    print('装入背包物品的最优价值为(The optimal value for backpack items is):', bestp)
    print('背包内所载物品如下(The items contained in the backpack are):', x)
```

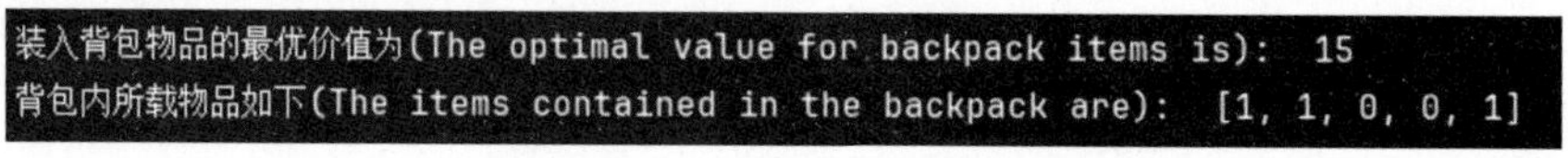

```
装入背包物品的最优价值为(The optimal value for backpack items is): 15
背包内所载物品如下(The items contained in the backpack are): [1, 1, 0, 0, 1]
```

图 4.4　0/1 背包程序运行结果

Figure 4.4　0/1 Result of the knapsack program

4.5　最优二叉查找树问题（Optimal Binary Search Tree Problem）

4.5.1　问题分析（Problem Analysis）

1. 基本概念（Basic Conception）

1）二叉查找树（Binary Search Tree）

给定由 n 个关键字组成的有序序列 $S=\{s_1, s_2, \cdots, s_n\}$，现在要用这些关键字建立一棵二叉查找树 T。对于每个关键字 s_i，其相应的查找概率为 p_i。由于在 S 中可能不存在对于某些值的检索，因此在二叉查找树中设置 $n+1$ 个虚节点 $e_0, e_1, \cdots, e_n$ 来表示不在 S 中的那些值，其中，e_0 表示小于 s_1 的所有值，e_n 表示大于 s_n 的所有值，对于 $i=1, 2, \cdots, n-1$，e_i 表示位于 s_i 与 s_{i+1} 之间的所有值。

Given an ordered sequence of n keywords $S=\{s_1, s_2, \cdots, s_n\}$, and now build a binary search tree T with these keywords. For each keyword s_i, its corresponding search probability is p_i. Since a search for some values may not exist in S, so set $n+1$ virtual nodes $e_0, e_1, \cdots, e_n$ to represent values that are not in S in the binary search tree, where e_0 represents all values less than s_1, e_n represents all values greater than s_n, for $i=1, 2, \cdots, n-1$, e_i represents all values between s_i and s_{i+1}.

每个虚节点 e_i 对应一个查找概率 q_i。在构建的二叉查找树中，s_i 为实节点（内部节点），e_i 表示虚节点（叶子节点）。每次检索要么成功，即检索到实节点 s_i，要么不成功，即检索到虚节点 e_i，因此 $\sum_{i=1}^{n} p_i + \sum_{i=0}^{n} q_i = 1$。显然，对于同一个关键字的集合，二叉查找树的形态会由于插入顺序的不同而不同。

Each virtual node e_i corresponds to a search probability q_i. In the constructed binary search tree, s_i is the real node (inner node) and e_i represents the virtual node (leaf node). Each search is either successful, that is, the real node s_i is searched, or unsuccessful, that is, the virtual node e_i is searched, so $\sum_{i=1}^{n} p_i + \sum_{i=0}^{n} q_i = 1$. Obviously, for the same set of keywords, the shape of the binary search tree will vary depending on the insertion order.

2) 平均比较次数(Average Number of Comparisons)

如何来衡量不同二叉查找树的查找效率呢？通常采用平均比较次数作为衡量的标准。设在表示 $S=\{s_1, s_2, \cdots, s_n\}$ 的二叉查找树 T 中，元素 s_i 的节点深度为 $c_i(1 \leqslant i \leqslant n)$，查找概率为 p_i，虚节点为 $\{e_0, e_1, \cdots, e_n\}$，$e_j$ 的节点深度为 d_j，查找概率为 $q_j(0 \leqslant j \leqslant n)$。

How to measure the search efficiency of different binary search trees? The average number of comparisons is usually used as a measure. Suppose there is a binary search tree T representing $S=\{s_1, s_2, \cdots, s_n\}$, the node depth of element s_i is $c_i(1 \leqslant i \leqslant n)$, the search probability is p_i, and the virtual node is $\{e_0, e_1, \cdots, e_n\}$, the node depth of e_j is d_j and the search probability is $q_j(0 \leqslant j \leqslant n)$.

平均比较次数通常被定义如下。

The average number of comparisons is usually defined as follows

$$C = \sum_{i=1}^{n} p_i(1 + c_i) + \sum_{j=0}^{n} q_j d_j \tag{4-8}$$

3) 最优二叉查找树(Optimal Binary Search Tree)

最优二叉查找树是在所有表示有序序列 S 的二叉查找树中，具有最小平均比较次数的二叉查找树。

The optimal binary search tree is the one with the smallest average number of comparisons among all binary search trees representing an ordered sequence S.

2. 建立最优价值的递归关系式(Establish the Recursive Relationship of the Optimal Value)

设 $C[i][j]$ 表示二叉查找树的平均比较次数，则最优价值的递归定义式如下。

Let $C[i][j]$ represent the average number of comparisons in a binary search tree, then the optimal value is recursively defined as follows.

当 $i \leqslant j$ 时(When $i \leqslant j$)，

$$C(i, j) = w_{ij} + \min_{i \leqslant k \leqslant j}\{C(i, k-1) + C(k+1, j)\} \tag{4-9}$$

其中(Thereinto)：

$$w_{ij} = w_{i(j-1)} + p_j + q_j$$

初始时(Initially)，

$$C(i, i-1) = 0, w_{i(i-1)} = q_{i-1} \quad (1 \leqslant i \leqslant n) \tag{4-10}$$

4.5.2 算法设计与描述(Algorithm Design and Description)

1. 设计合适的数据结构(Design the Right Data Structure)

设有序序列 $S=\{s_1, \cdots, s_n\}$，数组 $s[n]$ 存储序列 S 中的元素；数组 $p[n]$ 存储序列 S 中

相应元素的查找概率；二维数组 $C[n+1][n+1]$ 中的 $C[i][j]$ 表示二叉查找树 $T(i,j)$ 的平均比较次数；二维数组 $R[n+1][n+1]$ 中的 $R[i][j]$ 表示二叉查找树 $T(i,j)$ 中作为根节点的元素在序列 S 中的位置。数组 $q[n]$ 存储虚节点 $e_0,e_1,\cdots,e_n$ 的查找概率。为了提高效率，不是每次计算 $C(i,j)$ 时都计算 w_{ij} 的值，而是把这些值存储在二维数组 $W[i][j]$ 中。

Suppose there is an ordered sequence $S=\{s_1, \cdots, s_n\}$, array $s[n]$ stores elements in sequence S; array $p[n]$ stores the search probability of the corresponding element in sequence S; In the two-dimensional array $C[n+1][n+1]$, $C[i][j]$ represents the average number of comparisons in the binary search tree $T(i, j)$; $R[i][j]$ in the two-dimensional array $R[n+1][n+1]$ represents the position of the element as the root node in the sequence S in the binary search tree $T(i, j)$. Array $q[n]$ stores virtual nodes e_0, $e_1, \cdots, e_n$ search probability. For efficiency, instead of calculating the values of w_{ij} every time $C(i, j)$ is computed, these values are stored in the two-dimensional array $W[i][j]$.

2. 初始化（Initialization）

设置 $C[i][i-1]=0$；$W[i][i-1]=q_{i-1}$，其中 $l\leq i\leq n+1$。

Set $C[i][i-1]=0$. $W[i][i-1]=q_{i-1}$, where $l\leq i\leq n+1$.

3. 循环阶段（Cycle Stage）

采用自底向上的方式逐步计算最优价值，记录最优决策。

The bottom-up method is used to calculate the optimal value step by step and record the optimal decision.

（1）字符集规模为 1 时，即 $S_{ij}=\{s_i\}$，$i=1,2,\cdots,n$，且 $j=i$，显然这种规模的子问题有 n 个，即首先要构造出 n 棵最优二叉查找树 $T(1,1),T(2,2),\cdots,T(n,n)$。依据递归式，很容易求得 $W[i][i]$ 和 $C[i][i]$。同时，对于所构造的 n 棵最优二叉查找树，它们的根分别记为 $R[1][1]=1, R[2][2]=2,\cdots, R[n][n]=n$。

When the character set size is 1, that is, $S_{ij}=\{s_i\}$, $i=1, 2, \cdots, n, and\ j=i$, obviously there are n subproblems of this scale, that is, n optimal binary search trees must be constructed first $T(1, 1), T(2, 2), \cdots, T(n, n)$. According to the recursion, it is easy to find $W[i][i]$ and $C[i][i]$. Meanwhile, for the constructed n optimal binary search trees, their roots are denoted as $R[1][1]=1, R[2][2]=2, \cdots, R[n][n]=n$.

（2）字符集规模为 2 时，即 $S_{ij}=\{s_i,s_j\}$，$i=1,2,\cdots,n-1$ 且 $j=i+1$，显然这种规模的子问题有 $n-1$ 个，即要构造出 $n-1$ 棵最优二叉查找树 $T(1,2),T(2,3),\cdots,T(n-1,n)$。依据式(4-4)，求得 $W[i][j]$，然后分别在整数 $i,i+1,i+2,\cdots,j$ 中选择适当的 k 值，使得 $C(i,j)$ 最小，树的根记为 $R[i][j]=k$。

When the character set size is 2, $S_{ij}=\{s_i, s_j\}$, $i=1, 2, \cdots, n-1$ and $j=i+1$, obviously there are $n-1$ subproblems of this scale, that is, to construct an $n-1$ optimal binary search tree $T(1, 2), T(2, 3), \cdots, T(n-1, n)$. According to the formula (4-4), $W[i][j]$ is obtained, and then in the integers $i, i+1, i+2, \cdots$, choose an appropriate value of k in j such that $C(i, j)$ is minimal and the root of the tree is denoted $R[i][j]=k$.

（3）以此类推,构造出字符集 S_{ij} 中含三个字符的最优二叉查找树、含 4 个字符的最优二叉查找树,直到字符集规模为 n 时,即 $S_{1n}=\{s_1,s_2,\cdots,s_n\}$,显然这种规模的子问题有 1 个,即要构造出 1 棵最优二叉查找树 $T(1,n)$。依据式(4-4),求得 $W[i][j]$,然后在整数 1, 2,$\cdots$,n 中选择适当的 k 值,使得 $C(i,j)$ 最小。同时,记录该树的根 $R[1][n]=k$。

In this way, the optimal binary search tree with 3 characters and the optimal binary search tree with 4 characters in the character set S_{ij} are constructed until the character set size is n, that is, $S_{1n}=\{s_1, s_2, \cdots, s_n\}$, obviously there is 1 subproblem of this scale, that is, to construct an optimal binary search tree $T(1, n)$. According to the formula (4-4), $W[i][j]$ is obtained, and then in the integer 1, 2, $\cdots$, n, choose an appropriate value of k, so that $C(i, j)$ is minimal. At the same time, record the root of the tree $R[1][n]=k$.

4. 最优解的构造(Construction of Optimal Solutions)

从 $R[i][j]$ 中保存的最优二叉查找子树 $T(i,j)$ 的根节点信息,可构造出问题的最优解,当 $R[1][n]=k$ 时,元素 s_k 即为所求的最优二叉查找树的根节点。此时,需要计算两个子问题:求左子树 $T(1,k-1)$ 和右子树 $T(k+1,n)$ 的根节点信息。若 $R[1][k-1]=i$,则元素 s_i 即为 $T(1,k-1)$ 的根节点元素。以此类推,将很容易由 R 中记录的信息构造出问题的最优解。

The optimal solution of the problem can be constructed from the root node information of the optimal binary search subtree $T(i, j)$ stored in $R[i][j]$. When $R[1][n]=k$, element s_k is the root node of the optimal binary search tree. In this case, two subproblems need to be computed: the root node information of the left subtree $T(1, k-1)$ and the right subtree $T(k+1, n)$. If $R[1][k-1]=i$, then s_i is the root element of $T(1, k-1)$. By analogy, it will be easy to construct an optimal solution to the problem from the information recorded in R.

4.5.3 算法实现(Algorithm Implementation)

1. 实例构造(Instance Construction)

假设 5 个有序元素的集合为 $\{s_1,s_2,s_3,s_4,s_5\}$,查找概率 $p=<p_1,p_2,p_3,p_4,p_5>=<0.15,0.1,0.05,0.1,0.2>$;叶节点元素 $\{e_0,e_1,e_2,e_3,e_4,e_5\}$,查找概率 $q=<q_0,q_1,q_2,q_3,q_4,q_5>=<0.05,0.1,0.05,0.05,0.05,0.1>$。试构造这 5 个有序元素的最优二叉查找树。

Suppose there is a set of 5 ordered element $\{s_1, s_2, s_3, s_4, s_5\}$, and the search probability $p=<p_1, p_2, p_3, p_4, p_5>=<0.15, 0.1, 0.05, 0.1, 0.2>$; Leaf node element $\{e_0, e_1, e_2, e_3, e_4, e_5\}$, find the probability $q=<q_0, q_1, q_2, q_3, q_4, q_5>=<0.05, 0.1, 0.05, 0.05, 0.05, 0.1>$. Try to construct an optimal binary search tree for these 5 ordered elements.

（1）令 $C[i][i-1]=0$;$W[i][i-1]=q_{i-1}$;$1\leq i\leq 5$,如表 4.15 所示。

Let $C[i][i-1]=0$; $W[i][i-1]=q_{i-1}$; $1\leq i\leq 5$, as shown in Table 4.15.

表 4.15　初始化 $W[i][i-1]$ 和 $C[i][i-1]$

Table 4.15　Initializing $W[i][i-1]$ and $C[i][i-1]$

W	0	1	2	3	4	5
1	0.05					
2		0.1				
3			0.05			
4				0.05		
5					0.05	
6						0.1

C	0	1	2	3	4	5
1	0					
2		0				
3			0			
4				0		
5					0	
6						0

（2）字符集规模为 1 时，即构造 5 棵二叉查找树 $T(i,j)$，此时 $1 \leqslant i \leqslant 5$ 且 $i=j$。当 $i=1$ 时，$W[1][1]=W[1][0]+p_1+q_1=0.3$；$C[1][1]=W[1][1]+C[1][0]+C[2][1]=0.3$；$R[1][1]=1$；同理，可求出 i 取值为 2、3、4、5 时的 $W[i][i]$、$C[i][i]$ 和 $R[i][i]$。结果如表 4.16 所示。

When the character set size is 1, that is, five binary search trees $T(i, j)$ are constructed, and then $1 \leqslant i \leqslant 5$ and $i=j$. When $i=1$, $W[1][1]=W[1][0]+p_1+q_1=0.3$; $C[1][1]=W[1][1]+C[1][0]+C[2][1]=0.3$; $R[1][1]=1$; Similarly, we can find $W[i][i]$, $C[i][i]$, and $R[i][i]$ when i is 2, 3, 4, 5. The results are shown in Table 4.16.

表 4.16　i 取值为 1，2，3，4，5 时 $W[i][i]$、$C[i][i]$ 和 $R[i][i]$ 的值

Table 4.16　Values of $W[i][i]$, $C[i][i]$ and $R[i][i]$ when the value of i is 1, 2, 3, 4, 5

W	0	1	2	3	4	5
1	0.05	0.3				
2		0.1	0.25			
3			0.05	0.15		
4				0.05	0.20	
5					0.05	0.35
6						0.1

C	0	1	2	3	4	5
1	0	0.3				
2		0	0.25			
3			0	0.15		
4				0	0.20	
5					0	0.35
6						0

R	0	1	2	3	4	5
1		1				
2			2			
3				3		
4					4	
5						5

（3）字符集规模为 2 时，即构造 4 棵二叉查找树 $T(i,j)$，此时 $1 \leqslant i \leqslant 4$ 且 $j-i=1$。

When the character set size is 2, that is, four binary search trees $T(i, j)$ are constructed, where $1 \leqslant i \leqslant 4$ and $j-i=1$.

当 $i=1$ 时，此时 $j=2$，且 $1 \leqslant k \leqslant 2$。$W[1][2]=W[1][1]+p_2+q_2=0.3+0.1+0.05=0.45$。

If $i=1$, then $j=2$ and $1 \leqslant k \leqslant 2$. $W[1][2]=W[1][1]+p_2+q_2=0.3+0.1+0.05=0.45$.

当 $k=1$ 时，$C[1][2]=W[1][2]+C[1][0]+C[2][2]=0.7$。

When $k=1$, $C[1][2]=W[1][2]+C[1][0]+C[2][2]=0.7$.

当 $k=2$ 时，$C[1][2]=W[1][2]+C[1][1]+C[3][2]=0.75$。由此可得，$C[1][2]=0.7$，

$R[1][2]=1$。同理可求出 i 取值为 $2,3,4$ 时的 $W[i][j]$、$C[i][j]$ 和 $R[i][j]$。结果如表 4.17 所示。

When $k=2$, $C[1][2]=W[1][2]+C[1][1]+C[3][2]=0.75$. It follows that $C[1][2]=0.7$, $R[1][2]=1$. Similarly, $W[i][j]$, $C[i][j]$ and $R[i][j]$ can be obtained when i is 2, 3, 4. The results are shown in Table 4.17.

表 4.17　i 取值为 1，2，3，4 时 $W[i][j]$、$C[i][j]$ 和 $R[i][j]$ 的值

Table 4.17　Values of $W[i][j]$, $C[i][j]$ and $R[i][j]$ when the value of i is 1, 2, $3,4$

W	0	1	2	3	4	5
1	0.05	0.3	0.45			
2		0.1	0.25	0.35		
3			0.05	0.15	0.3	
4				0.05	0.20	0.5
5					0.05	0.35
6						0.1

C	0	1	2	3	4	5
1	0	0.3	0.7			
2		0	0.25	0.5		
3			0	0.15	0.45	
4				0	0.20	0.7
5					0	0.35
6						0

R	0	1	2	3	4	5
1		1	1			
2			2	2		
3				3	4	
4					4	5
5						5

（4）以此类推，直至字符集规模为 5 时，即构造一棵二叉查找树 $T(i,j)$，此时 $i=1$ 且 $j-i=4$。

And so on, until the character set size is 5, that is, a binary search tree $T(i, j)$ is constructed, where $i=1$ and $j-i=4$.

当 $i=1$ 时，此时 $j=5$，且 $1\leqslant k\leqslant 5$。$W[1][5]=W[1][4]+p_5+q_5=0.7+0.2+0.1=1.0$。

If $i=1$, then $j=5$ and $1\leqslant k\leqslant 5$. $W[1][5]=W[1][4]+p_5+q_5=0.7+0.2+0.1=1.0$.

当 $k=1$ 时，

when $k=1$

$C[1][5]=W[1][5]+C[1][0]+C[2][5]=1.0+0+1.65=2.65$。

当 $k=2$ 时，

when $k=2$

$C[1][5]=W[1][5]+C[1][1]+C[3][5]=1.0+0.3+1.05=2.35$。

当 $k=3$ 时，

when $k=3$

$C[1][5]=W[1][5]+C[1][2]+C[4][5]=1.0+0.7+0.7=2.4$。

当 $k=4$ 时，

when $k=4$

$C[1][5]=W[1][5]+C[1][3]+C[5][5]=1.0+1.0+0.35=2.35$。

当 $k=5$ 时，

when $k=5$

$C[1][5]=W[1][5]+C[1][4]+C[6][5]=1.0+1.45+0=2.45$。

由此可得，$C[1][5]=2.35$，$R[1][5]=2$，结果如表 4.18 所示。

Thus, $C[1][5]=2.35$, $R[1][5]=2$, the results are shown in Table 4.18.

表 4.18　i 取值为 1 时 $W[i][j]$、$C[i][j]$ 和 $R[i][j]$ 的值

Table 4.18　Values of $W[i][j]$, $C[i][j]$ and $R[i][j]$ when the value of i is 1

W	0	1	2	3	4	5
1	0.05	0.3	0.45	0.55	0.7	1.0
2		0.1	0.25	0.35	0.5	0.8
3			0.05	0.15	0.3	0.6
4				0.05	0.20	0.5
5					0.05	0.35
6						0.1

C	0	1	2	3	4	5
1	0	0.3	0.7	1.0	1.45	2.35
2		0	0.25	0.5	0.95	1.65
3			0	0.15	0.45	1.05
4				0	0.20	0.7
5					0	0.35
6						0

R	0	1	2	3	4	5
1		1	1	2	2	2
2			2	2	2	4
3				3	4	5
4					4	5
5						5

（5）由表 4.18 可知，构造 5 个有序元素的最优二叉查找树的最优价值为 2.35。从 $R[i][j]$ 中保存的最优二叉查找子树 $T(i,j)$ 的根节点信息，构造出问题的最优解。

As can be seen from Table 4.18, the optimal value of constructing an optimal binary search tree with 5 ordered elements is 2.35. Based on the information of the root node of the subtree $T(i, j)$ saved in $R[i][j]$, the optimal solution of the problem is constructed.

由于 $R[1][5]=2$，即 $k=2$，所以 s_2 是最优二叉查找树 $T(1,5)$ 的根节点。此时，分别求出 $T(1,5)$ 的左子树 $T(1,1)$ 和右子树 $T(3,5)$ 的根节点。由于 $R[1][1]=1$，则其左子树 $T(1,1)$ 的根节点为 s_1；由于 $R[3][5]=5$，则右子树 $T(3,5)$ 的根节点为 s_5。下一步，求出子树 $T(3,5)$ 的左子树的根节点，由于 $R[3][4]=4$，故 $T(3,4)$ 的根节点为 s_4；则 s_3 为 s_4 的左孩子。由此构造出如图 4.5 所示的最优二叉查找树。

Since $R[1][5]=2$, this is, $k=2$, s_2 is the root of the optimal binary search tree $T(1, 5)$. In this case, the root nodes of the left subtree $T(1, 1)$ and the right subtree $T(3, 5)$ of $T(1, 5)$ are found respectively. Since $R[1][1]=1$, the root node of its left subtree $T(1, 1)$ is s_1; since $R[3][5]=5$, the root node of the right subtree $T(3, 5)$ is s_5. Next, find the root node of the left subtree of the subtree $T(3, 5)$. Since $R[3][4]=4$, the root node of $T(3, 4)$ is s_4; s_3 is the left child of s_4. Thus, the optimal binary search tree as shown in Figure 4.5 is constructed.

2. 代码实现（Code Implementation）

定义一个 $optimal\_bst()$ 函数计算最优价值 c。接收实节点概率 p、虚节点概率 q 和问题规模 n，输出最优价值 c 和相关决策 R。定义 $best\_solution()$ 函数构造最优解。接收相关决策 R、子问题 s_i，…，s_j 及有序序列 s，输出各级子问题的根。

Define an $optimal\_bst()$ function to compute the optimal value c. The real node probability p, virtual node probability q and problem size n are received, Output the optimal value c and related decision R. Define the $best\_solution()$ function to construct the optimal solution. Receive related decision R, subproblem s_i, …, s_j and ordered sequence s, outputs the root of each level of subproblem.

在 $main()$ 函数中，给定规模为 5 的有序序列 s、实节点的查找概率 p、虚节点的查找概

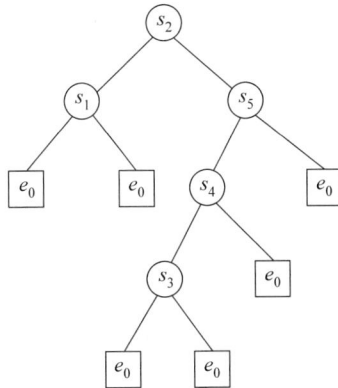

图 4.5 构造完成的最优二叉查找树

Figure 4.5 Constructed the optimal binary search tree

率 q,调用 *optimal_bst*( ) 函数得到该实例的最优价值,调用 *best_solution*( ) 函数构造最优解,并将结果打印输出到显示器上。程序运行结果如图 4.6 所示,代码如下。

In the *main* ( ) function, given the ordered sequence s of scale 5, the search probability p of the real node and the search probability q of the virtual node, the *optimal_bst*( ) function is called to obtain the optimal value of the instance, the *best_solution*( ) function is called to construct the optimal solution, and the result is printed out to the display. The result of running the program is shown in Figure 4.6, and the code is as follows.

```
def optimal_bst(p, q, n):
    #初始化 C、W、R 矩阵(Initialize the C, W, and R matrices)
    C = [[0 for _ in range(n+2)] for _ in range(n+2)]
    W = [[0 for _ in range(n+2)] for _ in range(n+2)]
    R = [[0 for _ in range(n+2)] for _ in range(n+2)]

    #设置初始条件(Set initial condition)
    for i in range(1, n +2):
        C[i][i-1] = 0
        W[i][i-1] = q[i-1]

    #自底向上计算(Calculate from bottom up)
    for length in range(1, n +1):          #遍历所有可能的子树长度(Traverse all possible subtree lengths)
        for i in range(1, n - length +2):  #确定子树的起始位置(Determine the starting position of the subtree)
            j = i +length -1
            W[i][j] = W[i][j-1] +p[j-1] +q[j]
            C[i][j] = float('inf')
            for r in range(i, j +1):
                t = C[i][r-1] +C[r+1][j] +W[i][j]
                if t < C[i][j]:
                    C[i][j] = t
                    R[i][j] = r
```

续

```
    return C[1][n], R

def best_solution(R, s, i, j):
    if i <= j:
        k = R[i][j]
        print("The root of s" +str(i) +" to s" +str(j) +" is:" +s[k -1])
        print("s" +str(i) +" : s" +str(j) +"的根是" +s[k -1])
        best_solution(R, s, i, k-1)          #左子树(Left subtree)
        best_solution(R, s, k+1, j)          #右子树(Right subtree)

def main():
    s = ['s1', 's2', 's3', 's4', 's5']
    p = [0.15, 0.1, 0.05, 0.1, 0.2]
    q = [0.05, 0.1, 0.05, 0.05, 0.05, 0.1]
    n = len(s)

    c, R = optimal_bst(p, q, n)
    print(f"最优价值为(The optimal value is): {c:.2f}")
    print("构造的最优解为(The optimal solution of the construction is):")
    best_solution(R, s, 1, n)

if __name__ == "__main__":
    main()
```

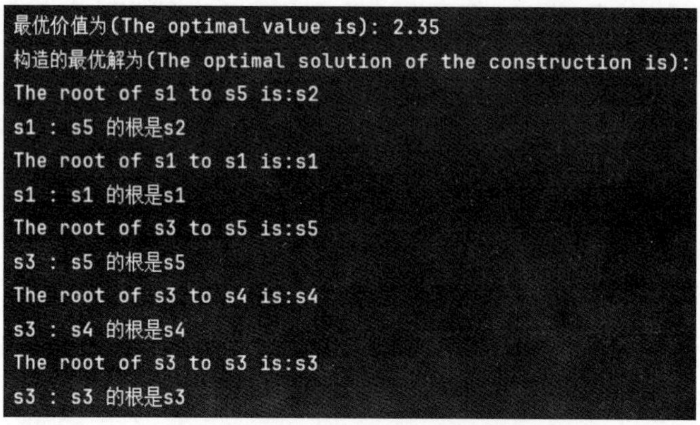

```
最优价值为(The optimal value is): 2.35
构造的最优解为(The optimal solution of the construction is):
The root of s1 to s5 is:s2
s1 : s5 的根是s2
The root of s1 to s1 is:s1
s1 : s1 的根是s1
The root of s3 to s5 is:s5
s3 : s5 的根是s5
The root of s3 to s4 is:s4
s3 : s4 的根是s4
The root of s3 to s3 is:s3
s3 : s3 的根是s3
```

图 4.6　最优二叉查找树程序运行结果

Figure 4.6　Result of the optimal binary search tree program

习题 4(Exercises Four)

1. 简述动态规划的基本思想。

Describe the basic idea of dynamic programming.

2. 简述动态规划算法解最优化问题的步骤。

Describe the steps of using dynamic programming algorithms to solve optimization problems.

3. 解释动态规划中的"最优子结构"概念,并给出一个实例说明。

Explain the concept of "optimal substructure" in dynamic programming and give an example to illustrate it.

4. 什么是"重叠子问题"？为什么避免重复计算这些子问题是动态规划方法高效的关键？

What is the "overlapping subproblem"? Why is avoiding recomputing these subproblems key to the efficiency of dynamic programming methods?

5. 斐波那契数列问题:请利用动态规划算法求出斐波那契数列的第 10 个数。斐波那契数列的定义为: $F(0) = 0, F(1) = 1, F(n) = F(n-1) + F(n-2)(n \geqslant 2)$。

Fibonacci sequence problem: Find the tenth digit of the Fibonacci sequence using dynamic programming. The Fibonacci sequence is defined as $F(0) = 0$, $F(1) = 1$, $F(n) = F(n-1) + F(n-2)(n \geqslant 2)$.

6. 最长递增子序列问题:给定一个整数序列 $\{10, 9, 2, 5, 3, 7, 101, 18\}$,找到其中最长的递增子序列的长度并输出。子序列可以不是连续的。

The longest increasing subsequence problem: Given an integer sequence $\{10, 9, 2, 5, 3, 7, 101, 18\}$, find the length of the longest increasing subsequence and output it. Subsequences can be non-continuous.

7. 0/1 背包问题:给定一个容量为 10 的背包和 4 件物品,每件物品的重量和价值分别为<2,3,4,5>和<3,4,5,6>,要求选取一些物品放入背包中,使得背包中物品的总价值最大,但不能超过背包的容量,输出最大总价值。

0/1 Backpack problem: Given a backpack with a capacity of 10 and 4 items, the weight and value of each item are <2, 3, 4, 5> and <3, 4, 5, 6>, it is required to select some items to put into the backpack, so that the total value of the items in the backpack is maximized, but cannot exceed the capacity of the backpack, Output the maximum total value.

第5章

回溯算法

Chapter 5　Backtracking Algorithm

回溯(Backtracking)算法是一种在解空间中搜索可行解或最优解的方法。它在包含问题的所有解的解空间树中,按照深度优先的策略,从根节点出发搜索解空间树。算法搜索至任一节点时,总是先判断该节点是否肯定不包含问题的解。如果肯定不包含,则跳过对以该节点为根的子树的系统搜索,逐层向其祖先节点回溯。否则,进入该子树,继续按深度优先的策略进行搜索。

Backtracking algorithm is a method of searching for feasible or optimal solutions in a solution space. In the solution space tree containing all the solutions of the problem, it searches the solution space tree from the root node according to depth-first strategy. When the algorithm searches for any node, it always checks whether the node does not contain the solution to the problem. If it does not contain, the systematic search for the subtree rooted by this node is skipped and the ancestor node is traced back layer by layer. Otherwise, enter the subtree and continue searching according to the depth-first strategy.

回溯算法在求解问题的所有解时,要回溯到根,且根节点的所有子树都已被搜索才结束。而在用来求问题的任一解时,只要搜索到问题的一个解就可以结束。这种以深度优先的方式系统地搜索问题的解的算法称为回溯算法,它适用于求解一些组合数较大的问题。

When the backtracking algorithm is used to solve all the solutions of the problem, it must be traced back to the root, and all the subtrees of the root node have been searched before it finishes. When used to find any solution to the problem, the search is terminated if it has found a solution to the problem. This algorithm, which systematically searches for the solution of the problem in a depth-first manner, is called backtracking algorithm, and it is suitable for solving some problems with many combinations.

5.1　概述(Overview)

5.1.1　基本思想(Basic Idea)

1. 算法框架(Algorithm Framework)

用回溯算法解决问题时,首先应明确搜索范围,即问题所有可能解的组成范围。这个范围越小越好,且至少包含问题的一个(最优)解。为了定义搜索范围,需要明确以下 4 个方面。

When using backtracking method to solve a problem, the first is to define the search scope, that is, the composition range of all possible solutions of the problem. The smaller the range, the better, and it contains at least one (optimal) solution to the problem. To define the scope of the search, you need to clarify the following four aspects.

1) 问题解的形式(The Form of the Solution to the Problem)

回溯算法希望问题的解能够表示成一个 n 元组$(x_1, x_2, \cdots, x_n)$的形式。

Backtracking hopes that the solution to the problem can be expressed as an n-tuple of $(x_1, x_2, \cdots, x_n)$.

2) 显约束(Explicit Constraint)

对分量 $x_i(i=1,2,\cdots,n)$ 的取值范围限定。

Limit the value range of the component $x_i(i=1, 2\cdots, n)$.

3) 隐约束(Implicit Constraint)

为满足问题的解而对不同分量之间施加的约束。

Constraints imposed between the different components to satisfy the solution of the problem.

4) 解空间(Solution Space)

对于问题的一个实例,解向量满足显约束的所有 n 元组构成了该实例的一个解空间。

For an instance of a problem, all n-tuples of the solution vector satisfying an explicit constraint form a solution space for that instance.

2. 搜索问题的解空间树(Search the Solution Space Tree of the Problem)

在搜索的过程中,需要了解以下三个名词。

In the process of searching, you need to know the following three terms.

1) 扩展节点(The Extension Node)

一个正在生成孩子的节点称为扩展节点。

A child being generated node called an extension node.

2) 活节点(The active Node)

一个自身已生成但其孩子还没有全部生成的节点称为活节点。

A node that has generated itself but has not yet generated all its children is called an active node.

3) 死节点(The Dead Node)

一个所有孩子已经生成的节点称为死节点。

A node whose all children have been generated is called a dead node.

3. 搜索思想(Search Idea)

从根开始,以深度优先搜索的方式进行搜索。根节点是活节点并且是当前的扩展节点。在搜索过程中,当前的扩展节点沿纵深方向移向一个新节点,判断该新节点是否满足隐约束。如果满足,那么新节点成为活节点,并且成为当前的扩展节点,继续深一层的搜索;如果不满足,那么换到该新节点的兄弟节点(扩展节点的其他分支)继续搜索;如果新节点没有兄弟节点,或其兄弟节点已全部搜索完毕,那么扩展节点成为死节点,搜索回溯到其父节点处继续进行。搜索过程继续直到找到问题的解或根节点变成死节点为止。

Start at the root and search in a depth-first way. The root node is an active node and is the current extension node. In the search process, the current extension node moves to a new node along the depth direction, and whether the new node satisfies the implicit constraint is judged. If it does, the new node becomes an active node, and becomes the current extension node, and the search continues at a deeper level. If not, then change to the sibling node of the new node (other branches of the extension node) to continue the search; If the new node has no siblings, or all its siblings have been searched, then the extension node becomes a dead node and the search continues back to the parent node. The search process continues until a solution to the problem is found or the root node becomes a dead node.

从回溯算法的搜索思想可知,搜索开始之前必须确定问题的隐约束。隐约束一般是考查解空间结构中的节点是否有可能得到问题的可行解或最优解。如果不可能得到问题的可行解或最优解,就不用沿着该节点的分支继续搜索了,需要换到该节点的兄弟节点或回到上一层节点。

From the search idea of backtracking method, the implicit constraint of the problem must be determined before the search begins. Implicit constraint is generally to check whether the nodes in the solution space structure are possible to obtain a feasible or an optimal solution to the problem. If it is impossible to obtain a feasible or optimal solution to the problem, it is not necessary to continue searching along the branch of the node, but to change to the sibling node of the node or return to the node of the previous layer.

在深度优先搜索的过程中,不满足隐约束的分支被剪掉,只沿着满足隐约束的分支搜索问题的解,从而避免了无效搜索,加快了搜索速度。因此,隐约束又称为剪枝函数。隐约束(剪枝函数)一般有两种:一是判断是否能够得到可行解的隐约束,称之为约束条件(约束函数);二是判断是否有可能得到最优解的隐约束,称之为限界条件(限界函数)。可见,回溯算法是一种具有约束函数或限界函数的深度优先搜索方法。

In the process of depth-first search, the branches that do not satisfy the implicit constraint are cut off, and only the branches that satisfy the implicit constraint are searched, thus avoiding the invalid search, and speeding up the search speed. Therefore, implicit constraints are also called pruning functions. Gnerally, there are two types of implicit constraints (also known as pruning functions): one is the constraint function, which determines whether a feasible solution can be obtained under the given implicit constraints; the other is the bound function, which evaluates whether it is possible to reach an optimal solution under the implicit constraints. So backtracking is a depth-first search method with constraint function or bound function.

总之,回溯算法求解的基本步骤包括以下三部分。

In short, the basic steps of backtracking algorithm solution include the following three parts.

（1）针对所给问题,定义问题的解空间。

For the given problem, the solution space of the problem was defined.

（2）确定易于搜索的解空间结构(找出适当的剪枝函数)。

Determining the structure of the solution space that is easy to search（find appropriate pruning function）.

（3）以深度优先方式搜索解空间,并在搜索过程中用剪枝函数避免无效搜索。

The solution space is searched in a depth-first manner, and a pruning function is used to avoid invalid searches.

5.1.2　算法的适用条件(Applicable Conditions of the Algorithm)

回溯算法是一种基于深度优先搜索的穷举式求解方法,通常用于解决组合优化问题,其适用条件包括如下。

Backtracking algorithm is an exhaustive solution method based on depth-first search, which is usually used to solve combinatorial optimization problems. Its application conditions include the following.

（1）问题具有多个解,回溯算法适用于那些具有多个解的问题,因为它可以穷尽所有可能的解空间。

Problems have multiple solutions, and the backtracking algorithm is suitable for those with multiple solutions because it can exhaust all possible solution spaces.

（2）问题可以分解为阶段,问题的解可以根据问题规模分解为一系列的阶段,每个阶段决策一个部分解,而后面的阶段依赖于前面的决策结果。

The problem can be decomposed into stages, and the solution of the problem can be decomposed into a series of stages according to the scale of the problem. Each stage determines a partial solution, and the subsequent stages depend on the results of the previous decisions.

（3）问题可以用树结构表示,问题的解空间可以用树结构表示,其中树的每个节点代表一个可能的解,树的分支代表每个决策点的选择。

The problem can be represented by a tree structure, and the solution space of the problem can be represented by a tree structure, where each node of the tree represents a possible solution, and the branch of the tree represents the choice of each decision point.

（4）存在可行解的部分集合。对于组合优化问题,存在一些部分解是可行的,可以通过递归的方式向下扩展解空间,直到找到最终的解或者确定无解。

There is a partial set of feasible solutions. For combinatorial optimization problems, some partial solutions are feasible, and the solution space can be recursively expanded downward until the final solution is found or no solution is determined.

（5）可以通过剪枝进行优化。在搜索过程中,可以通过一些约束条件或者启发式方法来剪枝,减少搜索空间,提高搜索效率。

It can be optimized by pruning. In the search process, some constraints or heuristic methods can be used to prune to reduce the search space and improve the search efficiency.

总体来说,回溯算法适用于那些具有多个解、可以分解为阶段、可以用树结构表示、存在可行解的部分集合,并且可以通过剪枝进行优化的问题。

In general, backtracking algorithms are suitable for problems that have multiple solutions, can be decomposed into stages, can be represented by a tree structure, there exists a partial set of feasible solutions, and can be optimized by pruning.

5.1.3　算法的效率估计（Efficiency Estimation of Algorithm）

首先,需要了解问题的复杂度。包括问题的规模、搜索空间的大小以及解空间的结构。通常,问题的复杂度越高,回溯算法的效率就越低。对于给定的问题,需要估计其搜索空间的大小,搜索空间的大小取决于问题的规模和约束条件,较大的搜索空间意味着更多的搜索和可能的组合,这可能会导致回溯算法的效率降低。

First, you need to understand the complexity of the problem. It includes the size of the problem, the size of the search space and the structure of the solution space. In general, the more complex the problem, the less efficient the backtracking algorithm is. For a given problem, the size of its search space needs to be estimated, and the size of the search space depends on the size and constraints of the problem, a larger search space means more searches and possible combinations, which may lead to lower efficiency of the backtracking algorithm.

另外,还可以通过剪枝和其他优化技术来提高回溯算法的效率。某些问题可能具有特定的优化方法,可以根据问题的特性进行调整。这可能包括启发式方法、领域知识的利用以及问题特定的剪枝策略。

In addition, the efficiency of the backtracking algorithm can be improved by pruning and other optimization techniques. Some problems may have specific optimization methods that can be adjusted based on the characteristics of the problem. This may include heuristics, the use of domain knowledge, and problem-specific pruning strategies.

总体来说,回溯算法的效率估计涉及对问题本身的理解、搜索空间的大小、剪枝和优化方法的使用以及算法实现的质量。在实际应用中,需要综合考虑这些因素来评估回溯算法的效率,并根据需要进行调整和优化。

In general, the efficiency estimation of backtracking algorithms involves the understanding of the problem itself, the size of the search space, the use of pruning and optimization methods, and the quality of the algorithm implementation. In practical applications, it is necessary to consider these factors comprehensively to evaluate the efficiency of the backtracking algorithm, and adjust and optimize it according to the needs.

5.2　0/1 背包问题（0/1 Knapsack Problem）

给定 n 件物品和一个背包。物品 i 的重量是 w_i，其价值为 v_i，背包的容量为 W。一件物品要么全部装入背包，要么全部不装入背包，不允许部分装入。装入背包的物品的总重量不超过背包的容量。

Given n items and a backpack. The weight of item i is w_i, its value is v_i, and the capacity of the backpack is W. Either all an item is loaded into the backpack, or all is not loaded into the backpack, and partial loading is not allowed. The total weight of the items loaded into the backpack does not exceed the capacity of the backpack.

5.2.1　基本思想（Basic Idea）

1. 定义问题的解空间（Define the Solution Space of the Problem）

0/1 背包问题是要将物品装入背包，并且物品有且只有两种状态。因此，可以用变量 x_i 表示第 i 件物品是否被装入背包的行为，如果用"0"表示不被装入背包，用"1"表示装入背包，则 x_i 的取值为 0 或 1。该问题解的形式是一个 n 元组，且每个分量的取值为 0 或 1。由此可得，问题的解空间为 $(x_1, x_2, \cdots, x_n)$，其中，$x_i = 0$ 或 $1(i = 1, 2, \cdots, n)$。

The 0/1 backpack problem is to pack an item into a backpack, and the item has and only has two states. Therefore, the variable x_i can be used to represent the behavior of whether the item i is loaded into the backpack, and if "0" is used to indicate that it is not loaded into the backpack, and "1" is used to indicate that it is loaded into the backpack, the value of x_i is 0 or 1. The solution to the problem is a tuple of n, and each component takes the value 0 or 1. It follows that the solution space of the problem is $(x_1, x_2, \cdots, x_n)$, where $x_i = 0$ or $1(i = 1, 2, \cdots, n)$.

2. 确定解空间的组织结构（Determine the Organization Structure of the Solution Space）

问题的解空间描述了 2^n 种可能的解，也可以说是 n 个元素组成的集合的所有子集个数。可见，问题的解空间树为子集树。采用一棵满二叉树将解空间有效地组织起来，解空间树的深度为问题的规模 n。以 $n = 2$ 为例，解空间树结构如图 5.1 所示，其中，根节点为起始节点，表示当前状态为空包，还未放入物品。

The solution space of the problem describes 2^n possible solutions, or the number of subsets of a set of n elements. Thus, the solution space tree of the problem is a subset tree. A full binary tree is used to organize the solution space effectively, the depth of the solution space tree is size n of the problem. Taking $n = 2$ as an example, the structure of the solution space tree is shown in Figure 5.1, where the root node is the start node, indicating that the current state is an empty packet and no items have been placed in it yet.

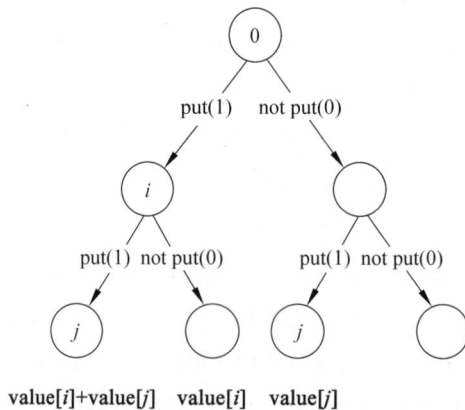

图 5.1　$n=2$ 时背包问题的解空间树结构

Figure 5.1　Tree structure of the solution space of the knapsack problem for $n=2$

3. 搜索解空间（Search Solution Space）

1）判断是否需要约束条件（Determine Whether Constraints are Required）

0/1 背包问题的解空间包含 2^n 个可能的解，会存在某种或某些物品无法装入背包的情况。因此，需要设置约束条件来判断所有可能的解描述的装入背包的物品总重量是否超出背包的容量。如果超出，就为不可行解；否则，视为可行解。搜索过程将不再搜索那些导致不可行解的节点。约束条件的形式化描述如下。

The solution space of the 0/1 knapsack problem contains 2^n possible solutions, and there will be cases where one or more items will not fit in the knapsack. Therefore, it is necessary to set constraints to judge whether the total weight of the items loaded into the backpack described by all possible solutions exceeds the capacity of the backpack, if so, it is not feasible. Otherwise, it is considered a feasible solution. The search process will no longer search for nodes that lead to infeasible solutions. The formal description of a constraint is as follows.

$$\sum_{i=1}^{n} w_i x_i \leqslant W$$

2）判断是否需要限界条件（Determine Whether a Limiting Condition is Required）

0/1 背包问题的可行解可能不止一个，问题的目标是找一个所描述的装入背包的物品总价值最大的可行解，即最优解。因此需要设置限界条件来加速找出该最优解的速度。

There may be more than one feasible solution to the 0/1 knapsack problem, and the goal of the problem is to find a feasible solution with the maximum total value of the items in the knapsack described, that is, the optimal solution. Therefore, it is necessary to set a limiting condition to accelerate the speed of finding the optimal solution.

根据解空间的组织结构可知，任何一个中间节点 z（中间状态）均表示从根节点到该中间节点的分支所代表的行为已经确定，从 z 到其子孙节点的分支的行为是不确定的。也就是说，如果 z 在解空间树中所处的层次是 t，从第 1 件物品到第 $t-1$ 件物品的状态已经确定，接下来要确定第 t 件物品的状态。无论沿着 z 的哪一个分支进行扩展，第 t 件物品的状态就确定了。那么，从第 t+1 件物品到第 n 件物品的状态还不确定。

According to the organization structure of the solution space, any intermediate node z (intermediate state) means that the behavior represented by the branch from the root node to the intermediate node has been determined, and the behavior of the branch from z to its descendant node is uncertain. That is, if the level of z in the solution space tree is t, the state from item 1 to item $t-1$ has been determined, and then state determine the state of item t. No matter which branch of z is extended, the state of item t is determined. So, the state from item t+1 to item n is uncertain.

这样，可以根据前 t 件物品的状态确定当前已装入背包的物品的总价值，用 cp 表示。第 t+1 件物品到第 n 件物品的总价值用 rp 表示，则 cp+rp 是所有从根出发的路径中经过中间节点 z 的可行解的价值上界。如果价值上界小于或等于当前搜索到的最优解描述的装入背包的物品总价值(用 $bestp$ 表示，初始值为 0)，就说明从中间节点 z 继续向子孙节点搜索不可能得到一个比当前更优的可行解，没有继续搜索的必要；反之，则继续向 z 的子孙节点搜索。因此，限界条件可描述如下。

In this way, the total value of the items currently loaded into the backpack can be determined according to the state of the former t items, expressed by cp. The total value of item t+1 to item n is represented by rp, then cp+rp is the upper bound of the value of all feasible solutions that pass through the middle node z in the path from the root. If the upper bound on the value is less than or equal to the total value of the best solution found so far (denoted as $bestp$, which is initialized to 0), it indicates that continuing to explore the descendants of node z cannot lead to a better feasible solution. Therefore, further search along this branch is unnecessary. Otherwise, the search continues to the descendant node of z. Therefore, the limiting condition can be described as follows.

$$cp + rp > bestp$$

5.2.2　算法设计与描述(Algorithm Design and Description)

从根节点开始，以深度优先的方式进行搜索。根节点首先成为活节点，也是当前的扩展节点。由于子集树中约定左分支上的值为"1"，因此沿着扩展节点的左分支扩展，则代表装入物品，此时，需要判断是否能够装入该物品，即判断约束条件成立与否。如果成立，就进入左孩子节点，左孩子节点成为活节点，并且是当前的扩展节点，继续向纵深节点扩展；如果不成立，就剪掉扩展节点的左分支，沿着其右分支扩展。右分支代表物品不装入背包，肯定存在可行解。但是沿着右分支扩展有没有可能得到最优解呢？这需要由限界条件来判断。如果限界条件满足，说明有可能导致最优解，即进入右分支，右孩子节点成为活节点，并成为当前的扩展节点，继续向纵深节点扩展；如果不满足限界条件，则剪掉扩展节点的右分支，开始向最近的活节点回溯。搜索过程直到所有活节点变成死节点后结束。

Start at the root node and search in a depth-first manner. The root node first becomes an active node and is the current extension node. Since the value on the left branch of the subset tree is agreed to be "1", extending along the left branch of the extension node represents loading the item. At this time, it is necessary to determine whether the item can be loaded, that is, whether the constraint condition is established or not. If it is established, the left child node is entered, and the left child node becomes an active node, and is the current extension node, and continues to expand to the depth node. If not, cut off the left branch of the extension node and extend along its right branch. The right branch means that the item does not put into the backpack, so there must be a possible solution. But is it possible to extend along the right branch to get an optimal solution? This needs to be judged by the boundary condition. If the boundary condition is satisfied, it indicates that the optimal solution may be caused, that is, the right branch is entered, and the right child node becomes the active node, and becomes the current extension node, and continues to expand to the depth node. If the limit condition is not satisfied, the right branch of the extension node is pruned, and the algorithm backtracks to the nearest live node. The search process ends when all active nodes become dead.

下面以具体问题为例说明上述算法设计，若有三件物品，重量分别为 3、4、5，价值分别为 6、7、8，背包容量为 8，圆圈内的数字为当前背包的物品价值总量。画出解空间树，如图 5.2 所示。

To illustrate the design of the above algorithm with a specific problem, if there are three items with weights of 3, 4, and 5, values of 6, 7, and 8, and a backpack capacity is 8, the number inside the circle is the total value of the items in the current backpack. Draw the solution space tree, as shown in Figure 5.2.

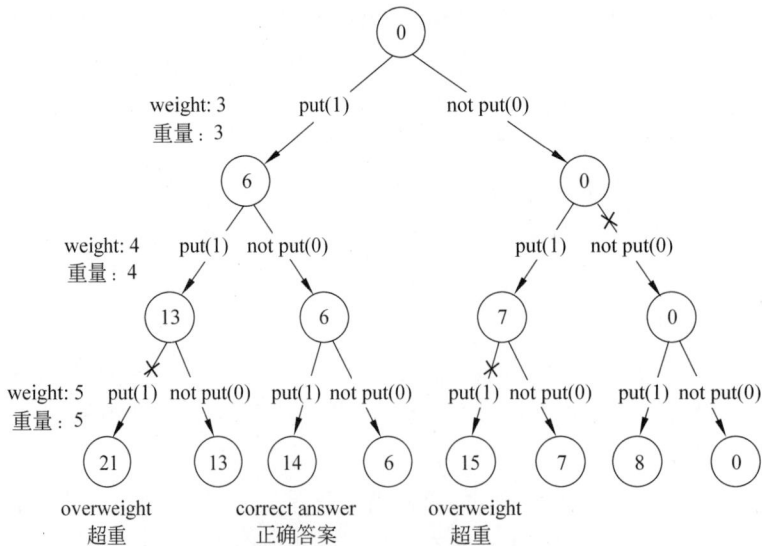

图 5.2　三件物品的解空间树

Figure 5.2　Solution space tree for the three items

5.2.3 算法实现(Algorithm Implementation)

首先定义一个 *backtrack*( )递归函数,用于求解深度优先搜索问题的解,接收当前扩展节点在子集树中所处的层次 *t*,根节点所处的层次为 0(考虑到数组下标从 0 开始,这里将节点的层次记为 0)。搜索过程中记录最优解 *bestp* 和最优价值 *bestx*。

First of all, a *backtrack* ( ) recursive function is defined to solve the depth-first search problem, accepting the level *t* of the current extension node in the subset tree, and the level of the root node is 0 (considering that the array subscript starts from 0, the node's level is marked as 0 here). The best solution *bestp* and the best value *bestx* were recorded in the search process.

在 *main*( )函数中,提供物品的重量 *w*、物品的价值 *v* 和背包的容量 *W*,并做初始化工作,调用 *backtrack*( )函数,最后将结果打印输出到显示器上。程序运行结果如图 5.3 所示。其代码如下。

In the *main*( ) function, provide the weight of the item *w*, the value of the item *v*, and the capacity of the backpack *W*, and do the initialization work, call the *backtrack*( ) function, and finally print the result to the display. Figure 5.3 shows the program running result. The code is as follows.

```python
def backtrack(t):
    global bestp, cw, cp, x, bestx, rp
    if t >= n:
        if bestp < cp:
            bestp = cp
            bestx = x[:]
    else:
        if cw +w[t] <= W:
            x[t] = 1
            cw += w[t]
            cp += v[t]
            rp -= v[t]
            backtrack(t +1)
            cw -= w[t]
            cp -= v[t]
            rp += v[t]
        if cp +rp > bestp:
            x[t] = 0
            backtrack(t +1)

if __name__ == "__main__":
    bestp = 0            #当前最优价值(The current optimal value)
    cw = 0               #当前装入背包的重量(The weight currently loaded into the backpack)
    cp = 0               #当前装入背包的价值(The value of the current pack)
    bestx = None         #记录当前最优解(Record the current optimal solution)
```

续

```
n = 5                        # 物品个数(Number of items)
W = 10                       #背包容量(Backpack capacity)
w = [2, 2, 6, 5, 4]          #物品重量(Item weight)
v = [6, 3, 5, 4, 6]          #物品价值(Item value)
x = [0 for i in range(n)]    #当前可行解(Current feasible solution)
rp = 0                       #剩余物品的价值(The value of surplus item)
for i in range(len(v)):
    rp += v[i]
backtrack(0)                 #从根节点开始搜索(Search from the root node)
print("最优价值为(The optimal value is):", bestp)
print("背包内所载物品如下(The items contained in the backpack are):", bestx)
```

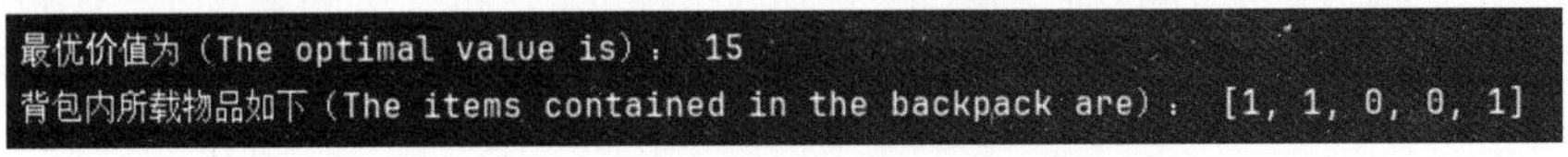

```
最优价值为 (The optimal value is): 15
背包内所载物品如下 (The items contained in the backpack are): [1, 1, 0, 0, 1]
```

图 5.3　0-1 背包回溯算法程序运行结果

Figure 5.3　0-1 Result of the knapsack backtracking program

5.3　n 皇后问题(n-queens Problem)

5.3.1　基本思想(Basic Idea)

n 皇后问题是寻找在 n×n 格的棋盘上放置 n 个皇后的方案,使得任意两个皇后不放在同一行或同一列或同一斜线上。用 n 元数组 x[1..n] 表示 n 皇后问题的解。其中,x[i] 表示皇后 i 放在棋盘的第 i 行的第 x[i] 列。由于不允许将第 2 个皇后放在同一列,所以解向量中的 x[i] 互不相同。两个皇后不能放在同一斜线上是问题的隐约束。对于一般的 n 皇后问题,这一隐约束条件可以转换成显约束的形式。将 n×n 格棋盘看作二维方阵,其行号从上到下依次编号为 1,2,…,n。如图 5.4 所示是 8 皇后问题的一个可行解。

The n-queens problem is to find a scheme for placing n queens on a chessboard of n×n cells such that no two queens are placed in the same row or column or on the same slash. The solution to the n-queen problem is represented by the n-tuple array x[1..n]. Where x[i] indicates that Queen i is placed in column x[i] of row i on the board. Since it is not allowed to place the second queen in the same column, the x[i] in the solution vector are different. The fact that two queens cannot be placed on the same slash is the implicit constraint of the problem. For the general n-queens problem, this implicit constraint can be expressed as an explicit constraint. The n×n chessboard is regarded as a two-dimensional square, and its row numbers are numbered from top to bottom as 1, 2, …, n. Figure 5.4 shows a feasible solution to the 8 queens problem.

设两个皇后的坐标分别为 (i,j) 和 (k,l)。若两个皇后在同一斜线上,那么这两个皇后的坐标连成的线的斜率为 1 或者 -1。从棋盘左上角到右下角的主对角线及其平行线(即斜率为 -1 的各斜线)上,两个下标值的差(行号 - 列号)值相等。同理,斜率为 1 的每一条斜线

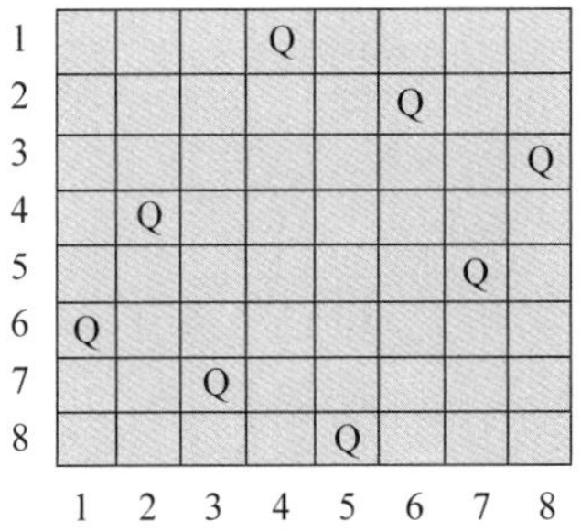

图 5.4 8 皇后问题的一个可行解

Figure 5.4 A feasible solution to the 8-queens problem

上，两个下标值的和（行号+列号）值相等，即

Let the coordinates of the two queens be (i, j) and (k, l). If two queens are on the same oblique line, then the coordinates of the two queens form a line with a slope of 1 or −1. The difference between the two subscripts（row number − column number）is equal on the main diagonal from the top left corner to the bottom right corner of the board and its parallel lines（i.e., each slash with a slope of −1）. Similarly, on each slash line with a slope of 1, the sum（row number +column number）of the two subscripts is equal, that is

$$\left.\begin{array}{l} \dfrac{i-k}{j-l}=1 \Rightarrow i-k=j-l \Leftrightarrow i-j=k-l \\[2mm] \dfrac{i-k}{j-l}=-1 \Rightarrow i-k=l-j \Leftrightarrow i+j=k+l \end{array}\right\} \Rightarrow |i-k|=|j-l|$$

因此，如果 $|i-k|=|j-l|$，则说明这两个皇后处于同一斜线上。以上两个皇后位于同一斜线上，问题的隐约束变成了显约束。

Therefore, if $|i-k|=|j-l|$, the two queens are on the same slash. The above two queens are located on the same slash line, and the implicit constraint of the problem becomes an explicit constraint.

5.3.2 算法设计与描述（Algorithm Design and Description）

用回溯算法解 n 皇后问题时，用完全 n 叉树表示解空间。可行性约束剪去不满足列和斜线约束的子树。以 3 皇后问题为例，解空间树结构如图 5.5 所示。

When solving the n-queens problem by backtracking, the solution space is represented by a complete n-tree. The feasibility constraint cuts out subtrees that do not satisfy the column and slash constraints. Taking the 3-queens problem as an example, the solution space tree structure is shown in Figure 5.5.

由图 5.5 的解空间树可知，三皇后问题是无解的，事实上只有 $n \geqslant 4$ 时，n 皇后问题才具有可行解，且随着问题规模的增大，可行解会越来越多。

As shown in the solution space tree of Figure 5.5, the 3-queens problem has no solution. In fact, the n-queens problem only has feasible solutions when $n \geqslant 4$, and the

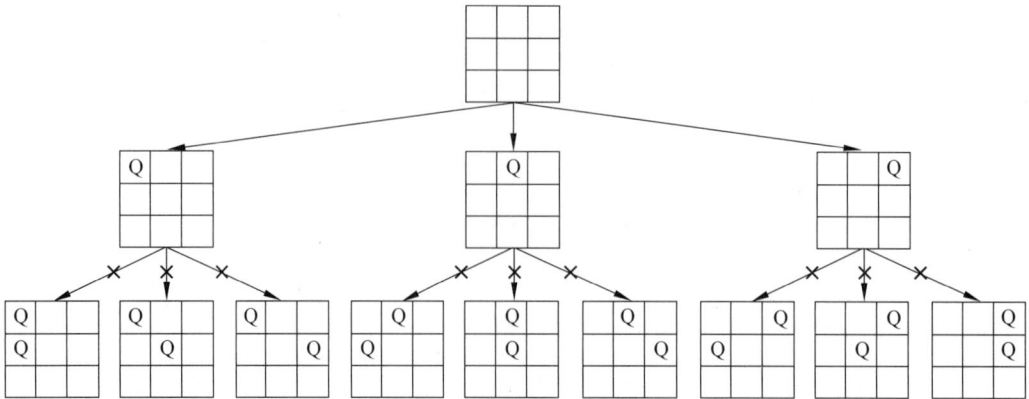

图 5.5　3 皇后问题的解空间树结构

Figure 5.5　Tree structure of the solution space of the 3-queens problem

number of feasible solutions increases with the problem size.

　　由于使用树结构解决 n 皇后问题时，皇后放置顺序是逐行放置的，所以不用考虑两个皇后是否会出现在同一行的情况。

　　Since the queen placement order is row-by-row when solving the n-queens problem using a tree structure, there is no need to consider whether two queens will appear on the same row.

　　当 $i=n$ 时，算法搜索至叶节点，得到一个新的放置方案，当前已找到的可行方案数 sum 增 1。

　　When $i=n$, the algorithm reaches a leaf node and obtains a new placement scheme; the count of feasible solutions, denoted as sum, is then incremented by 1.

　　当 $i<n$ 时，当前扩展节点 z 是解空间中的内部节点，该节点有 n 个儿子节点，对当前扩展节点 z 的每个儿子节点检查其可行性，并以深度优先的方式递归地对可行子树搜索，或剪去不可行子树。

　　When $i<n$, the current extension node z is an internal node in the solution space with n son nodes. The feasibility of each son of the current extension node z is checked, and the feasible subtree is searched recursively in a depth-first way, or the unfeasible subtree is cut.

　　以 4 皇后问题为例构建解空间树以说明上述算法设计，为了方便，用圆圈代替图 5.5 中的棋盘，红色圆圈代表当前节点与父节点或与根节点到该节点路径上的节点处于对角线位置，即不可行。在该图中也自动排除了任意两个节点可能会处于同一列的情况。圈内的数字代表当前皇后在棋盘中所处的坐标位置。4 皇后问题的解空间树如图 5.6 所示。

　　The 4-queens problem is used as an example to construct a solution space tree to illustrate the algorithm design described above. For convenience, circles are used instead of the chessboard in Figure 5.5, with the red circle representing the current node is in a diagonal position to the parent node or to a node on the path from the root node to that node, which means it is infeasible. The possibility that any two nodes may be in the same column is also automatically excluded in this figure. The number inside the circle represents the coordinate position of the current queen on the board. The solution space tree for the 4-queens problem is shown in Figure 5.6.

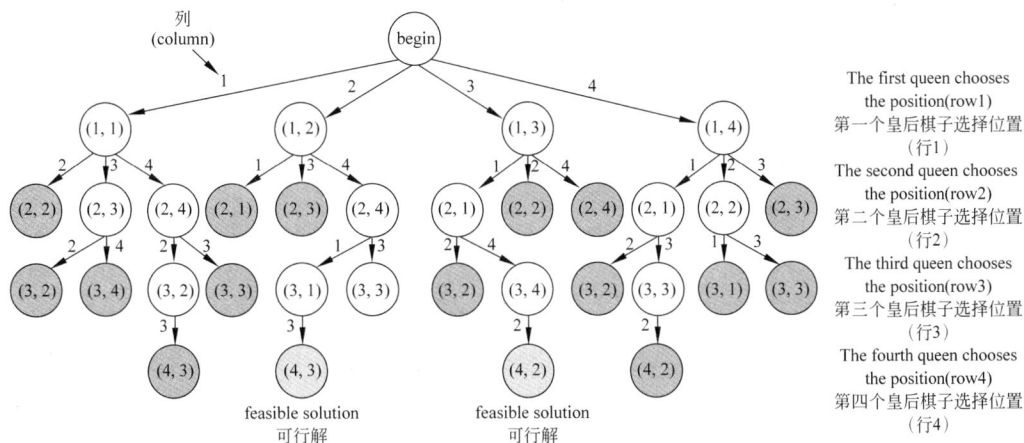

图 5.6 4 皇后问题的解空间树

Figure 5.6 Solution space tree for the 4-queens problem

通过图 5.6 得到的解空间树,从根节点到叶子节点,可以找到 4 皇后问题的两个可行解以及坐标路径,将其转换为棋盘形式得到 4 皇后问题的两个可行解,如图 5.7 所示。

With the solution space tree obtained in Figure 5.6, the two feasible solutions of the 4-queens problem as well as the coordinate paths can be found from the root node to the leaf nodes, which are transformed into the chessboard form to obtain the two feasible solutions of the 4-queens problem, as shown in Figure 5.7.

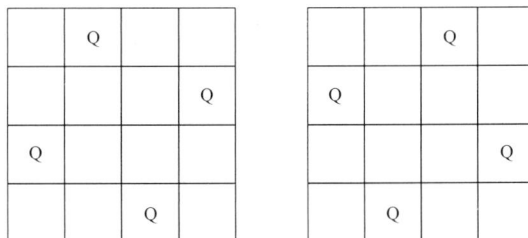

图 5.7 4 皇后问题的解空间树

Figure 5.7 Solution space tree of 4-queens problem

5.3.3 算法实现(Algorithm Implementation)

定义 *backtrack*( )递归函数,用于深度优先搜索问题的解,再定义 *is_valid*( )函数检查当前位置是否可以放置皇后,以 4 皇后为例,程序运行结果如图 5.8 所示。其代码如下。

The *backtrack*( ) recursive function is defined to search for the solution of the depth-first problem, and the *is_valid*( ) function is defined to check whether the current position can be placed. Take the 4 queens as an example, the results of the program are shown in Figure 5.8. The code is as follows.

```python
def NQueens(n):
    board = [['□'] * n for _ in range(n)]
    solutions = []

    def is_valid(row, col):
        #检查当前位置是否可以放置皇后(Check whether the current location can hold the queen)
        for i in range(row):
            if board[i][col] == 'Q':
                return False
            if col - (row - i) >= 0 and board[i][col - (row - i)] == 'Q':
                return False
            if col + (row - i) < n and board[i][col + (row - i)] == 'Q':
                return False
        return True

    def backtrack(row):
        if row == n:
            #找到一个解决方案(Find a solution)
            solutions.append([''.join(row) for row in board])
            return

        for col in range(n):
            if is_valid(row, col):
                board[row][col] = 'Q'
                backtrack(row + 1)
                board[row][col] = '□'

    backtrack(0)
    return solutions


n = 4
solutions = NQueens(n)
print('-' * 8)
for solution in solutions:
    print('\n'.join(solution))
    print('-' * 8)
```

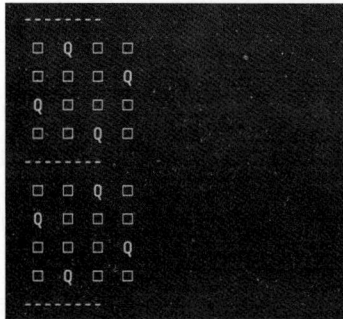

图 5.8　4 皇后程序运行结果

Figure 5.8　Result of the 4-queens program

5.4　最大团问题（Maximum Clique Problem）

5.4.1　基本思想（Basic Idea）

给定无向图 $G=(V, E)$。如果 $U \subseteq V$，且对任意 $u, v \in U$，有 $(u, v) \in E$，则称 U 是 G 的完全子图。G 的完全子图 U 是 G 的团，G 的最大团是指 G 中所含顶点数最多的团。即最大团问题就是找出无向图 G 中包含顶点个数最多的团。

Given an undirected graph $G=(V, E)$. If $U \subseteq V$, and for any $u, v \in U$, there is $(u, v) \in E$, then U is said to be a complete subgraph of G. The complete subgraph U of G is a clique of G, and the largest clique of G is the clique containing the largest number of vertexes in G. The maximum clique problem is to find the clique with the largest number of vertexes in an undirected graph G.

图 G 的最大团问题可以看作图 G 的顶点集 V 的子集选取问题。因此可以用子集树来表示问题的解空间。首先设最大团为一个空团，向其中加入一个顶点，然后依次考虑每个顶点。查看该顶点加入团之后是否仍然构成一个团，如果可以，考虑将该顶点加入团或者舍弃两种情况；如果不行，直接舍弃，然后递归判断下一顶点。对于无连接或者直接舍弃两种情况，在递归前，可采用剪枝策略来避免无效搜索。为了判断当前顶点加入团之后是否仍是一个团，只需要考虑该顶点和团中顶点是否都有连接。

The maximal clique problem of graph G can be regarded as the problem of subset selection of vertex set V of graph G. Therefore, a subset tree can be used to represent the solution space of the problem. First set the maximum clique as an empty clique, add a vertex to it, and then consider each vertex in turn. Check whether the vertex still forms a group after joining the group, and if possible, consider adding the vertex to the group or dropping both cases; If not, directly discard, and then recursively determine the next vertex. For both cases of no connection or direct discarding, pruning strategies can be used before recursion to avoid invalid searches. To determine whether the current vertex is still a clique after joining the clique, it is only necessary to consider whether the vertex and the vertex in the clique are connected.

5.4.2　算法设计与描述（Algorithm Design and Description）

最大团问题的搜索和 0/1 背包问题的搜索相似，只是进行判断的约束条件和限界条件不同而已。

The search for the maximum clique problem is similar to the search for the 0/1 knapsack problem, except that the constraints and bounds of the judgment are different.

从根节点开始，以深度优先的方式进行搜索。根节点首先成为活节点，也是当前的扩展节点。沿着扩展节点的左分支扩展，需要判断加入该节点之后是否能够成团。如果成立，就进入左孩子节点，左孩子节点成为活节点，并且是当前的扩展节点，继续向纵深节点扩展；如果不成立，就剪掉扩展节点的左分支，沿着其右分支扩展。右分支代表该节点不加入团，肯定有可能存在可行解。但是沿着右分支扩展有没有可能得到最优解呢？这一点需要由限界条件来判断。如果限界条件满足，说明有可能导致最优解，即进入右分支，右孩子节点成为活节点，并成为当前的扩展节点，继续向纵深节点扩展；如果不满足限界条件，则剪掉扩展节点的右分支，开始向最近的活节点回溯。搜索过程直到所有活节点变成死节点后结束。

Starting from the root node, the search is performed in a depth-first manner. The root node first becomes a living node and is also the current extension node. To expand along the left branch of the expansion node, it is necessary to determine whether it can form a clique after joining the node. If so, it enters the left child node, and the left child node becomes a living node, which is the current expansion node and continues to expand to the depth node. If not, cut off the left branch of the extension node and extend along its right branch. The right branch means that the node does not join the clique, and there must be a possible solution. But is it possible to extend along the right branch to get an optimal solution? This needs to be judged by the boundary condition. If the boundary condition is satisfied, it indicates that the optimal solution may be caused, that is, the right branch is entered, and the right child node becomes the living node, and becomes the current extension node, and continues to expand to the depth node. If the limit condition is not satisfied, the right branch of the extension node is pruned, and the algorithm backtracks to the nearest live node. The search process ends when all living nodes become dead.

5.4.3　算法实现（Algorithm Implementation）

首先定义一个 *place*( )函数，用于判断指定的点是否能加入最大团集合中。该函数接收待判定的点，返回 *True* 或 *False*，*True* 表示指定点能放入最大团集合，*False* 表示指定点不能放入最大团集合。再定义一个递归深度优先搜索的 *backtrack*( )函数，搜索最优解。

First define a *place*( ) function that determines whether the specified node can be added to the maximal clique set. This function receives the node to be determined and returns *True* or *False*. *True* means the specified node can fit into the maximum clique set, *False* means the specified node cannot fit into the maximum clique set. Then define a recursive depth-first search function *backtrack*( ) to search for the optimal solution.

在 *main*( )函数中，用邻接矩阵存储给定的图 *G*，调用 *backtrack*( )函数，求最优解和最优价值，最后将结果打印输出到显示器上。程序运行结果如图 5.9 所示。其代码如下。

In the *main*( ) function, use the adjacency matrix to store the given graph *G*, call the *backtrack*( ) function, find the optimal solution and the optimal value, and finally print the result to the display. Figure 5.9 shows the program running result. The code is as follows.

```
def place(t):
    global x
    global a
```

```
        OK = True
        for j in range(t):
            if x[j] and a[t][j] ==0:
                OK = False
                break
        return OK


def backtrack(t):
    global cn, bestn, n, bestx, x
    if t ==n:
        if cn > bestn:
            bestx = x[:]
            bestn = cn
        return

    if place(t):
        x[t] = 1
        cn +=1
        backtrack(t + 1)
        cn -=1

    if cn + n - t - 1 > bestn:
        x[t] = 0
        backtrack(t + 1)

if __name__ =="__main__":
    a = [[0, 1, 1, 0, 0], [1, 0, 1, 1, 1], [1, 1, 0, 1, 1], [0, 1, 1, 0, 1], [0, 1, 1, 1, 0]]
    n = len(a)
    x = [1 for i in range(n)]
    bestx = None
    bestn = cn = 0
    backtrack(0)
    print("最大团点个数(Maximum number of dots): ", bestn)
    print("最大团为(Maximum clique is): ", bestx)
```

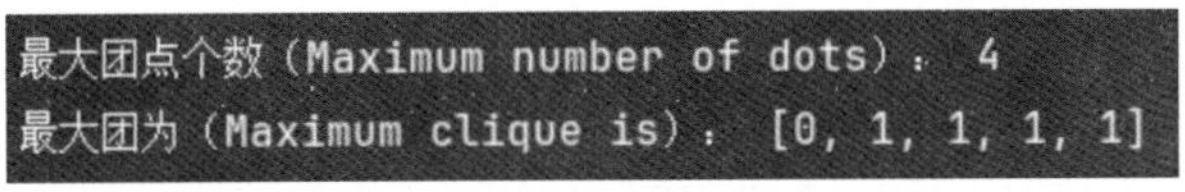

```
最大团点个数(Maximum number of dots): 4
最大团为(Maximum clique is): [0, 1, 1, 1, 1]
```

图 5.9　程序运行结果

Figure 5.9　Result of the program running

5.5　图的 m 着色问题(The M-Coloring Problem of Graph)

给定无向连通图 $G = (V, E)$ 和 m 种不同的颜色。用这些颜色为图 G 的各顶点着色,每个顶点着一种颜色。如果有一种着色法使 G 中有边相连的两个顶点着不同颜色,则称这个图是 m 可着色的。图的 m 着色问题是对于给定图 G 和 m 种颜色,找出所有不同的着色方法。

Given an undirected connected graph $G = (V, E)$ and m different colors. Use these colors to color the vertexes of the graph G, each vertex having a color. If there is a coloring method so that the two vertexes connected by edges in G are of different colors, the graph is said to be colorable in m. The m coloring problem of graphs is to find all the different coloring methods for a given graph G and m colors.

5.5.1 基本思想（Basic Idea）

该问题中每个顶点所着的颜色均有 m 种选择，n 个顶点所着颜色的一个组合是一个可能的解。从给定的已知条件来看，无向连通图 G 中假设有 n 个顶点，它肯定至少有 $n-1$ 条边，有边相连的两个顶点所着颜色不相同，n 个顶点所着颜色的所有组合中必然存在不是着色问题方案的组合，因此需要设置约束条件。而针对所有可行解，不存在可行解优劣的问题。所以，不需要设置限界条件。

There are m possible colors for each vertex in this problem, and any combination of colors for n vertices is a possible solution. From the given known conditions, if there are n vertexes in an undirected connected graph G, it must have at least $n-1$ edges, and the two vertexes connected by the edge have different colors, in all combinations of the colors of n vertexes, there must be combinations that are not solutions to the coloring problem, so it is necessary to set constraints. However, for all feasible solutions, there is no question of whether the feasible solution is good or bad. Therefore, there is no need to set a boundary condition.

1. 定义问题的解空间（Define the Solution Space of the Problem）

图的 m 着色问题的解空间形式为 $(x_1, x_2, \cdots, x_n)$，分量 $x_i(i=1,2,\cdots,n)$ 表示第 i 个顶点着第 x_i 号颜色。m 种颜色的色号组成的集合为 $S=\{1,2,\cdots,m\}$，$x_i \in S, i=1,2,\cdots,n$。

The solution space of the m coloring problem of a graph is $(x_1, x_2, \cdots, x_n)$, component $x_i(i=1, 2, \cdots, n)$ represents the i vertex with the x_i color. The set of color numbers of m colors is $S=\{1, 2, \cdots, m\}$, $x_i \in S, i=1, 2, \cdots, n$.

2. 确定解空间的组织结构（Determine the Organization Structure of the Solution Space）

问题的解空间组织结构是一棵满 m 叉树，树的深度为 n。

The solution space structure of the problem is a full m-fork tree with a depth of n.

3. 搜索解空间（Search Solution Space）

（1）设置约束条件。当前顶点要和前面已确定颜色且有边相连的顶点所着颜色不相同。假设当前扩展节点所在的层次为 t，则下一步扩展就是要判断第 t 个顶点着什么颜色，第 t 个顶点所着的颜色要与已经确定所着颜色的第 $1 \sim (t-1)$ 个顶点中与其边相连的颜色不相同。约束函数代码描述如下。

Set constraint condition. The current vertex must have a color different from the previous vertex that has been determined and connected by an edge. Assuming that the current level of the expansion node is t, the next step of the expansion is to determine what color the t vertex is, and the color of the t vertex should be different from the color connected with the edge of the first to $(t-1)$ vertex that has been determined to be the color. The constraint function code is described below.

```
def ok(k):
    for j in range(1, k):
        if a[k][j] ==1 and x[j] ==x[k]:
            return False
    return True
```

（2）无须设置限界条件。

There is no need to set a boundary condition.

5.5.2 算法设计与描述（Algorithm Design and Description）

扩展节点沿着某个分支扩展时需要判断约束条件。如果满足，就进入深一层继续搜索；如果不满足，就将扩展生成的节点剪掉。搜索到叶子节点时，找到一种着色方案。搜索过程直到全部活节点变成死节点为止。

When the extension node expands along a branch, it needs to determine the constraints, and if they are satisfied, it goes into a deeper layer to continue the search. If not, the nodes generated by the extension are clipped off. When the leaf node is searched, a coloring scheme is found. The search process continues until all living nodes become dead.

图的深度优先遍历需要借助邻接矩阵和邻接表，以确定图的深度优先遍历顶点顺序。以如图5.10所示的图为例，尝试确定它的遍历（上色）顺序。

Depth-first traversal of a graph requires the help of adjacency matrix and adjacency table to determine the depth-first traversal vertex order of the graph. Take the graph shown in Figure 5.10 as an example and try to determine its traversal (coloring) order.

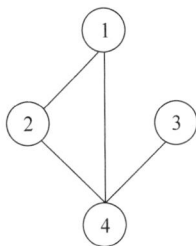

图 5.10　示例图

Figure 5.10　Example graph

构建该图的邻接矩阵如下。

The adjacency matrix for constructing this graph is as follows.

$$a = \begin{bmatrix} 0 & 1 & 0 & 1 \\ 1 & 0 & 0 & 1 \\ 0 & 0 & 0 & 1 \\ 1 & 1 & 1 & 0 \end{bmatrix}$$

构建该图的邻接表，如图 5.11 所示。

The adjacency list for building this graph is shown in Figure 5.11.

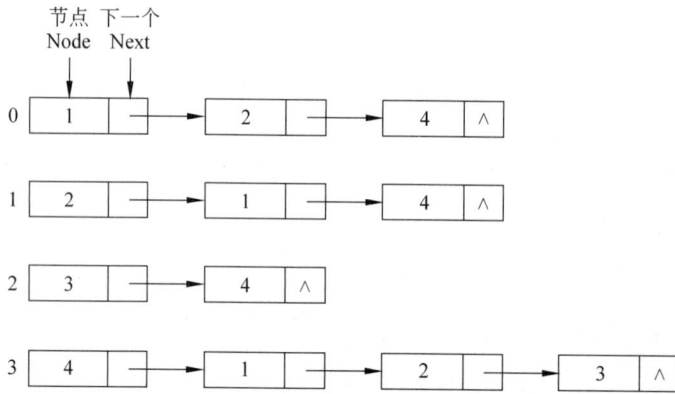

图 5.11　图 5.10 的邻接表

Figure 5.11　The adjacency table of Figure 5.10

从节点 1 开始，根据图 5.11 的邻接表，对该图进行深度优先遍历，得到该图的遍历（上色）顺序为 1、2、4、3。

Starting from node 1, the graph is traversed depth-first according to the adjacency table in Figure 5.11, and the traversal (coloring) order of the graph is 1, 2, 4, and 3.

至少使用三种颜色才能满足该图的上色要求，该图的解空间树如图 5.12 所示。

At least three colors can be used to satisfy the coloring requirements for this graph, which has the solution space tree shown in Figure 5.12.

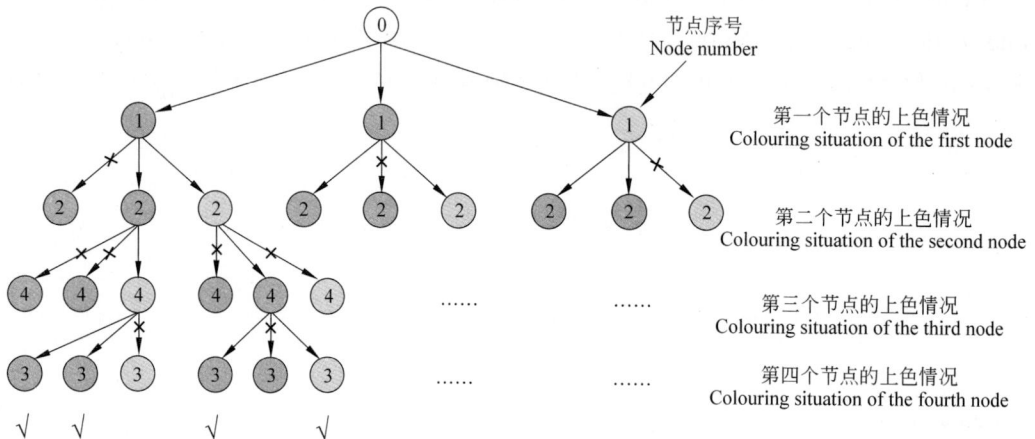

图 5.12　例图着色问题的解空间树

Figure 5.12　Solution space tree of the example graph coloring problem

由解空间树可知，该图共有 12 种着色方案。

According to the solution space tree, there are 12 coloring schemes in the graph.

5.5.3 算法实现(Algorithm Implementation)

选用邻接矩阵 a 存储图 G,用二维列表存储所有的着色方案,用一维列表存储当前着色方案。

The adjacency matrix a is selected to store the graph G, all the coloring schemes are stored in a two-dimensional list, and the current coloring schemes are stored in a one-dimensional list.

首先定义 $ok()$ 函数,接收待着色的第 k 号顶点编号,输出是否都能为该点着相应的颜色 $x[k]$。$True$ 表示 k 号顶点能着 $x[k]$ 号色,$False$ 表示 k 号顶点不能着 $x[k]$ 号色。再定义一个深度优先搜索的 $backtrack()$ 函数,搜索所有可能的着色方案,并统计着色方案数。

First define the $ok()$ function to receive the number of the k vertex to be colored, and output whether the corresponding color $x[k]$ can be used for that point. $True$ indicates that the vertex of k can be colored $x[k]$, $False$ indicates that the vertex of k cannot be colored $x[k]$. Then define a depth-first search $backtrack()$ function to search for all possible coloring schemes and count the number of coloring schemes.

在 $main()$ 函数中,用邻接矩阵 a 存储无向图,牺牲 0 行 0 列位置的存储单元,下标从 1 开始有效。初始化相关辅助变量,调用 $backtrack()$ 函数,求所有可能的着色方案,最后将结果打印输出到显示器上。输出结果中,0 号存储单元数据无效,从下标 1 开始有效。程序运行结果如图 5.13 所示。其代码如下。

In the $main()$ function, the undirected graph is stored with the adjacency matrix a, sacrificing the storage unit at 0 rows and 0 columns, and the subscript is valid from 1. Initialize the relevant auxiliary variables, call the $backtrack()$ function, find all possible coloring schemes, and finally print out the results to the display. In the output result, the metadata of storage unit 0 is invalid. The metadata is valid from subscript 1. Figure 5.13 shows the program running result. The code is as follows.

```python
def ok(k):
    for j in range(1, k):
        if a[k][j] == 1 and x[j] == x[k]:
            return False
    return True


def backtrack(t):
    global colors, x, sum1
    if t > n:                        #搜索到叶子节点
        sum1 += 1
        colors.append(x[:])
    else:
        for i in range(1, m + 1):    #搜索中间节点的每一个分支,即尝试着任何一种颜色
            x[t] = i
```

续

```
            if ok(t):
                backtrack(t + 1)

if _ _name_ _ == "_ _main_ _":
    a = [[0, 0, 0, 0, 0, 0], [0, 0, 1, 1, 0, 0], [0, 1, 0, 1, 1, 1], [0, 1, 1, 0, 1, 0], [0, 0, 1, 1, 0, 1],
        [0, 0, 1, 0, 1, 0]]
    n = len(a) - 1
    m = 3
    sum1 = 0
    colors = []
    x = [0 for i in range(n + 1)]
    backtrack(1)
    for i in range(len(colors)):
        print(colors[i])
    print("共有:" + str(sum1) + "种着色方案")
    print("There are: " + str(sum1) + " coloring schemes")
```

```
[0, 1, 2, 3, 1, 3]
[0, 1, 3, 2, 1, 2]
[0, 2, 1, 3, 2, 3]
[0, 2, 3, 1, 2, 1]
[0, 3, 1, 2, 3, 2]
[0, 3, 2, 1, 3, 1]
共有:6种着色方案
There are: 6 coloring schemes
```

图 5.13　图的 m 着色问题程序运行结果

Figure 5.13　Result of m coloring problem program running

习题 5（Exercises Five）

1. 什么是回溯算法？

What is backtracking algorithm?

2. 什么是剪枝函数？

What is the pruning function?

3. 回溯算法解题的步骤是什么？

What are the steps of backtracking algorithm?

4. 简述回溯算法的基本思想。

Briefly describe the basic idea of backtracking algorithm.

5. 组合求和问题：给定一个候选数组 *candidates* 和一个目标数 *target*，找出 *candidates* 中所有可以使数字和为 *target* 的唯一组合。*candidates* 中的每个数字在每个组合中只能使用一次。

Combinatorial summation problem: Given an array of *candidates* and a target number *target*, find all the unique combinations of *candidates* that can make the number sum the *target*. Each number in the *candidates* can only be used once in each combination.

6. 全排列问题：给定一个不含重复数字的数组 *nums*，返回其所有可能的全排列。

Full permutation problem: Given an array *nums* with no duplicate numbers, return all possible full permutations.

7. 子集问题：给定一组不含重复元素的整数 *nums*，返回所有可能的子集（幂集）。

Subset problem: Given a set of integers *nums* with no repeating elements, return all possible subsets（power sets）.

第6章

分支限界算法

Chapter 6 Branch and Bound Algorithm

6.1 概述(Overview)

分支限界算法类似于回溯算法,也是一种在问题的解空间树中搜索问题解的算法,常以宽度优先或以最小耗费(最大收益)优先的方式搜索问题的解空间树。

Branch and bound algorithm is similar to backtracking algorithm, and it is also an algorithm to search the solution of a problem in the solution space tree. It usually searches the solution space tree in the way of width first or minimum cost (maximum benefit) first.

分支限界算法首先将根节点加入活节点表(用于存放活节点的数据结构),接着从活节点表中取出根节点,使其成为当前扩展节点,一次性生成其所有的子节点,判断子节点应该舍弃(导致不可行解或非最优解)或是保留(其余节点)。再从活节点表中取出一个活节点当作当前扩展节点,重复上述扩展过程,直到找到所需的解或活节点表为空。因此,每个活节点最多只有一次机会成为扩展节点。

The branch and bound algorithm firstly add the root node to the living node table (the data structure used to store the living nodes), then extracts the root node from the living node table to make it the current expansion node, generates all its child nodes at once, and determines whether the child nodes should be discarded (resulting in infeasible or non-optimal solutions) or kept (the rest of the nodes). Then a living node is taken from the living node table as the current expansion node, and the above expansion process is repeated until the required solution is found or the living node table is empty. Thus, each living node has at most one chance to become an extension node.

可见,分支限界算法搜索过程的关键在于判断子节点的舍弃或保留。因此,在搜索之前要设定子节点的判断标准,与回溯算法搜索过程中用到的约束条件和限界条件的含义相同。

The key to the search process of the branch and bound algorithm is to determine whether to discard or retain the child nodes. Therefore, the judgment criteria for child nodes should be set before searching, which has the same meaning as the constraints and bounds used in backtracking search.

活节点的实现通常有两种方法,一是先进先出队列,二是优先级队列,它们对应的分支限界算法分别称为队列式分支限界算法和优先队列式分支限界算法。

There are usually two methods to implement living nodes, one is FIFO (First in First Out) queue, the other is priority queue, and their corresponding branch and bound methods are called queue branch and bound algorithm and priority queue branch and bound algorithm respectively.

队列式分支限界算法按照队列先进先出(FIFO)原则选取下一个节点作为当前扩展节点。优先队列式分支限界算法按照规定的优先级选取队列中优先级最高的节点作为当前扩展节点。优先队列一般用二叉堆来实现,即最大堆实现最大优先队列,体现最大效益优先;最小堆实现最小优先队列,体现最小费用优先。

The queue branch and bound algorithm selects the next node as the current extension node according to the queue FIFO principle. The priority queue branch and bound algorithm selects the node with the highest priority in the queue as the current extension node according to the specified priority. The priority queue is generally implemented using binary heaps, that is, the maximum heap implements the maximum priority queue, reflecting the maximum benefit priority, and the minimum heap implements the minimum priority queue, reflecting the minimum cost priority.

分支限界算法的一般解题步骤如下。

The general steps of branch and bound algorithm are as follows.

(1) 定义问题的解空间。

Define the solution space of the problem.

(2) 确定问题的解空间组织结构(树或图)。

Determine the solution space organization structure of the problem (tree or graph).

(3) 搜索解空间。搜索前要定义判断标准(约束函数或限界函数),如果选用优先队列分支限界算法,就必须确定优先级。

To search the solution space, the criteria (constraint function or bound function) should be defined before the search. If the priority queue branch and bound algorithm is selected, the priority must be determined.

6.2　0/1 背包问题(0/1 Knapsack Problem)

6.2.1　基本思想(Basic Idea)

1. 问题描述(Problem Description)

分别用队列式分支限界算法和优先队列式分支限界算法求解 0/1 背包问题,其中,物资件数为 n,物资权重为 $w_1, w_2, \cdots, w_n$,物资价值为 $v_1, v_2, \cdots, v_n$,背包总容量为 W。

The 0/1 knapsack problem is solved by the queue branch and bound algorithm and the priority queue branch and bound algorithm, respectively, where the number of items is n and the weight of items is w_1, w_2, $\cdots$, w_n, the material value is v_1, v_2, $\cdots$, v_n, the total capacity of the backpack is W.

2. 求解分析（Solution Analysis）

1）定义解空间（Define the Solution Space）

该问题实例的解空间为$(x_1, x_2, \cdots, x_n)$，解空间内元素取值为 0（不装入）或 1（部分或完整装入）。

The solution space of this problem instance is $(x_1, x_2, \cdots, x_n)$, the value of the element in the solution space is 0 (not loaded) or 1 (partially or completely loaded).

2）确定问题的组织结构（Determine the Organization Structure）

解空间是一棵子集树，深度为 n。

Determine the organization structure of the problem solution space. The solution space is a subset tree with depth n.

3）搜索解空间（Search Solution Space）

根据不同的搜索方法，定义合适的约束条件和限界条件并开始搜索。初始时，将根节点放入活节点表中。搜索过程，从活节点表中取出一个活节点作为当前的扩展节点，一次性生成扩展节点的所有子节点，判断是否满足约束条件和限界条件，决定保留或抛弃。找到问题的解或活节点表为空时，搜索停止。

According to different search methods, appropriate constraints and boundary conditions are defined and the search is started. Initially, the root node is placed in the living node table. In the search process, a living node is selected from the living node table as the current extension node, all the child nodes of the extension node are generated at one time, and then determine whether the constraints and boundary conditions are satisfied, and decide to keep or discard. The search stops when the solution to the problem is found or the living node table is empty.

6.2.2　算法设计（Algorithm Design）

1. 队列式分支限界算法（Queued Branch and Bound Algorithm）

定义约束条件为 $\sum_{i=1}^{n} w_i x_i \leqslant C$，限界条件为 $cp+r'p>bestp$。其中，cp 表示当前已装入背包的物品总价值，初始值为 0；$r'p$ 表示剩余物品装入剩余背包容量的最优价值，初始值为所有物品的价值之和；$bestp$ 表示当前最优解，初始值为 0，当 $cp>bestp$ 时，更新 $bestp$ 为 cp。

The constraint condition is $\sum_{i=1}^{n} w_i x_i \leqslant C$, the limiting condition is $cp+r'p>bestp$, cp represents the total value of the items currently loaded into the backpack, and the initial value is 0; $r'p$ represents the optimal value of remaining items loaded into the remaining backpack capacity, and the initial value is the sum of the values of all items; $bestp$ represents the current optimal solution, the initial value is 0, when $cp > bestp$, update $bestp$ to cp.

算法伪代码如下。（The algorithm pseudocode is as follows.）

算法 6.1 队列式分支限界算法解决 0/1 背包问题（Algorithm 6.1 Algorithm queued branch bound solves the 0/1 knapsack problem）

输入：背包的容量（Input：Capacity of backpack）

输出：最优价值的 *bestp* 及最优解叶子节点编号 *best*（Output：*bestp* of the best value and the number of the optimal leaf node *best*）

1.   que←空队列（empty queue）

2.   node←Node(0, 0, 1)   //根节点,cw 为 0,cp 为 0,根节点编号为 1（For the root node, cw is 0, cp is //0, and the root node number is 1）

3.   insert(que, node)   //将 node 节点插入队列（Insert the node into the queue）

4.   while(! que, empty( ))do   //当队列 *que* 非空时,循环（When the queue *que* is not empty, it loops）

5.   current_node←delete(que)   //从队列中取出队首元素（Removes the first element from the queue）

6.   depth←current_node   //节点的深度（Depth of node）

7.   if depth $==n$ then   //叶子表示找到了一个比当前解更好的一个解,并记录（The leaf indicates //that it has found a better solution than the current one and records it）

8.   bestp←current_node.cp

9.   best←current_node.id

10.   else   //如果该扩展节点不是叶子节点（If the extension node is not a leaf node）

11.   if current_node.cw + goods[depth][1]≤capacity then //判断约束条件（Judgment constraint）

12.   if current_node.cp + goods[depth][2]>bestp then

13.   bestp ← current_node.cp + goods[depth][2]

14.   best ← current_node.id $*$ 2 //记录当前最优的节点编号（Record the optimal node number）

15.   end if

16.   alive_node←Node(current_node.cp+godds[depth][2]

17.   current_node.cw+godds[depth][1], current_node.id $*$ 2)

18.   insert(que, alive_node)   //插入队列 *que*（Insert queue *que*）

19.   end if

20.   up←bound(current_node)   //计算当前节点的价值上界（Calculate the upper bound of the value of //the current node）

21.   if up>bestp then

22.   alive_node←Node(current_node.cp, current_node.cw, current_node.id $*$ 2+1)

23.   insert(que, alive_node)   //插入队列 *que*（Insert queue *que*）

24.   end if

25.   end if

26. end while

27. return bestp, beat

2. 优先队列式分支限界算法（Priority Queue Branch and Bound Algorithm）

优先级定义为活节点代表的部分解所描述的装入背包的物品价值上界,该价值上界越大,优先级越高。活节点的价值上界 $up=cp+r'p$,cp 代表活节点的价值,$r'p$ 代表剩余物品装满背包剩余容量的最优质的价值。

Priority is defined as the upper bound of the value of the item to be loaded into the backpack described by the partial solution represented by the living node. The greater the upper bound of the value, the higher the priority. Value of living nodes upper bound $up=cp+r'p$, which cp represent the value of active nodes, and $r'p$ represent remaining items of active nodes fill the remaining capacity of the backpack with the best value.

约束条件与队列式分支限界算法相同,限界条件为 $up=cp+r'p>bestp$。算法伪代码如下。

The constraint condition is the same as that of queue branch and bound algorithm, and the limit condition is $up = cp + r'p > bestp$. The algorithm pseudocode is as follows.

算法 6.2　优先队列式分支限界算法解决 0/1 背包问题（Algorithm 6.2　Priority queue branch bound algorithm solves the 0/1 knapsack problem）

输入：背包的容量（Input：Capacity of backpack）
输出：问题的最优价值 bestp 及最优解叶子节点编号 best（Output：the optimal value of the problem *bestp* and the optimal solution leaf node number *best*）

```
1.   heap←[]                    //初始化空堆(Initialize the empty heap)
2.   node←Node(0, 0, 1)   //Node 为节点类,node 为根节点(Node is the node class and node is the root node)
3.   node.up←bound(node)//bound 计算节点的价值上界(bound computes the upper bound of a node's value)
4.   insert(heap, node)        //将根节点插入堆(Insert the root node into the heap)
5.   while(heap 非空) do
6.     current_node←delete(que)        //取堆顶元素(Take the top element)
7.     depth←current_node              //节点的深度(Depth of node)
8.     if depth ==n then   //叶子表示找到了一个比当前解更好的一个解,并记录(The leaf indicates that it has
                           //found a better solution than the current one and records it)
9.        bestp←current_node.cp
10.       best←current_node.id
11.    else                 //如果该扩展节点不是叶子节点(If the extension node is not a leaf node)
12.      if current_node.cw + goods[depth][1]≤capacity then        //判断约束条件(Judgment constraint)
13.       if current_node.cp + goods[depth][2]>bestp then
14.         bestp ← current_node.cp + goods[depth][2]
15.         best ← current_node.id * 2        //记录当前最优的节点编号(Record the optimal node number)
16.       end if
17.       alive_node←Node(current_node.cp + godds[depth][2]
18.       current_node.cw+godds[depth][1], current_node.id * 2)
19.       insert(heap, alive_node)        //将活节点插入堆(Insert the living node into the heap)
20.      end if
21.      up←bound(current_node)          //计算当前节点的价值上界(Calculate the upper bound of the
                                        //value of the current node)
22.      if up>bestp then
23.        alive_node←Node(current_node.cp, current_node.cw, current_node.id *2+1)
24.        insert(heap, alive_node)
25.      end if
26.    end if
27. end while
28. return bestp, beat        //返回最优价值 beatp 和最优解的子节点的编号 best(Return the best value beatp and the
                            //number of the child of the best solution best)
```

6.2.3　算法实现（Algorithm Implementation）

分别用队列式分支限界算法和优先队列式分支限界算法求解 0/1 背包问题,其中,物资件数为 $n=4$,物资权重为 $w = [3,5,2,1]$,物资价值为 $v=[9,10,7,4]$,背包总容量为 $W=7$。

Queue branch and bound algorithm and priority queue branch and bound algorithm are used to solve the 0/1 knapsack problem, respectively, where the number of items is $n=5$, the weight of materials is $w = [3, 5, 2, 1]$, the value of items is $v=[9, 10, 7, 4]$, and the total capacity of knapsack is $W=7$.

1. 队列式分支限界算法（Queued Branch and Bound Algorithm）

1）实例构造（Instance Construction）

初始时，将根节点 A 插入活节点表中，节点 A 是唯一的活节点。从活节点表中取出 A，节点 A 是当前的扩展节点，如图 6.1 所示。

Initially, the root node A is inserted into the living node table, and node A is the only living node. Take A from the living node table, node A is the current extension node. This is shown in Figure 6.1.

一次性生成 A 的两个孩子节点 B 和 C，节点 B 满足约束条件，因为满足 $cp>bestp$，故将 $bestp$ 改写为节点 B 的 cp，即 $bestp=9$；对于节点 C，由于 $cp=0$，$rp=21$，$bestp=9$，满足限界条件，依次将 B 和 C 插入活节点表，如图 6.2 所示。（图中深节点表示死节点，已不在活节点中。节点旁括号内的数据表示背包的剩余容量和已装入背包的物品价值。）

Two children nodes B and C of A are generated at one time, and node B satisfies the constraint condition. Since it satisfies $cp>bestp$, $bestp$ is rewritten as the cp of node B, that is $bestp=9$. For node C, B and C are successively inserted into the living node table since $cp=0$, $rp=21$, $bestp=9$ and the bound condition is satisfied, as shown in Figure 6.2. (In the figure, a deep node represents a dead node, which is no longer a living node. The data in parentheses next to the node represents the remaining capacity of the knapsack and the value of the items already in the knapsack.)

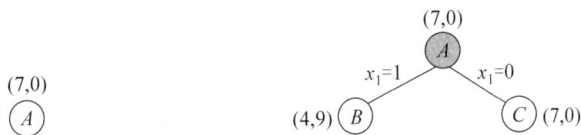

图 6.1　搜索过程（1）

Figure 6.1　Searching process（1）

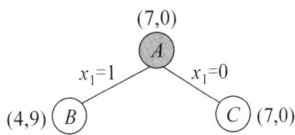

图 6.2　搜索过程（2）

Figure 6.2　Searching process（2）

再从活节点表中取出一个活节点 B 作为当前的扩展节点，一次性生成 B 的两个孩子节点，左孩子节点不满足约束条件（超出背包总重量），舍弃；对于右孩子节点 D，由于 $cp=9$，$rp=11$，$bestp=9$，满足限界条件将节点 D 保存到活节点表中，如图 6.3 所示。

Then, a living node B was taken from the living node table as the current expansion node, and two child nodes of B were generated at once. The left child node did not meet the constraints (exceeding the total weight of the knapsack) and was discarded. For the right child node D, since $cp=9$, $rp=11$, $bestp=9$, the bound condition is satisfied to save node D into the living node table. This is shown in Figure 6.3.

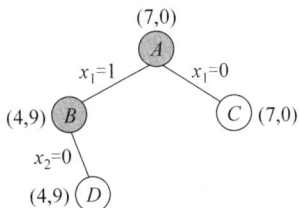

图 6.3　搜索过程（3）

Figure 6.3　Searching process（3）

从活节点表中取出 C，节点 C 是当前的扩展节点，一次性生成它的两个孩子节点 E 和 F，节点 E 满足约束条件，因为 E 节点的 $cp=10$，满足 $cp>bestp$，则 $bestp=10$。节点 F 中 $cp=0$，$rp=11$，$bestp=10$，满足限界条件。依次将 E 和 F 插入活节点表中，如图 6.4 所示。

Take C from the living node table, node C is the current expansion node, and generate its two children E and F at once, node E satisfies the constraint, since the node E has $cp=10$ and satisfies $cp>bestp$, then $bestp=10$. In node F, $cp=0$, $rp=11$, $bestp=10$, which satisfies the bound condition. Insert E and F sequentially into the living node table. This is shown in Figure 6.4.

从活节点表中取出 D，节点 D 是当前的扩展节点，一次性生成它的两个孩子节点 G 和 H。节点 G 满足约束条件，将 G 插入活节点表，因为 G 节点的 $cp=16$，满足 $cp>bestp$，则 $bestp=16$。由于 $cp=9$，$rp=4$，$bestp=16$，节点 H 不满足限界条件 $cp+rp>bestp$，舍弃，如图 6.5 所示。

Take D from the living node table, where node D is the current extension node, and generate its two children G and H at once. The node G satisfies the constraint, and G is inserted into the living node table, since $cp=16$ of the node G satisfies $cp>bestp$, then $bestp=16$. Since $cp=9$, $rp=4$ and $bestp=16$, node H does not satisfy the bound condition $cp+rp>bestp$ and is discarded, as shown in Figure 6.5.

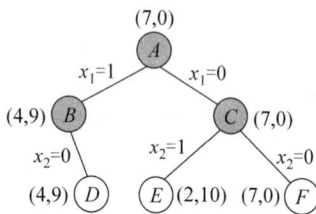

图 6.4　搜索过程（4）

Figure 6.4　Searching process （4）

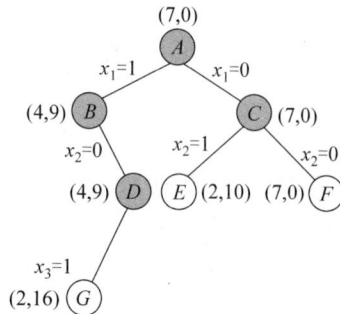

图 6.5　搜索过程（5）

Figure 6.5　Searching process （5）

从活节点表中取出 E，节点 E 是当前的扩展节点，一次性生成它的两个孩子节点 I 和 J，节点 I 满足约束条件。将其插入活节点表，因为 I 节点的 $cp=17$，满足 $cp>bestp$，则 $bestp=17$。对于节点 J，由于 $cp=10$，$rp=4$，$bestp=17$，不满足限界条件 $cp+rp>bestp$，舍弃，如图 6.6 所示。

Take E from the living node table, node E is the current extension node, generate its two children I and J at once, and node I satisfies the constraints. Insert it into the living node table, since $cp=17$ at node I satisfies $cp>bestp$, then $bestp=17$. For node J, since $cp=10$, $rp=4$ and $bestp=17$, the bound condition $cp+rp>bestp$ is not satisfied and is discarded, as shown in Figure 6.6.

从活节点表中取出 F，节点 F 是当前的扩展节点，一次性生成它的两个孩子节点 K 和 L。由于节点 K 的 $cp=7$，$rp=4$，$bestp=17$，不满足限界条件 $cp+rp>bestp$，舍弃。由于节点

L 的 $cp=0$，$rp=4$，$bestp=17$，不满足限界条件 $cp+rp>bestp$，舍弃，如图 6.7 所示。

Take F from the living node table, node F is the current extension node, and generate its two children K and L at once. Since $cp=7$, $rp=4$ and $bestp=17$ for node K, the bound condition $cp+rp>bestp$ is not satisfied and is discarded. Since $cp=0$, $rp=4$ and $bestp=17$ for node L, the bound condition $cp+rp>bestp$ is not satisfied and is discarded, as shown in Figure 6.7.

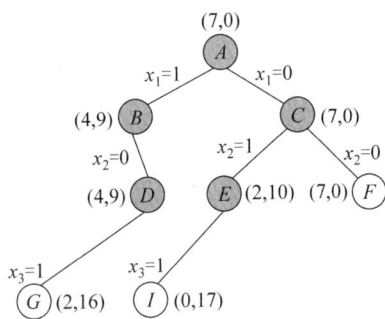

图 6.6　搜索过程（6）

Figure 6.6　Searching process（6）

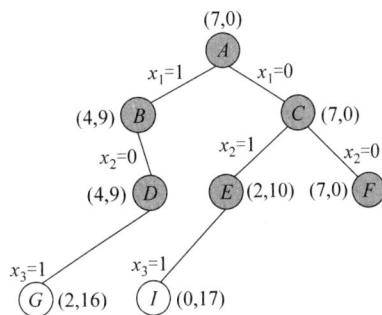

图 6.7　搜索过程（7）

Figure 6.7　Searching process（7）

从活节点表中取出 G，节点 G 是当前的扩展节点，一次性生成它的两个孩子节点 M 和 N，左孩子节点 M 满足约束条件且已经是叶子节点，此时找到了当前最优解，修改 $bestp=20$，右孩子节点 N 不满足限界条件 $cp+rp>bestp$，舍弃，如图 6.8 所示。

Take G from the living node table, node G is the current expansion node, and its two children M and N are generated at once. The left child M satisfies the constraints and is already a leaf node, and the current optimal solution is found at this time, so $bestp=20$ is modified, and the right child N does not satisfy the bound condition $cp+rp>bestp$, and is discarded, as shown in Figure 6.8.

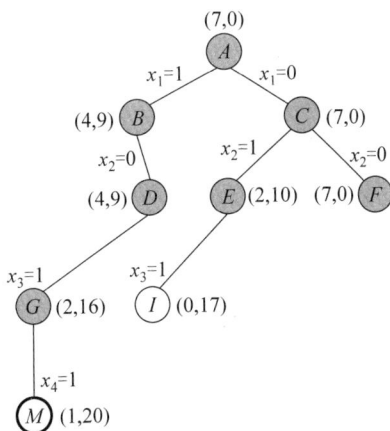

图 6.8　搜索过程（8）

Figure 6.8　Searching process（8）

从活节点表中取出活节点 I，它扩展生成的孩子节点不满足约束条件或限界条件，舍弃，如图 6.9 所示。

Take out the living node I from the living node table. The child node generated by its expansion does not meet the constraints or boundary conditions, and discard it, as shown in Figure 6.9.

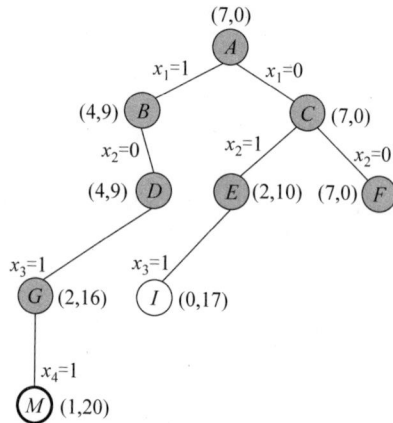

図 6.9 搜索过程（9）

Figure 6.9 Searching process (9)

此时活节点表为空，算法结束，找到了问题的最优解，即从根节点 A 到叶子节点 M 的路径（粗线条表示的路径）(1，0，1，1)，最优价值为 20，如图 6.10 所示。

At this point, the living node table is empty, and the algorithm finishes, finding the optimal solution to the problem, which is the path (1, 0, 1, 1) from the root node A to the leaf node M (the path represented by the thick line), and the optimal value is 20, as shown in Figure 6.10.

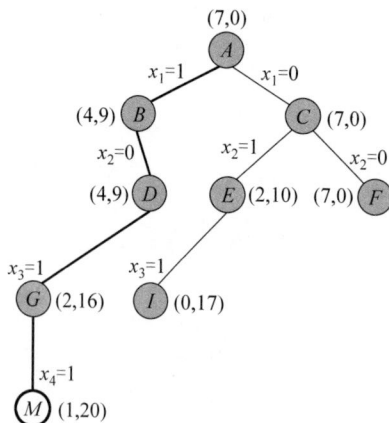

図 6.10 搜索过程（10）

Figure 6.10 Searching process (10)

2）代码实现（Code Implementation）

Python 实现需要导入依赖包 *math* 和 *queue*。

The *Python* implementation needs to import the dependency packages *math* and *queue*.

首先定义一个 *Node* 类，用于描述树节点，该类包含节点的价值 *cp*、质量 *cw* 和编号 *id* 三个字段。其中，编号 *id* 是指该节点在完全二叉解空间树中的编号，其根节点编号为 1，左孩子节点为 2，右孩子节点为 3。以此类推，编号为 *i* 的节点，其左孩子节点编号为 2*i*，右孩子节点编号为 2*i*+1。

First, a *Node* class is defined to describe tree nodes, which contains the value *cp*, quality *cw*, and number *id* fields of the node. Where *id* refers to the number of the node in the complete binary solution space tree, the root node number is 1, the left child node is 2, and the right child node is 3. Similarly, for a node numbered *i*, the left child node is numbered 2*i* and the right child node is numbered 2*i*+1.

然后定义如下函数。

Then define the following function。

（1）*bound*（）函数。用于求当前节点的价值上限，即 *cp*+*r′p*，*r′p* 为剩余物品装入剩余背包容量能装入的最优价值。

Function *bound*（） is used to find the upper limit of the value of the current node，that is，*cp*+*r′p*，*r′p* is the optimal value that the remaining items can be loaded into the remaining backpack capacity.

（2）*queque_branch*（）函数，用于搜索问题的最优价值，并记录最优解叶子节点的编号，该函数接收背包的容量 *capacity*，输出最优价值 *bestp* 和最优解叶子节点编号 *best*。

Function *queque_branch*（） is used to search the optimal value of the problem and record the number of the leaf node of the optimal solution. The function receives the *capacity* of the backpack and outputs the best value *bestp* and the number of the leaf node of the optimal solution *best*.

（3）*get_bestx*（）函数，用于构造最优解，从最优解叶子节点编号 *best* 出发，如果节点编号能被 2 整除，说明是左孩子节点，对应物品装入背包，否则是右孩子节点（不装入背包），从叶子节点回到根节点，就能够构造出问题的最优解。

Function *get_bestx*（） is used to construct the optimal solution, starting from the optimal solution of the leaf node number *best*, if the node number can be divisible by 2, it is a left child node, and the corresponding items are loaded into the backpack, otherwise it is a right child node（not loaded into the backpack），from the leaf node back to the root node, it can be constructed the optimal solution of the problem.

最后是定义入口 *main*（）函数。其中，初始化 0/1 背包问题的一个实例，将物品按照单位重量的价值降序排列，然后调用得到最优解和最优价值，最后打印输出。程序代码如下。代码运行结果如图 6.11 所示。

The last is to define the entry function *main*（）. It initializes an example of the 0/1 knapsack problem, sort the items in descending order according to the value per unit weight, and then calls to get the optimal solution and the optimal value, and finally prints the output. The program code is as follows. The result of the code running is shown in Figure 6.11.

```
#队列式分支限界算法(Queued branch bound algorithm)
import math
import queue
class Node：
    def __init__(self, cp, cw, myid)：
        self.cp = cp
        self.cw = cw
        self.id = myid
def bound(node)：
    global c, goods
    cleft = c -node.cw
    b = node.cp
    i=int(math.log2(node.id)) + 1   #round(math.log2(node.id))求出节点深度,节点深度+1 为所处的层次数,剩余
                                    #物品为层次数….n+1(Figure out the node depth, node depth + 1 is the
                                    #number of layers, the remaining items are the number of layers…. n+1)
    while(i<n and goods[i][1]<=cleft)：
        cleft -=goods[i][1]
        b +=goods[i][2]
        i +=1
    if(i<n)：
        b +=goods[i][2]/goods[i][1] * cleft;
    return b
def queue_branch(capacity)：
    global goods, n
    bestp = 0
    best = 0
    que = queue.Queue()
    node = Node(0, 0, 1)   #根节点,当前重量 cw 为 0,当前价值 cp 为 0,根节点编号 id 为 1(The current weight
                          #cw is 0, the current value cp is 0, and the number id of the root node is 1)
    que.put(node)
    while(not que.empty())：
        current_node = que.get()
        depth = int(math.log2(current_node.id))
        if depth ==n：      #叶子表示找到了比当前解更好的一个解,记录之(The leaf indicates that a
                          #better solution has been found than the current one, which is recorded)
            bestp = current_node.cp
            best = current_node.id
        else：
            if current_node.cw + goods[depth][1] <=capacity：
                if(current_node.cp + goods[depth][2] > bestp)：
                    bestp = current_node.cp + goods[depth][2]
                    best = current_node.id * 2   #记录当前最优的节点编号(Record the optimal node number)
                alive_node = Node(current_node.cp + goods[depth][2], current_node.cw + goods[depth][1],
                current_node.id *2)
                    que.put(alive_node)
            up = bound(current_node)
            if up > bestp：
```

```
                alive_node = Node(current_node.cp, current_node.cw, current_node.id * 2 + 1)
                que.put(alive_node)
        return bestp, best
def get_bestx(best):
    global n, goods
    bestx = [0 for i in range(n)]
    i = best
    depth = int(math.log2(best))
    while i>1:
        depth = int(math.log2(i))
        s, y = divmod(i, 2) #s为商,是父节点的编号,y为余数,余数是0,则为左孩子,记录1,反之,记录0(s is the
                            #quotient, is the number of the parent node, y is the remainder, if the remainder is 0,
                            #then is the left child, record 1, otherwise, record 0)
        if y == 0:
            bestx[goods[depth-1][0]] = 1
        else:
            bestx[goods[depth-1][0]] = 0
        i = s
    return bestx


if __name__ == '__main__':
    n = 5 #问题规模(Problem scale)
    c = 10#背包容量(Backpack capacity)
    goods = [[0, 2, 6], [1, 2, 3], [2, 6, 5], [3, 5, 4], [4, 4, 6]]
    goods.sort(key = lambda x:x[2]/x[1], reverse = True)
    bestp, best = queue_branch(c)
    bestx = get_bestx(best)
    print("最大价值为(The maximum value is): ", bestp)
    print("最优解为(The optimal solution is): ", bestx)
```

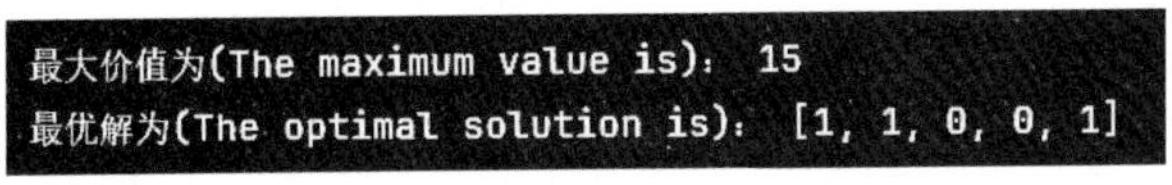

```
最大价值为(The maximum value is): 15
最优解为(The optimal solution is): [1, 1, 0, 0, 1]
```

图 6.11　队列式分支限界算法解决背包问题的运行结果

Figure 6.11　Result of the queued branch bound algorithm to solve knapsack problem

2. 优先队列式分支限界算法(Priority Queue Branch and Bound Algorithm)

1) 实例构造(Instance Construction)

初始时,将根节点 A 插入活节点表,节点 A 是唯一的活节点,如图 6.12 所示。

Initially, the root node A is inserted into the living node table, and node A is the only living node. As shown in Figure 6.12.

从活节点表中取出 A,节点 A 是当前的扩展节点,一次性生成它的两个孩子节点 B 和 C,节点 B 满足约束条件,将节点 B 插入活节点表,B 节点的 $up = 9+4+7+1/5\times10 = 22$,修改 $bestp = cp = 9$;节点 C 的 $up = 0+4+7+4/5\times10 = 19$,满足限界条件,将 C 插入活节点表,如图 6.13 所示。

Take A from the living node, node A is the current expansion node, generate its two children nodes B and C at once, node B meets the constraints, insert node B into the living node, node B's $up=9+4+7+1/5\times10=22$, modify $bestp=cp=9$; The $up=0+4+7+4/5\times10=19$ of node C satisfies the bound condition, and C is inserted into the living node. This is shown in Figure 6.13.

图 6.12　搜索过程（1）

Figure 6.12　Searching process（1）

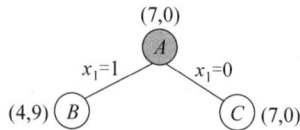

图 6.13　搜索过程（2）

Figure 6.13　Searching process（2）

从活节点表中取出一个优先级最高的活节点 B 作为当前的扩展节点，一次性生成 B 的两个孩子节点，左孩子节点不满足约束条件，舍弃；右孩子节点 D 的 $up=9+4+7=20$，满足限界条件，将节点 D 保存到活节点表中，如图 6.14 所示。

A living node B with the highest priority was taken from the living node table as the current expansion node, and two child nodes of B were generated at once. The left child node did not meet the constraints and was discarded. The right child node D has $up=9+4+7=20$, which satisfies the bound condition and saves node D into the living node table, as shown in Figure 6.14.

从活节点表中取出优先级最高的活节点 D，节点 D 是当前的扩展节点，一次性生成它的两个孩子节点 E 和 F，节点 E 满足约束条件，其 $up=16+4=20$，修改 $bestp=cp=16$，将 E 插入活节点表；节点 F 的 $up=9+4=11$，不满足限界条件 $up=cp+r'p>bestp$，舍弃，如图 6.15 所示。

Take out the living node D with the highest priority from the living node table, node D is the current expansion node, generate its two children nodes E and F at once, node E satisfies the constraints, $up=16+4=20$, modify $bestp=cp=16$, insert E into the living node table. Node F has $up=9+4=11$, which does not satisfy the bound condition $up=cp+r'p>bestp$ and is discarded, as shown in Figure 6.15.

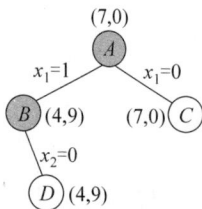

图 6.14　搜索过程（3）

Figure 6.14　Searching process（3）

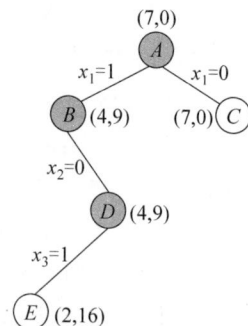

图 6.15　搜索过程（4）

Figure 6.15　Searching process（4）

从活节点表中取出优先级最高的活节点 E，节点 E 是当前的扩展节点，一次性生成它的两个孩子节点，左孩子节点 G 满足约束条件且是叶子节点，G 的 $up=20$，修改 $bestp=20$，将 G 插入活节点表；右孩子节点的 $up=16$，不满足限界条件 $up=cp+r'p>bestp$，舍弃，如图 6.16 所示。

Take out the living node E with the highest priority from the living node table, node E is the current expansion node, generate its two child nodes at once, the left child node G satisfies the constraints and is a leaf node, G's $up=20$, modify $bestp=20$, and insert G into the living node table. The right child node has $up=16$, which does not satisfy the bound condition $up=cp+r'p>bestp$ and is discarded, as shown in Figure 6.16.

从活节点表取出优先级最高的活节点 G，由于 G 已经是叶子节点，搜索结束，找到了问题的最优解，即从根节点到叶节点的路径 $(1, 0, 1, 1)$，$bestp=20$，如图 6.17 所示。

Take out the living node G with the highest priority from the living node table. Since G is already a leaf node, the search ends and the optimal solution of the problem is found, that is, the path $(1, 0, 1, 1)$ from the root node to the leaf node, $bestp=20$, as shown in Figure 6.17.

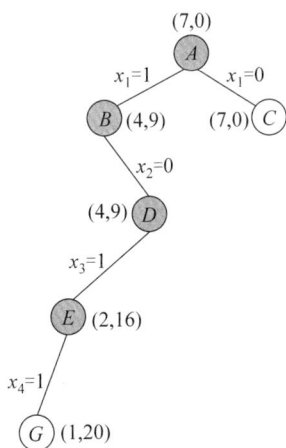

图 6.16 搜索过程（5）
Figure 6.16 Searching process （5）

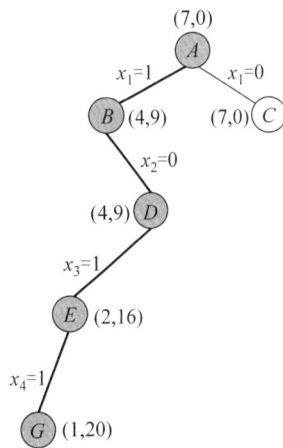

图 6.17 搜索过程（6）
Figure 6.17 Searching process （6）

2）代码实现（Code Implementation）

优先队列算法的代码实现思路与队列式分支限界算法大致类似，主要区别是关于类 $Node$ 的定义。设计了 4 个字段，分别是节点的价值上界 up、价值 cp、质量 cw 和编号 id。类 $Node$ 重载了大于方法 $\_\_gt\_\_()$ 和等于方法 $\_\_eq\_\_()$，定义类中参与字段的字段。程序代码如下。代码运行结果如图 6.18 所示。

The code implementation of the priority queue algorithm is like that of the queued branch and bound algorithm, with the main difference being the definition of the $Node$ class. Four fields are designed, which are value upper bound up, value cp, quality cw and number id. Class $Node$ overrides methods greater than $\_\_gt\_\_()$ and equal to $\_\_eq\_\_()$, defining fields that participate in fields in the class. The program code is as follows. The result of the code running is shown in Figure 6.18.

```python
#优先队列式(Priority queue)
import math
import heapq
class Node：
    def __init__(self, cp, cw, myid, up)：
        self.up = up                        #价值上界,根据价值上界构造极大堆(The maximum heap is constructed
                                            #according to the upper bound of value)
        self.cp = cp
        self.cw = cw
        self.id = myid
    def __gt__(self, other)：
        return self.up > other.up
    def __eq__(self, other)：
        if(other == None)：
            return False
        if(not isinstance(other, HeapNode))：
            return False
        return self.up == other.up

def bound(node)：
    global c, goods
    cleft = c - node.cw
    b = node.cp
    i = int(math.log2(node.id)) + 1    #round(math.log2(node.id))求出节点深度,节点深度+1为所处的层次数,
                                        #剩余物品为层次数…·.n+1(Figure out the node depth, node depth + 1 is the
                                        #number of layers, the remaining items are the number of layers…·. n+1)
    while(i<n and goods[i][1]<=cleft)：
        cleft -=goods[i][1]
        b +=goods[i][2]
        i +=1
    if(i<n)：
        b +=goods[i][2]/goods[i][1] * cleft;
    return b
def first_queue_branch(capacity)：
    global goods, n
    bestp=0
    best=0
    heap = []
    node = Node(0, 0, 1, 0)    #根节点,当前重量 cw 为 0, 当前价值 cp 为 0,根节点编号 id 为 1(The current weight
                               #cw is 0, the current value cp is 0, and the number id of the root node is 1)
    node.up = bound(node)
    heapq.heappush(heap, node)
    while(len(heap)>0)：
        current_node = heapq.heappop(heap)
        depth = int(math.log2(current_node.id))
        if depth ==n：           #叶子表示找到了比当前解更好的一个解,记录之(The leaf indicates that a better
                                 #solution has been found than the current one, which is recorded)
            bestp = current_node.cp
            best = current_node.id
        else：
            if current_node.cw + goods[depth][1] <=capacity：
                if(current_node.cp + goods[depth][2]> bestp)：
                    bestp = current_node.cp + goods[depth][2]
                    best = current_node.id * 2    #记录当前最优的节点编号(Record the optimal node number)
```

续

```
                    alive_node = Node(current_node.cp +goods[depth][2], current_node.cw + goods[depth][1],
                    current_node.id * 2, current_node.up)
                    heapq.heappush(heap, alive_node)
                up = bound(current_node)
                if up > bestp:
                    alive_node = Node(current_node.cp, current_node.cw, current_node.id * 2 + 1, up)
                    heapq.heappush(heap, alive_node)
    return bestp, best
def get_bestx(best):
    global n, goods
    bestx = [0 for i in range(n)]
    i = best
    depth = int(math.log2(best))
    while i>1:
        depth = int(math.log2(i))
        s, y = divmod(i, 2) #s 为商,是父节点的编号,y 为余数,余数是 0,则为左孩子,记录 1,反之,记录 0(s is the
                            #quotient, is the number of the parent node, y is the remainder, if the remainder is 0,
                            #then is the left child, record 1, otherwise, record 0)
        if y ==0:
            bestx[goods[depth-1][0]] =1
        else:
            bestx[goods[depth-1][0]] =0
        i = s
    return bestx


if __name__ =='__main__':
    n =5                            #问题规模(Problem scale)
    c = 20                          #背包容量(Backpack capacity)
    goods = [[0, 5, 7], [1, 4, 9], [2, 8, 15], [3, 12, 21], [4, 13, 24]]
                                    #物品编号,物品重量,物品价值(Item number, item weight, item value)
    goods.sort(key = lambda x:x[2]/x[1], reverse = True)
    bestp, best = first_queue_branch(c)
    bestx = get_bestx(best)
    print("最大价值为(The maximum value is): ", bestp)
    print("最优解为(The optimal solution is): ", bestx)
```

```
最大价值为(The maximum value is): 36
最优解为(The optimal solution is): [0, 0, 1, 1, 0]
```

图 6.18　优先队列分支限界算法实现背包问题的程序运行结果

Figure 6.18　Result of the priority queue branch bound algorithm to implement knapsack problem

6.3　旅行商问题(Traveling Salesman Problem)

6.3.1　基本思想(Basic Idea)

1. 问题描述(Problem Description)

某售货员要到 n 个城市去推销商品,已知各城市之间的路程(或旅费)。他要选定一条从驻地出发,经过每个城市一次,最后回到驻地的路线,使总的路程(或总旅费)最小。以

$n=4$ 为例分析。

A salesperson wants to go to n cities to sell goods, and the distance between the cities is known. He chose a route from the station, through each city once, and back to the station, to minimize the total distance (or total cost). Take $n=4$ as an example.

2. 求解分析（Solution Analysis）

（1）问题的解空间为 (x_1, x_2, x_3, x_4)，其中，令 $S=\{1,2,3,4\}$，$x_1=1$，$x_2 \in S-\{x_1\}$，$x_3 \in S-\{x_1, x_2\}$，$x_4 \in S-\{x_1, x_2, x_3\}$。

The solution space of the problem is (x_1, x_2, x_3, x_4), which makes the $S=\{1, 2, 3, 4\}$, the $x_1=1$, $x_2 \in S-\{x_1\}$, $x_3 \in S-\{x_1, x_2\}$, $x_4 \in S-\{x_1, x_2, x_3\}$.

（2）解空间的组织结构是一棵深度为 4 的排列树。

The organization structure of the solution space is a permutation tree with a depth of 4.

（3）搜索的约束条件 $g[i][j]! = \infty$，其中，$1 \leq i \leq 4$，$1 \leq j \leq 4$，g 是该图的邻接矩阵；限界条件设置为 $cl<bestl$，其中，cl 表示当前已经走的路径长度，初始值为 0，$bestl$ 表示当前最短路径长度，初始值为 ∞。

The search constraint $g[i][j]! = \infty$, where $1 \leq i \leq 4, 1 \leq j \leq 4$, and g is the adjacency matrix of the graph; The bound condition is set to $cl<bestl$, where cl represents the length of the current path that has been taken and the initial value is 0, and $bestl$ represents the length of the current shortest path and the initial value is ∞.

6.3.2　算法设计与描述（Algorithm Design and Description）

1. 队列式分支限界算法（Queued Branch Bound Algorithm）

用先进先出的队列存储活节点，当活节点表不空，循环地从表中取出一个活节点并一次性扩展所有子节点，判断约束条件和限界条件，决定保留和舍弃，直到活节点表为空或找到所需解为止。算法伪代码如下。

When the living node table is not empty, a living node is taken from the table and all its children are expanded at once cyclically. The constraints and bound conditions are determined, and the decision is made to keep or discard until the living node table is empty or the required solution is found. The algorithm pseudocode is as follows.

算法 6.3　**Traveling(a, start, g_n)**

输入：无向连通图 G 的临界矩阵 a，旅行商处出发地 start，城市数量 g_n（Input：critical matrix a of the undirected connected graph G，start from the traveler，number of cities g_n）

输出：最短旅行路径和最短路径长度（Output：Shortest travel path and shortest path length）

1.　heap←空堆（empty heap）
2.　node←第二层的节点（Layer 2 node）
3.　将 node 节点插入堆 heap 中（Insert the node into the heap）
4.　while（heap 非空）do　　　　//堆不空就循环（Cycle if the heap is not empty）
5.　　current_node←取出队首节点（Remove the first node of the queue）

续

6.   level←current_node 在解空间中所处的层次（The level of the solution space）

7.   cl←current_node 的当前路径长度（Current path length）

8.   if depth = =g_n then //叶子节点的父节点，表示找到了一个比当前更好的解并记录（The parent of a leaf
       //node indicates that a better solution is found and recorded）

9.   if（第 $n-1$ 个城市到第 n 个城市能走通、第 n 个城市到出发地城市能走通并且走的总路径长度小于 $best1$ 或 $best1 = =NoEdge$）then（The total path length from the $n-1$ city to the n city can go through, and the n city to the departure city can go through, and the total path length is less than $best1$ or $best1 = =NoEdge$）

10.   bestx←当前节点表示的解（The solution represented by the current node）

11.   beat1←回到出发地的总路径长度（The total path length back to the origin）

12.   end if

13. else

14.   for j←level to g_n then //扩展当前节点的所有分支（Extend all branches of the current node）

15.   if 满足约束条件和限界条件（Satisfy the constraint condition and limit condition）

16.   记录分支上的数据（Record the data on the branch）

17.   插入子节点到堆中（Insert child nodes into the heap）

18.   将子节点上的数据退回（Return the data on the child node）

19.   end if

20.   end for

21. end if

22. end while

23. return bestx best1

2. 优先队列式分支限界算法（Priority Queue Branch and Bound Algorithm）

用堆结构存储活节点，算法的优先级定义为当前节点的路径长度 cl，长度越短优先级越高。当活节点表不空，循环地从堆中取出活节点并一次性扩展所有子节点，判断并取舍，直到活节点表空或找到所需解为止。算法伪代码如下。

A heap structure is used to store living nodes, and the priority of the algorithm is defined as the path length cl of the current node, the shorter the length, the higher priority. When the living node table is not empty, a living node is taken from the heap and all its children are expanded at once cyclically, and the decision is made to keep or discard until the living node table is empty or the required solution is found. The pseudocode for the algorithm is as follows.

算法 6.4 Traveling（a，start，g_n）

输入：无向连通图 G 的临界矩阵 a，旅行商处出发地 $start$，城市数量 g_n（Input：critical matrix a of the undirected connected graph G, $start$ from the traveler, number of cities g_n）

输出：最短旅行路径和最短路径长度（Output：Shortest travel path and shortest path length）

1. que←空队列（Empty heap）

2. node←第二层的节点（Layer 2 node）

3. 将 node 节点插入队列 que 中（Insert the node into the queue que）

4. while（que 非空）do // que 队列不空就循环（The que queue loops if it is not empty）

5. current_node←取出队首节点（Remove the first node of the queue）

6. level←current_node 在解空间中所处的层次（The level of the solution space）

续

7.　cl←current_node 的当前路径长度（Current path length）

8.　if depth ==g_n then　//叶子节点的父节点，表示找到了一个比当前更好的解并记录（The parent of a leaf
　　　　//node indicates that a better solution is found and recorded）

9.　if（第 $n-1$ 个城市到第 n 个城市能走通、第 n 个城市到出发地城市能走通并且走的总路径长度小于 $best1$ 或 $best1 == NoEdge$）then（The total path length from the $n-1$ city to the n city can go through, and the n city to the departure city can go through, and the total path length is less than $best1$ or $best1 == NoEdge$）

10.　　bestx←当前节点表示的解（The solution represented by the current node）

11.　　beat1←回到出发地的总路径长度（The total path length back to the origin）

12.　end if

13.　else

14.　for j←level to g_n then　//扩展当前节点的所有分支（Extend all branches of the current node）

15.　　if 满足约束条件和限界条件（Satisfy the constraint condition and limit condition）

16.　　　记录分支上的数据（Record the data on the branch）

17.　　　插入子节点到队列 que（Insert a child node to queue que）

18.　　　将分支上的数据退回（Return the data on the branch）

19.　　end if

20.　end for

21.　end if

22. end while

23. return bestx best1

6.3.3　算法实现（Algorithm Implementation）

某售货员要到 4 个城市去推销商品，已知各城市之间的路程（或旅费），如图 6.19 所示。他要选定一条从驻地出发，经过每个城市一次，最后回到驻地的路线，使总的路程（或总旅费）最小。

A salesperson wants to go to four cities to sell goods, and the distance（or travel cost）between the cities is known, as shown in Figure 6.19. He wants to choose a route from the station, through each city once, and back to the station, so that the total distance（or total travel cost）is minimized.

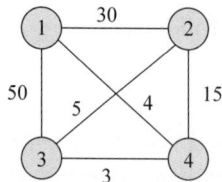

图 6.19　分支限界算法的旅行商问题

Figure 6.19　Traveling salesperson problem of branch and bound algorithm

用队列式分支限界算法和优先队列式分支限界算法分别解决该旅行商问题。

Queue branch and bound algorithm and priority queue branch and bound algorithm are used to solve the traveling salesman problem respectively.

1. 队列式分支限界算法(Queued Branch and Bound Algorithm)

1) 实例构造(Instance Construction)

初始时,由于 x 的取值是确定的,所以从根节点 A_0 的孩子节点 A 开始搜索即可。将节点 A 插入活节点表,节点 A 是活节点并且是当前的扩展节点,如图 6.20 所示。

Initially, since the value of x is certain, it is sufficient to start the search from the child node A of the root node A_0. Insert node A into the living node table. Node A is the living and current extension node, as shown in Figure 6.20.

从活节点表中取出活节点 A 作为当前的扩展节点,一次性生成它的三个孩子节点 B、C、D,均满足约束条件和限界条件,依次插入活节点表,节点 A 变成死节点,如图 6.21 所示。

The living node A is taken out from the living node table as the current expansion node, and its three child nodes B, C and D are generated once, all of which meet the constraints and bound conditions, and inserted into the living node table in turn, and node A becomes a dead node, as shown in Figure 6.21.

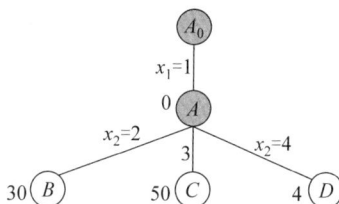

图 6.20 搜索过程(1)

Figure 6.20 Searching process (1)

图 6.21 搜索过程(2)

Figure 6.21 Searching process (2)

从活节点表中取出活节点 B 作为当前的扩展节点,一次性生成它的两个孩子节点 E、F,均满足约束条件和限界条件,依次插入活节点表,节点 B 变成死节点,如图 6.22 所示。

The living node B is taken out from the living node table as the current expansion node, and its two child nodes E and F are generated at one time, both of which meet the constraints and bound conditions, and inserted into the living node table in turn, node B becomes a dead node, as shown in Figure 6.22.

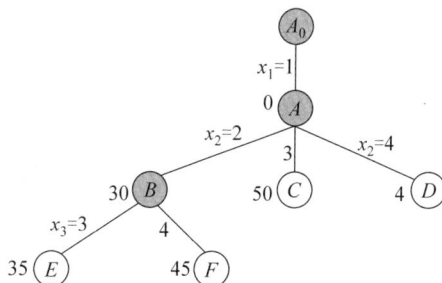

图 6.22 搜索过程(3)

Figure 6.22 Searching process (3)

从活节点表中取出活节点 C 作为当前的扩展节点,一次性生成它的两个孩子节点 G、H,均满足约束条件和限界条件,依次插入活节点表,节点 C 变成死节点,如图 6.23 所示。

The living node C is taken out from the living node table as the current expansion node, and its two child nodes G and H are generated at one time, both of which meet the constraints and bound conditions, and inserted into the living node table in turn, node C becomes a dead node, as shown in Figure 6.23.

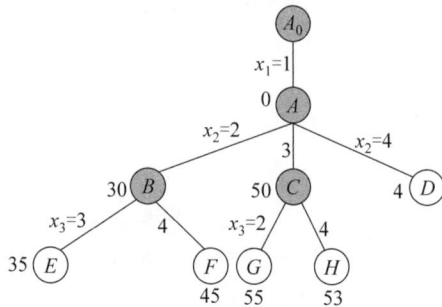

图 6.23　搜索过程（4）

Figure 6.23　Searching process (4)

从活节点表中取出活节点 D 作为当前的扩展节点，一次性生成它的两个孩子节点 I、J，均满足约束条件和限界条件，依次插入活节点表，节点 D 变成死节点，如图 6.24 所示。

The living node D is taken out from the living node table as the current expansion node, and its two child nodes I and J are generated at one time, both of which meet the constraints and bound conditions, and inserted into the living node table in turn, node D becomes a dead node, as shown in Figure 6.24.

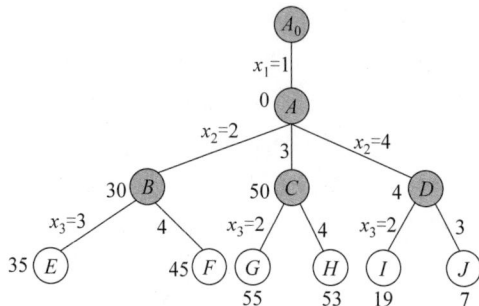

图 6.24　搜索过程（5）

Figure 6.24　Searching process (5)

从活节点表中取出活节点 E 作为当前的扩展节点，一次性生成它的一个孩子节点 K，满足约束条件和限界条件，节点 K 已经是叶子节点，且顶点 4 与城市 1 有边相连，说明已找到一个当前最优解，记录该节点，最短路径长度为 42，修改 $bestl=42$，如图 6.25 所示。

Take the living node E from the living node table as the current expansion node, generate one of its child node K at once, meet the constraints and bound conditions, node K is already a leaf node, and vertex 4 is connected to the city 1 by an edge, indicating that a current optimal solution has been found, record the node, the shortest path length is 42, modify $bestl=42$. This is shown in Figure 6.25.

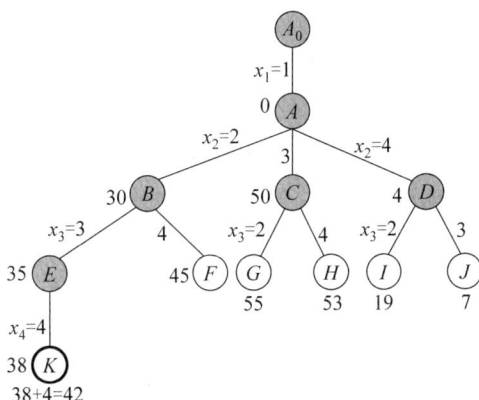

图 6.25　搜索过程(6)

Figure 6.25　Searching process (6)

　　从活节点表中依次取出活节点 F、G、H、I、J,一次性生成它们的孩子节点,均不满足限界条件 $cl<bestl$,舍弃,这些节点变成死节点。此时,活节点表为空,算法结束,找到的最优解是从根节点到叶子节点区的路径(1,2,3,4),路径长度为 42,如图 6.26 所示。

　　The living nodes F, G, H, I, J are taken out from the living node table in turn, and their children's nodes are generated at once. None of them satisfies the bound condition $cl<bestl$, so they are discarded and become dead nodes. At this point, the living node table is empty, and the algorithm ends. The optimal solution path (1, 2, 3, 4) is found from the root node to the leaf node, and the path length is 42, as shown in Figure 6.26.

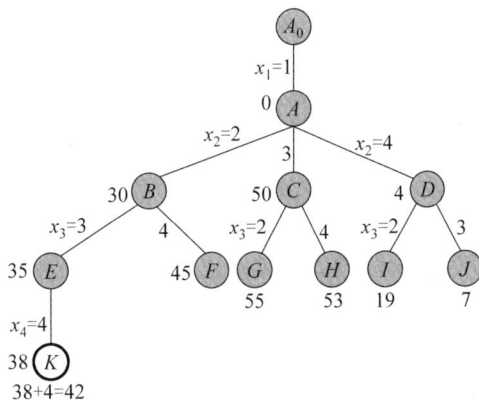

图 6.26　搜索过程(7)

Figure 6.26　Searching process (7)

2) 代码实现(Code Implementation)

Python 程序需要导入依赖包 *math*、*queue* 和 *sys*。

Python programs need to import dependency packages *math*, *queue*, and *sys*.

　　首先定义一个类 *Node*,用于描述树节点。该类有当前路径长度 *cl*、当前节点在解空间树中的层次 *level* 和当前节点代表的部分解 *x* 这三个字段。

　　First, defining a class *Node* that describes the tree nodes. This class has three fields: the current path length cl, the *level* represents the hierarchical level of current node in the solution space tree and x represents the partial solution of the current node.

然后定义一个 *traveling*( )函数，用于搜索最优旅行路径，并记录最短路径长度。该函数接收图的邻接矩阵 a、出发地 $start$ 和城市数量 g_n，输出最优旅行路线和最短路径长度。

Then define a *traveling*( ) function that searches for the optimal travel path and records the shortest path length. This function receives the adjacency matrix a of the graph, the *start* of origin and the number of cities g_n, and outputs the optimal travel route and the shortest path length.

最后定义入口 *main*( )函数，其中初始化旅行商问题的实例，为对应城市编号，程序设计的二维列表 a 中第 0 行和第 0 列为无效数据。调用函数得到问题的最优解和最优价值（其中，0 号存储单元为无效数据）。程序代码如下。程序的运行结果如图 6.27 所示。

Finally, the entry function *main*( ) is defined, where the instance of the traveling salesperson problem is initialized, and the row 0 and column 0 in the two-dimensional list a designed by the program are invalid data for the corresponding city number. Call the function to obtain the optimal solution and the optimal value of the problem (where storage cell 0 is invalid). The program code is as follows. The result of the code running is shown in Figure 6.27.

```python
#队列式分支限界算法,限界条件为 cl<bestL(Queued branch bound method, the bound condition is cl<bestL)
import math
import queue
class Node:
    def __init__(self, cl, level, x):    #cl:当前路径长度,level:当前节点层次,g_n:问题规模(cl: current path
                                          #length, level: current node level, g_n: problem scale)
        self.cl = cl                      #当前路径长度(Current path length)
        self.level = level                #节点的层次(Hierarchy of nodes)
        self.x = x                        #部分解(partial solution)

def traveling(a, start, g_n):
    que = queue.Queue()
    node = Node(0, 2, [i for i in range(g_n+1)])
    que.put(node)
    bestx = None                          #最大价值(maximum value)
    bestl = NoEdge
    while(not que.empty()):
        current_node = que.get()
        level = current_node.level
        cl = current_node.cl
        if level == g_n:                  #叶子表示找到了比当前解更好的一个解,记录之(The leaf indicates that a
                                          #better solution has been found than the current one, which is recorded)
            if (a[current_node.x[g_n-1]][current_node.x[g_n]] !=NoEdge and  a[current_node.x[g_n]][1]
            !=NoEdge and (cl + a[current_node.x[g_n-1]][current_node.x[g_n]] + a[current_node.x[g_n]][1]
            < bestl or bestl ==NoEdge)):
                bestx = current_node.x[:]
                bestl = cl + a[current_node.x[g_n-1]][current_node.x[g_n]] + a[current_node.x[g_n]][1]
```

续

```
    else：
        for j in range(level, g_n+1)：
            if (a[current_node.x[level-1]][current_node.x[j]] != NoEdge and (cl < bestl or bestl ==
            NoEdge))：
                current_node.x[level], current_node.x[j] = current_node.x[j], current_node.x[level]
                que.put(Node(cl + a[current_node.x[level-1]][current_node.x[level]], level+1, current_
                node.x[:]))
                current_node.x[level], current_node.x[j] = current_node.x[j], current_node.x[level]
    return bestx, bestl

if __name__ == '__main__'：
    import sys
    NoEdge = sys.maxsize
    a = [[NoEdge, NoEdge, NoEdge, NoEdge, NoEdge, NoEdge], [NoEdge, NoEdge, 10, NoEdge, 4, 12],
    [NoEdge, 10, NoEdge, 15, 8, 5], [NoEdge, NoEdge, 15, NoEdge, 7, 30], [NoEdge, 4, 8, 7, NoEdge, 6],
    [NoEdge, 12, 5, 30, 6, NoEdge]]
    g_n = len(a) -1
    bestx, bestl = traveling(a, 1, g_n)
    print("最短路径长度为(The shortest path length is)：", bestl)
    print("最优旅行路线为(The optimal travel route is)：", bestx)
```

```
最短路径长度为(The shortest path length is)： 43
最优旅行路线为(The optimal travel route is)： [0, 1, 4, 3, 2, 5]
```

图 6.27 队列式分支限界算法解决旅行商问题程序结果

Figure 6.27 Result of queue branch limiting algorithm to solve traveling salesman problem

2. 优先队列式分支限界算法(Priority Queue Branch Bound Algorithm)

1) 实例构造(Instance Construction)

初始时,由于 x 的取值是确定的,所以从根节点 A_0 的孩子节点 A 开始搜索即可。将节点 A 插入活节点表,节点 A 是活节点并且是当前的扩展节点,如图 6.28 所示。

Initially, since the value of x is certain, it is sufficient to start the search from the child node A of the root node A_0. Insert node A into the living node table. Node A is the living node and the current extension node, as shown in Figure 6.28.

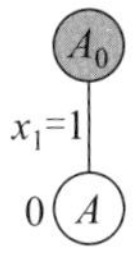

图 6.28 搜索过程(1)

Figure 6.28 Searching process (1)

从活节点表中取出活节点 A 作为当前的扩展节点,一次性生成它的三个孩子节点 B、C、D,均满足约束条件和限界条件,依次插入活节点表,节点 A 变成死节点,如图 6.29 所示。

The living node A is taken out from the living node table as the current expansion node, and its three child nodes B, C and D are generated once, all of which meet the constraints and bound conditions, and inserted into the living node table in turn, node A

becomes a dead node, as shown in Figure 6.29.

从活节点表中取出优先级最高的活节点 D 作为当前的扩展节点,一次生成它的两个孩子节点 E、F,均满足约束条件和限界条件,依次插入活节点表,节点 D 变成死节点,如图 6.30 所示。

Take out the living node D with the highest priority from the living node table as the current expansion node, generate its two child nodes E and F at one time, both meet the constraints and bound conditions, and insert them into the living node table in turn. Node D becomes a dead node, as shown in Figure 6.30.

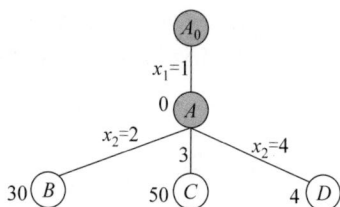

图 6.29　搜索过程(2)

Figure 6.29　Searching process (2)

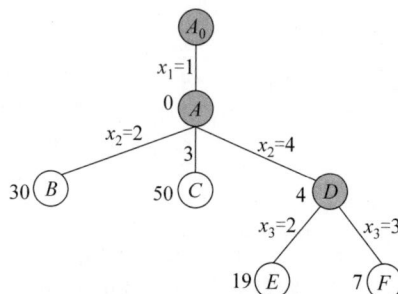

图 6.30　搜索过程(3)

Figure 6.30　Searching process (3)

从活节点表中取出优先级最高的活节点 F 作为当前的扩展节点,一次性生成它的一个孩子节点 G,满足约束条件和限界条件将 G 插入活节点表。由于节点 G 已经是叶子节点,此时找到了当前最优解,最短路径长度为 42,修改 $bestl$＝42,如图 6.31 所示。

The living node F with the highest priority is taken from the living node table as the current expansion node, and one of its child nodes G is generated at one time, and G is inserted into the living node table when the constraints and bound conditions are satisfied. Since node G is already a leaf node, the current optimal solution is found, the shortest path length is 42, so change $bestl$＝42, as shown in Figure 6.31.

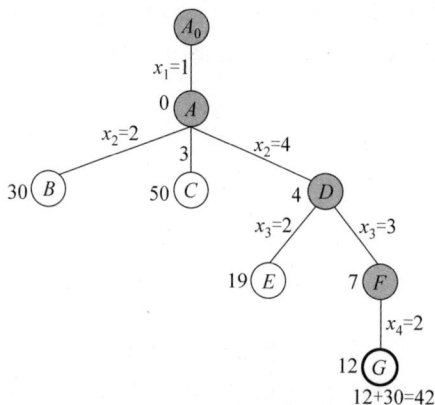

图 6.31　搜索过程(4)

Figure 6.31　Searching process (4)

从活节点表中取出优先级最高的活节点 E 作为当前的扩展节点，一次性生成它的一个孩子节点，不满足限界条件 $cl<bestl$，舍弃，节点 E 变成死节点，如图 6.32 所示。

Take out the living node E with the highest priority from the living node table as the current expansion node, generate one of its child nodes at one time, do not meet the bound condition $cl<bestl$, discard, and node E becomes a dead node, as shown in Figure 6.32.

从活节点表中取出优先级最高的活节点 B 作为当前的扩展节点，一次性生成它的两个孩子节点 H 和 I，节点 H 满足约束条件和限界条件，将其插入活节点表；节点 I 不满足限界条件 $cl<bestl$，舍弃。节点 B 变成死节点，如图 6.33 所示。

The living node B with the highest priority was taken from the living node table as the current expansion node, and its two child nodes H and I were generated at one time. Node H meets the constraints and bound conditions, and it was inserted into the living node table. Node I does not satisfy the bound condition $cl<bestl$ and is discarded. Node B becomes a dead node, as shown in Figure 6.33.

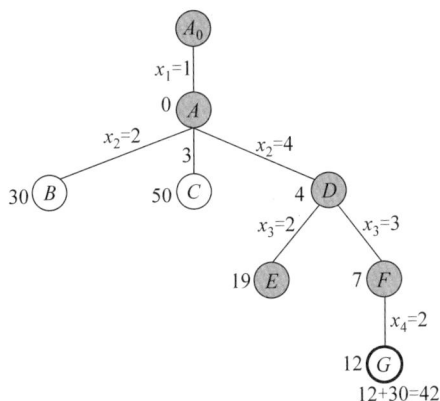

图 6.32 搜索过程(5)
Figure 6.32 Searching process (5)

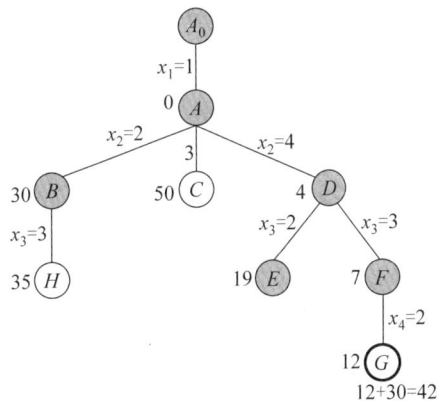

图 6.33 搜索过程(6)
Figure 6.33 Searching process (6)

从活节点表中取出优先级最高的活节点 H 作为当前的扩展节点，生成的孩子节点不满足限界条件 $cl<bestl$，舍弃，节点 H 变成死节点，再从活节点表中取出优先级最高的活节点 G 作为当前的扩展节点，G 已经是叶子节点，此时找到问题的最优解，算法结束，找到问题的最优解是从根节点 A_0 到叶子节点 G 的最短路径$(1,4,3,2)$，最短路径长度为 42，如图 6.34 所示。

The highest priority living node H is taken from the living node table as the current expansion node, and the generated child node does not meet the bound condition $cl<bestl$, so it is abandoned, and node H becomes a dead node. Then the highest priority living node G is taken from the living node table as the current expansion node, G is already a leaf node, currently, the optimal solution of the problem is found, and the algorithm ends. The shortest path $(1, 4, 3, 2)$ from the root node A_0 to the leaf node G with the shortest path length of 42, as shown in Figure 6.34.

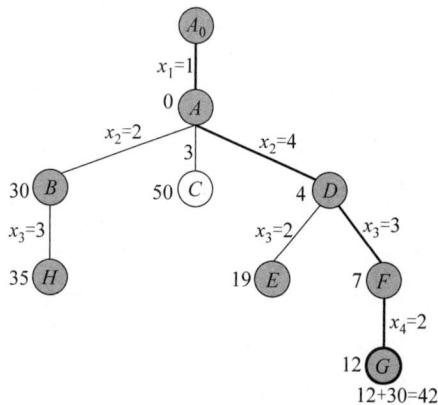

图 6.34　搜索过程（7）

Figure 6.34　Searching process (7)

2）代码实现（Code Implementation）

优先队列式分支限界算法可以不调用 *math* 依赖包。在定义类 *Node* 时，重载了 _ _lt_ _（ ）方法和 _ _eq_ _（ ）方法，定义参与比较的字段为 *cl*。其他部分与队列式分支限界算法的程序类似。算法运行结果如图 6.35 所示。程序代码如下。

The priority queue branch and bound algorithm can be used without calling the *math* package. When defining class *Node*, the _ _lt_ _( ) method and the _ _eq_ _( ) method are overridden, defining the field to be compared as *cl*. Other parts of the program are like queued branch and bound algorithms. The program code is as follows. The result of the code running is shown in Figure 6.35.

```
#优先队列式分支限界算法，限界条件为 cl<bestl
import heapq

class Node：
    def _ _init_ _(self, cl, level, x, g_n)：         # cl：当前路径长度,level：当前节点层次,g_n：问题规模
        self.cl = cl                                   #当前路径长度
        self.level = level                             #节点的层次
        self.x = x                                     #部分解
    def _ _lt_ _(self, other)：
        return self.cl < other.cl
    def _ _eq_ _(self, other)：
        if (other = =None)：
            return False
        if (not isinstance(other, HeapNode))：
            return False
        return self.cl = =other.cl
def traveling(a, start, g_n)：
    heap = [ ]
    bestx = None                                       #最大价值
    bestl = NoEdge
    node = Node(0, 2, [i for i in range(g_n + 1)])
    heapq.heappush(heap, node)
```

续

```
    while (len(heap) >0):
        current_node = heapq.heappop(heap)
        level = current_node.level
        cl = current_node.cl
        if level ==g_n:          #叶子表示找到了比当前解更好的一个解,记录之
            if (a[current_node.x[g_n -1]][current_node.x[g_n]] !=NoEdge and a[current_node.x[g_n]][1]
            !=NoEdge and (cl + a[current_node.x[g_n - 1]][current_node.x[g_n]] + a[current_node.x[g_n]]
            [1] < bestl or bestl ==NoEdge)):
                bestx = current_node.x[:]
                bestl = cl + a[current_node.x[g_n - 1]][current_node.x[g_n]] + a[current_node.x[g_n]][1]
        else:
            for j in range(level, g_n + 1):
                if (a[current_node.x[level - 1]][current_node.x[j]] !=NoEdge and (cl < bestl or bestl ==
                NoEdge)):
                    current_node.x[level], current_node.x[j] = current_node.x[j], current_node.x[level]
                    heapq.heappush(heap, Node(cl + a[current_node.x[level - 1]][current_node.x[level]],
                    level + 1, current_node.x[:]))
                    current_node.x[level], current_node.x[j] = current_node.x[j], current_node.x[level]
    return bestx, bestl
if __name__ =='__main__':
    import sys
    NoEdge = sys.maxsize
    a = [[NoEdge, NoEdge, NoEdge, NoEdge, NoEdge, NoEdge], [NoEdge, NoEdge, 10, NoEdge, 4, 12],
        [NoEdge, 10, NoEdge, 15, 8, 5], [NoEdge, NoEdge, 15, NoEdge, 7, 30], [NoEdge, 4, 8, 7, NoEdge,
        6], [NoEdge, 12, 5, 30, 6, NoEdge]]
    g_n = len(a) - 1
    bestx, bestl = traveling(a, 1, g_n)
    print("最短路径长度为(The shortest path length is): ", bestl)
    print("最优旅行路线为(The optimal travel route is): ", bestx)
```

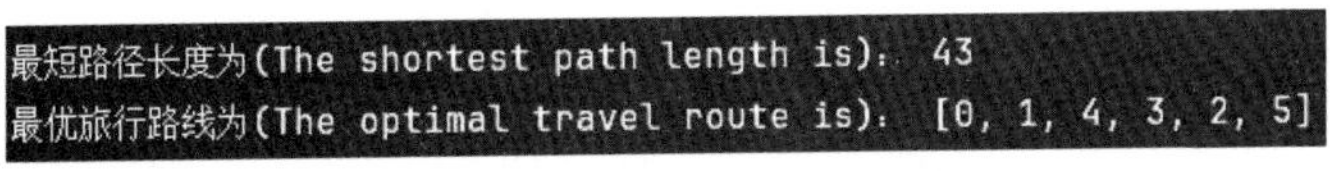

```
最短路径长度为(The shortest path length is): 43
最优旅行路线为(The optimal travel route is): [0, 1, 4, 3, 2, 5]
```

图 6.35　分支限界算法实现旅行商问题的程序运行结果

Figure 6.35　Result of the branch and bound algorithm to realize the traveling salesman problem

6.4　布线问题(Wiring Problem)

6.4.1　基本思想(Basic Idea)

1. 问题描述(Problem Description)

在 $N×M$ 的方格阵列中,指定一个方格的中点为 a,另一个方格的中点为 b,问题要求找出 a 到 b 的最短布线方案,如图 6.36 所示。布线时只能沿直线或直角,不能走斜线。黑色的单元格代表不可以通过的封锁方格。

There is an array of $N×M$ squares with midpoint a and midpoint b, the problem is to find the shortest routing from a to b, as shown in Figure 6.36. Wiring can only be along a straight line or right angle, not oblique line. The black cells represent blocked squares that cannot be passed.

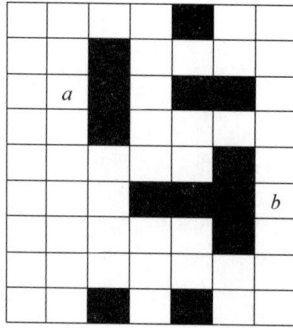

图 6.36　布线问题示意图

Figure 6.36　Schematic diagram of wiring problem

2. 问题分析（Problem Analysis）

将方格抽象为顶点，中心方格和相邻 4 个方向（上、下、左、右）能通过的方格用一条线连起来。这样，可以把问题的解空间定义为一张图。

The square is abstracted as a vertex, and the center square and the neighboring squares that can pass in four directions (up, down, left, and right) are connected by a line. In this way, the solution space of the problem can be defined as a graph.

该问题是特殊的最短路径问题，特殊之处在于用布线走过的方格数代表布线的长度，布线时每布一个方格，布线长度累加 1。由问题可知，从 a 点开始布线，只能朝上、下、左、右 4 个方向进行布线，并且遇到封锁方格、超出方格阵列边界和已布过线的方格都不能布线，把能布线的方格插入活节点表，然后从中取出一个继续扩展，搜索过程直到找到目标点或活节点表为空为止。

This problem is a special shortest path problem, which is special in that the number of squares passed by the wiring represents the length of the wiring, and the wiring length is accumulated by 1 for each square. It can be seen from the problem that the wiring from point a can only be routed in the four directions of up, down, left and right, and the blocked grid, beyond the grid array boundary and the grid that has been laid cannot be routed, insert the grid that can be routed into the living node table, and then take out a further expansion, the search process until the target node is found or the living node table is empty.

布线问题采用队列式分支限界算法。搜索从起点 a 开始，到终点 b 结束。约束条件为有边相连且未曾布线。

Queued branch and bound algorithm is used for wiring problems. The search starts at point a and ends at point b. The constraint is that there are edges connected and no wiring.

6.4.2　算法设计与描述（Algorithm Design and Description）

1. 布线关系（Wiring Relationship）

从 a 开始将其作为第一个扩展节点，沿 a 的上下左右 4 个方向的相邻节点扩展。判断约束条件是否成立,若成立,就放入活节点表中,并将方格标记为1。然后,从活节点列表中取出队首节点作为下一个扩展节点,并将与当前扩展节点相邻且未被标记过的方格记为2。以此类推,一直继续到算法搜索到目标方格或活节点表为空位置。目标方格里的数据表示布线长度。

Starting from a, it is taken as the first expansion node, and it is extended along the neighboring nodes of a in the four directions of up, down, left, and right. Check if the constraint is true, and if so, put it in the living node table and mark the square as 1. Then, the head of the line node is taken from the living node list as the next expansion node, and the square adjacent to the current expansion node that has not been marked is marked as 2. And so on, until the algorithm finds the target square or an empty position in living node table. The data in the target square indicates the wiring length.

算法过程中,不能布线的条件为封锁的方格、超出方格阵列的边界和已布过线的方格。方格阵列用二维数组表示,不同类型的方格用不同的数字表示,具体地,封锁的方格用-2表示,布过线的方格用大于或等于0的整数顺序表示,未曾布过线的方格用-1表示,边界方格外围加了"一堵墙",墙上的方格用数字-2表示,即边界不能布线。这样,约束条件就可以简单地表示为 $grid[i][j]=-1$。算法伪代码如下。

In the process of algorithm, the conditions that cannot be wired are blocked squares, squares beyond the bound of square array and squares already laid. The square array is represented by a two-dimensional array, and different types of squares are represented by different numbers, specifically, the blocked squares are represented by -2, the lined squares are represented by integers greater than or equal to 0, and the unlined squares are represented by -1, and "a wall" is added to the periphery of the boundary square, and the squares on the wall are represented by the number -2, that is, the boundary cannot be wired. In this way, the constraint can simply be expressed as $grid[i][j]=-1$. The algorithm pseudocode is as follows.

算法6.5 形成布线关系(Algorithm 6.5 Forms wiring relationships)

输入:起点 $start$、终点 $finish$(Input:starting from the $start$ and end point $finish$)

输出:搜索过程形成的方格阵列 $grid$,最短布线长度(Output:search process forms of the lattice $grid$ array, the shortest length of the wiring)

1.　que←空队列(empty queue)
2.　insert(que, start)　　　　　　　//起点入队列 que(Starting point into the queue que)
3.　while(True) do
4.　　here←队列 que 中取出一个活节点(Remove a living node from queue que)
5.　　for i←0 to 3 do　　　　　　　//沿着扩展节点的右、下、左、上4个方向扩展(Expand along the right,
　　　　　　　　　　　　　　　　　　//down, left, and up directions of the extension node)
6.　　　nbr←here 扩展的子节点(Extended child node)
7.　　　if(nbr 方格中的数字 ==-1) then//判断约束条件(Judgment constraint)
8.　　　　nbr 方格中的数字←当前扩展节点 $here$ 方格中的数字+1(Number + 1 in the $here$ square of the current expansion node)
9.　　　if nbr 是终点 then
10.　　　break　　　　　　　　　　//如果到达终点结束(If the end is reached)

续

```
11.      end if
12.      insert(que, nbr)    //将 nbr 插入队列(活节点表)que(Insert the nbr into the queue (active node table) que)
13.    end if
14.  end for
15.  if nbr 是终点 then
16.    break                 //完成布线(Finish wiring)
17.  end if
18.  if que 为空队列 then(If que is empty queue then)
19.    return                //返回,算法结束(Return, algorithm end)
20.  end for
21. end while
22. return grid
```

2. 最优解构造(Optimal Solution Construction)

构造最优解的过程从目标点开始逆向推理。沿着上下左右 4 个方向,判断如果某个方向方格里的数据比扩展节点方格里的数据小,就进入该方向方格,使其成为当前的扩展节点,以此类推,搜索过程一直持续到起点。算法伪代码如下。

The process of constructing the optimal solution starts from the targetnode and reasoning backward. Along with up, down, left, and right, and determine that if the data in one direction is smaller than the data in the extension node, it will enter the square in that direction and make it the current expansion node, and so on, the search process continues until the starting node. The pseudocode for the algorithm is as follows.

算法 6.6　布线问题构造最优解(Algorithm 6.6　Construct an optimal solution for wiring problems)

输入:搜索过程中得到的方格阵列 grid,起点 start,终点 finish(Input:The grid array obtained in the search process, starting point start, terminal point finish)

输出:布线方案 path 及布线长度 pathlen(Output:Routing scheme path and routing length pathlen)

```
1.  pathlen←目标点中的数字(The number in the target point)
2.  path←空队列                    //存放布线方案(Storage wiring scheme)
3.  here←finish
4.  for j←pathlen−1 to 0 do
5.    将 here 插入 path 中(Insert here into the path)
6.    for i←0 to 3 do              //沿 4 个方向扩展(Spread in four directions)
7.      nbr←扩展的子节点(Extended child node)
8.      if(nbr 方格中的数字 = =j) then    //回到上一层(Go back to the previous level)
9.        break
10.     end if
11.   end for
12.   here←nbr                     //往回推进(Push back)
13. end for
14. 将 start 插入 path 中 (Insert start into the path)
15. return path, pathlen
```

6.4.3　算法实现(Algorithm Implementation)

1. 实例构造(Instance Construction)

从 a 开始将其作为第一个扩展节点,沿 a 的上、下、左、右 4 个方向的相邻节点扩展。判断约束条件是否成立,如果成立,则放入活节点表,并将这些方格标记为1,如图6.37所示。

Starting from a, it is taken as the first expansion node, and it is extended along the neighboring nodes of a in the four directions of top, bottom, left and right. Determine whether the constraints hold, and if so, put into the living node and mark these squares as 1. This is shown in Figure 6.37.

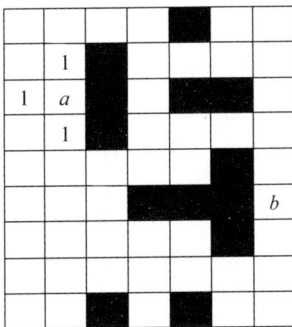

图 6.37　构造最优解过程(1)

Figure 6.37　Construct the optimal solution procedure (1)

接着从活节点队列中取出队首节点作为下一个扩展节点,并将与当前扩展节点相邻且未标记过的方格记为2,如图6.38所示。

Then, the head node is taken from the living node queue as the next expansion node, and the unmarked square adjacent to the current expansion node is marked as 2, as shown in Figure 6.38.

图 6.38　构造最优解过程(2)

Figure 6.38　Construct the optimal solution procedure (2)

以此类推,直到算法搜索到目标方格或活节点表为空为止。目标方格里的数据表示布线长度,如图6.39所示。

And so on, until the algorithm searches the target square or the living node list is empty. The data in the target square represents the wiring length, as shown in Figure 6.39.

构造最优解过程是从目标点 b 开始,沿着上、下、左、右4个方向。判断如果某个方向方格里的数据比扩展节点方格里的数据小1,则进入该方向方格,使其成为当前的扩展节点。

图 6.39 构造最优解过程（3）

Figure 6.39 Construct the optimal solution procedure (3)

以此类推，搜索过程一直持续到起始点 a，最后生成最短路径，如图 6.40 所示。

The process of constructing the optimal solution starts from the targetnode b and follows four directions: up, down, left and right. If the data in a direction square is 1 less than the data in the expansion node square, enter the direction square and make it the current expansion node. And so on, the search process continues until the starting node a, and finally the shortest path is generated, as shown in Figure 6.40.

图 6.40 构造最优解过程（4）

Figure 6.40 Construct the optimal solution procedure (4)

2. 代码实现（Code Implementation）

Python 程序需要导入依赖包 *queue*。

Python programs need to import the dependency package *queue*.

首先定义一个类 *Node*，类中定义了方格的行和列两个字段，用以表示方格的位置。

First, a class *Node* is defined in which has two fields of row and column, are defined to represent the position of the square.

然后定义两个函数：①*findpath*( )函数，用于搜索最短布线方案，接收搜索起点 *start* 和终点 *finish*，记录搜索过程中方格阵列中元素的变化；②*build_path*( )函数，用于构造最短布线方案，接收方格阵列、搜索的起点和终点，返回最短布线长度和最短布线方案。

And then two functions are defined. ①the *findpath*( ) function is used to search for the shortest wiring scheme. Receive the search *start* and *finish*, and record the changes of elements in the square array during the search process. The *build_path* ( ) function is used to construct the shortest wiring scheme, receive the square array, the starting node and end node of the search, and return the shortest wiring length and the shortest wiring scheme.

最后定义入口 *main* ( )函数,其中初始化一个 9×7 方格阵列,指定布线起点、终点及封锁方格。初始化搜索的 4 个方向为方格阵列周围添加的"围墙",然后调用函数得到最短的布线方案和布线长度。最后打印输出结果。程序运行结果如图 6.41 所示。程序代码如下。

Finally, define the entry function *main* ( ), which initializes a 9×7 square array, specifying the wiring start, end, and block square. Initializing the search for 4 directions adds a " wall" around the square array and then calls the function to get the shortest wiring scheme and wiring length. Finally print out the result. The program code is as follows. The result of the code running is shown in Figure 6.41.

```python
import queue
class Node:
    def _ _init_ _(self, row, col):
        self.row = row
        self.col = col
def findpath(start, finish):
    global grid, offset
    pathLen = 0
    if(start.row = =finish.row) and (start.col = =finish.col):  #起点与终点相同,不用布线(Same starting point and
                                                                #end point, no wiring)
        pathLen = 0
    # return
    here = start
    grid[start.row][start.col] =0
    que = queue.Queue()
    que.put(start)
    while(True):
        here = que.get()
        for i in range(4):      #沿着扩展节点的右、下、左、上 4 个方向扩展(Extend along the right, down, left and up
                                #directions of the extension node)
            nbr = Node(here.row + offset[i].row, here.col + offset[i].col)
            #print("nbr.row = " +str(nbr.row))
            #print("nbr.col = " + str(nbr.col))
            if(grid[nbr.row][nbr.col] = =-1):   #如果这个方格还没有扩展(If this square hasn't been expanded)
                grid[nbr.row][nbr.col] = grid[here.row][here.col] + 1
                if((nbr.row = =finish.row) and (nbr.col = =finish.col)):
                    break                       #如果到达终点结束(If the end is reached)
                que.put(nbr)                    #此邻节点放入队列(This neighbor node is put into the queue)
        if((nbr.row = =finish.row) and (nbr.col = =finish.col)):
            break                               #完成布线(Finish wiring)
        if que.empty():
            return
def build_path(grid, start, finish):
    global offset
    pathlen =grid[finish.row][finish.col]
    path = []
    here =finish
    for j in range(pathlen -1, -1, -1):
```

续

```
        path.insert(0, here)
        for i in range(4):                    #4 个方向扩展(Four-direction expansion)
            nbr = Node(here.row+offset[i].row, here.col+offset[i].col)
            if (grid[nbr.row][nbr.col] ==j):
                break
        here=nbr                              #往回推进(Push back)
    path.insert(0, start)
    return path , pathlen
if __name__ =="__main__":
    n = 9                                     #行数(Row number)
    m = 7                                     #列数(Column number)
    grid = [[ -1 for j in range(m+2) ]for i in range(n+2) ]
    grid[1][6] = -2
    grid[2][3] = -2
    grid[3][3] = -2
    grid[3][5] = -2
    grid[3][6] = -2
    grid[4][3] = -2
    grid[5][6] = -2
    grid[6][5] = -2
    grid[6][4] = -2
    grid[6][6] = -2
    grid[7][6] = -2
    grid[9][3] = -2
    grid[9][5] = -2
    offset = [Node(0, 1), Node(1, 0), Node(0, -1), Node(-1, 0)]
    start = Node(3, 2)
    finish = Node(6, 7)
    for i in range(n+2):             #方格阵列的上下“围墙”(The "wall" of the top and bottom of the grid array)
        grid[i][0] = -2
        grid[i][m+1] = -2
    for i in range(m+2):
        grid[0][i] = -2
        grid[n+1][i] = -2
    findpath(start, finish)
    path, pathlen = build_path(grid, start, finish)
    print("布线长度为(Wiring length is):", pathlen)
    print("布线方案为(The wiring scheme is):")
    for i in range(len(path)):
        print("path["+str(i)+"].row=" + str(path[i].row)+"      path["+str(i)+"].col=" + str(path[i].col))
```

```
布线长度为(Wiring length is): 10
布线方案为(The wiring scheme is):
path[0].row=3      path[0].col=2
path[1].row=4      path[1].col=2
path[2].row=5      path[2].col=2
path[3].row=5      path[3].col=3
path[4].row=5      path[4].col=4
path[5].row=5      path[5].col=5
path[6].row=4      path[6].col=5
path[7].row=4      path[7].col=6
path[8].row=4      path[8].col=7
path[9].row=5      path[9].col=7
path[10].row=6      path[10].col=7
```

图 6.41　分支限界算法解决布线问题的运行结果

Figure 6.41　Result of the branch and bound algorithm to solve wiring problem

习题 6(Exercises Six)

1. 简述队列式分支限界算法和优先队列式分支限界算法的区别。

Briefly introduces the difference between queue branchand bound algorithm and priority queue branch and bound algorithm.

2. 简述分支限界算法与回溯算法的区分。

Briefly introduces the distinction between branch and bound algorithm and backtracking algorithm.

3. 简述分支限界算法的搜索策略。

Briefly introduces the search strategy of branch and bound algorithm.

4. 给定背包容量 $W=20$,6 件物品的重量分别为 $(5,3,2,10,4,2)$,价值分别为 $(11,8,15,18,12,6)$。画出分支限界算法求解上述 0/1 背包问题的搜索空间。

Given the backpack capacity $W=20$, the weight of the 6 items is $(5, 3, 2, 10, 4, 2)$, and the value is $(11, 8, 15, 18, 12, 6)$. Draw the search space of the branch and bound algorithm for solving the 0/1 knapsack problem.

5. 设某一机器由 n 个部件组成,每一种部件都可以从 m 个不同的供应商处购得。设 w_{ij} 是从供应商 j 处购得的部件 i 的重量,c_{ij} 是相应的价格。试设计一个算法,给出总价格不超过 c 的最小重量机器设计。

Suppose a machine consists of n parts, each of which can be purchased from m different suppliers. Let w_{ij} be the weight of part i purchased from supplier j and c_{ij} be the corresponding price. Try to design an algorithm to give the minimum weight machine design whose total price does not exceed c.

6. 分配问题要求将 n 个任务分配给 n 个人,每个人完成任务的代价不同,要求分配的结果最优解空间。

The assignment problem requires n tasks to be assigned to n individuals, each with different costs to complete the task, and requires the resulting optimal solution space of the assignment.

第7章

图算法

Chapter 7　Graph Algorithm

　　图算法是图分析的工具之一。图算法提供了一种最有效的分析连接数据的方法,描述了如何处理图以发现一些定性或者定量的结论。图算法基于图论,利用节点之间的关系来推断复杂系统的结构和变化。可以使用这些算法来发现隐藏的信息,验证业务假设,并对行为进行预测。

　　Graph algorithm is one of the tools of graph analysis. Graph algorithms provide one of the most efficient ways to analyze connected data. They describe how to process graphs to find some qualitative conclusions. Graph algorithms are based on graph theory and use the relationships between vertexes to infer the structure and changes of complex systems. They can be used to discover hidden information, verify business assumptions, and make predictions about behavior.

7.1　概述(Overview)

　　图(Graph)是由若干给定的顶点及连接两顶点的边所构成的图形,这种图形通常用来描述某些事物之间的某种特定关系。顶点用于代表事物,连接两顶点的边则用于表示两个事物间具有某种关系。

　　A Graph is consisting of a number of given vertexes and edges connected two vertexes. This graph is usually used to describe a particular relationship between something. Vertexes are used to represent things, and the edges connecting two vertexes are used to indicate that two things have a relationship.

　　使用 $G=(V, E)$ 表示一个图,V 为顶点(Vertex),E 为边(Edge)。图中不允许没有顶点,但是可以没有边。

　　Use $G=(V, E)$ to represent a graph, where V is the Vertex and E is the Edge. The graph is not allowed to be without vertexes, but it can be without edges.

　　(1) 无向图。若顶点 v_i 到 v_j 之间的边没有方向,则称这条边为无向边,用无序偶对 (v_i, v_j) 来表示,如果图中任意两个顶点之间的边都是无向边,则称该图为无向图。

　　Undirected graph: If the edge between the vertexes v_i and v_j has no direction, then this edge is called an undirected edge, expressed by an unordered pair (v_i, v_j). If the edges between any two vertexes in the graph are all undirected, then the graph is called an undirected graph.

(2) 有向图。若从顶点 v_i 到 v_j 的边有方向，则称这条边为有向边，也称为弧(Arc)，用有序偶<v_i,v_j>来表示，v_i 称为弧尾(Tail)，v_j 称为弧头(Head)。如果图中任意两个顶点之间的边都是有向边，则称该图为有向图。

Directed graph. If the edge from vertex v_i to v_j has a direction, then this edge is called a directed edge, also known as arc, with ordered pair <v_i, v_j>, where v_i is called the tail and v_j is called the head. If the edges between any two vertexes in a graph are all directed edges, the graph is called a directed graph.

(3) 邻接/相邻。顶点 u 和 v，若存在边(u,v)，则称 u 和 v 是邻接或相邻的。

Adjacent/neighboring. The vertexes u and v, if there are edges (u, v), they are said to be adjacent or neighboring.

(4) 完全图。任意两个不同的节点都是邻接的简单图称为完全图。

Complete graph. If any two vertexes in one graph are adjacent, then the graph is called as a complete graph.

(5) 连通图(一般都是指无向图)。从顶点 v 到 m 有路径，就称顶点 v 和 m 连通。如果图中任意两顶点都连通，则该图为连通图。

Connected graph (generally refers to undirected graph). If there is a path from vertex v to m, then vertexes v and m are said to be connected. If any two vertexes in a graph are connected, the graph is connected.

(6) 关联。顶点 v 是边 e 的一个端点，则称 e 和 v 关联或相邻。

Association. If a vertex v is an endpoint of an edge e, then e and v are said to be associated or adjacent.

(7) 度。顶点相邻边的数目，常用 $deg(V)$ 或 $d(v)$ 表示。

Degree. The number of edges that are adjacent to one certain vertex v, usually represented by $deg(V)$ or $d(v)$.

(8) 入度。以该顶点为终点的边的数目。

Indegree. The number of edges ending at one certain vertex.

(9) 出度。以该顶点为起点的边的数目。

Outdegree. The number of edges starting at one certain vertex.

(10) 顶点的度。入度与出度之和。

Degree of the vertex. The sum of the indegree and the outdegree for one certain vertex.

7.2 图的表示(Representation of Graph)

对于图比较常用的表示有直接存边，邻接表，邻接矩阵，链式前向星。

The common representation of graph methods includes direct edge, adjacency list, adjacency matrix and chain forward star.

　　每种图的表示都有不同的用处，需要根据实际需求选择。比较标准和常规的表示方法是邻接表和邻接矩阵。这两种表示法都既可以表示无向图，也可以表示有向图。邻接链表通常用来表示稀疏图（边的条数 $|E|$ 远远小于 $|V^2|$ 的图），而邻接矩阵通常用来表示稠密图（边的条数 $|E|$ 接近 $|V^2|$ 的图）。另外，如果需要快速判断任意两个节点之间是否有边相连，可能也需要使用邻接矩阵表示法。

　　Each representation of graph method has a different purpose and needs to be selected according to actual needs. The standard and conventional representation methods are adjacency list and adjacency matrix. Both representations can represent undirected and directed graphs. Adjacency lists are often used to represent sparse graphs ($|E|$ is much smaller than $|V^2|$), while adjacency matrices are often used to represent dense graphs ($|E|$ is close to $|V^2|$). In addition, if it needs to quickly determine whether any two vertexes are connected by an edge, the adjacency matrix representation is also used.

7.2.1　直接存边（Direct Edge）

　　直接存边比较简单，使用一个数组来存边，数组中的每个元素都包含一条边的起点与终点（带权重的图还包含权重），或者使用多个数组分别存起点、终点和边权。

　　It is easier to store edges directly, using an array in which each element contains the start and end of an edge（weighted graphs also contain weights），or using multiple arrays to store the start, end, and edge weights separately.

　　由于直接存边的遍历效率低下，一般不用于遍历图。在最小生成树的 *Kruskal* 算法中，由于需要将边按权重排序，可以使用直接存边。有时候可能需要多次建图（如建一遍原图，建一遍反图），此时既可以使用多个其他数据结构来同时存储多张图，也可以将边直接存下来，需要重新建图时直接使用。

　　Because of the low efficiency of the direct edge traversal, it is generally not used for traversing graphs. In *Kruskal* algorithm of minimum spanning tree, direct edge storage can be used because of the operate of sorting edges by weight. Sometimes it may be necessary to build multiple diagrams（such as building the original diagram and the inverse diagram at one time），in which case it can either use multiple other data structures to store multiple diagrams at the same time, or can save the edge directly and use it straight when it need to rebuild the diagram.

7.2.2　邻接表（Adjacency List）

　　邻接表是图的一种链式存储结构。在邻接表中，对图中每个顶点建立一个单链表，第 i 个单链表中的节点表示依附于顶点 v_i 的边（对有向图是以顶点 v_i 为尾的弧）。每个节点由三个域组成，其中，邻接点域（*adjvex*）指示与顶点 v_i 邻接的点在图中的位置，链域（*nextarc*）指示下一条边或弧的节点，数据域（*info*）存储和边或弧相关的信息，如权值等。每个链表上附设一个表头节点。在表头节点中，除了设有链域（*firstarc*）指向链表中第一个节点之外，还设有存储顶点 v_i 的名或其他有关信息的数据域（*data*），如图 7.1 所示。

The adjacency list is a chained storage structure for graphs. In the adjacency list, a single linked list is created for each vertex in the graph, and the node in the i single linked list represents the edge attached to vertex v_i (a directed graph is an arc with vertex v_i as its tail). Each node is composed of three fields, of which the *adjvex* indicates the position of the point adjacent to vertex v_i in the graph, the *nextarc* indicates the node of the next edge or arc, and the *info* stores information related to the edge or arc, such as weights. Each linked list is attached to a table head node. In the table head node, in addition to having a *firstarc* pointing to the first node in the linked list, there is also a *data* storing the name of the vertex v_i or other relevant information. As shown in Figure 7.1.

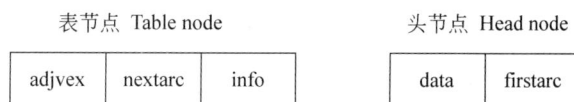

图 7.1 邻接表结构

Figure 7.1 Structure diagram of adjacency list

有向图 G_1 邻接表的表示如图 7.2 所示。

Figure 7.2 shows the adjacency table of directed graph G_1.

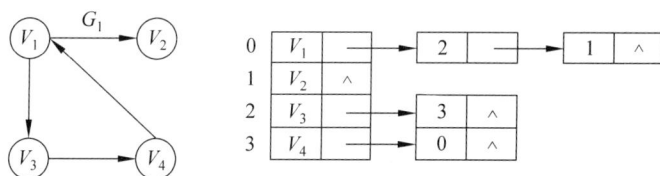

图 7.2 有向图邻接表

Figure 7.2 Directed graph adjacency table

无向图 G_2 邻接表的表示如图 7.3 所示。

The representation of the undirected graph G_2 adjacency table is shown in Figure 7.3.

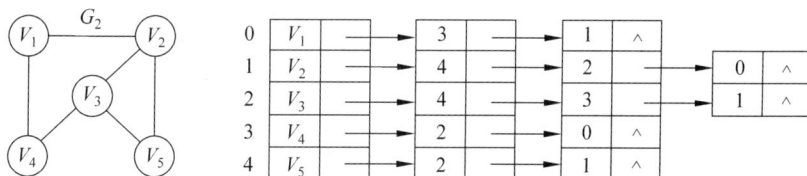

图 7.3 无向图邻接表

Figure 7.3 Undirected graph adjacency table

7.2.3 邻接矩阵(Adjacency Matrix)

图的邻接矩阵(Adjacency Matrix)存储方式是用两个数组来表示图。一个一维数组存储图中顶点信息,一个二维数组(称为邻接矩阵)存储图中的边或弧的信息。

The adjacency matrix of a graph is stored as two arrays representing the graph. A one-dimensional array stores information about the vertexes of the graph, and a two-dimensional array (called the adjacency matrix) stores information about the edges or arcs of the graph.

假设图 G 有 n 个顶点，则邻接矩阵 A 是一个 $n \times n$ 的方阵，定义如下。

Suppose that the graph G has n vertexes, then the adjacency matrix A is an $n \times n$ square matrix and is defined as follows.

$$A[i][j] = \begin{cases} 1, & (v_i, v_j) 或 <v_i, v_j> 是 E(G) 中的边 \\ 0, & (v_i, v_j) 或 <v_i, v_j> 不是 E(G) 中的边 \end{cases}$$

$$A[i][j] = \begin{cases} 1, & (v_i, v_j) \text{ or } <v_i, v_j> \text{ is an edge of } E(G) \\ 0, & (v_i, v_j) \text{ or } <v_i, v_j> \text{ is not an edge of } E(G) \end{cases}$$

无向图邻接矩阵的表示如图 7.4 所示。

The representation of the adjacency matrix of an undirected graph is shown in Figure 7.4.

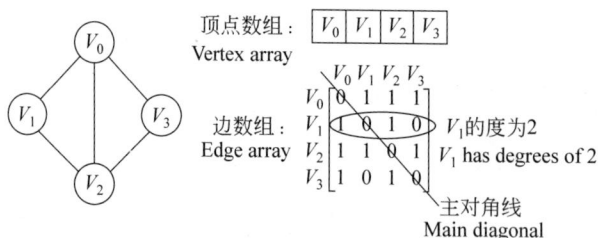

图 7.4　无向图邻接矩阵

Figure 7.4　Adjacency matrix of undirected graph

无向图的邻接矩阵一定是一个对称矩阵（即从矩阵的左上角到右下角的主对角线为轴，右上角的元与左下角相对应的元素全部是相等的）。因此，在实际存储邻接矩阵时只需存储上（或下）三角矩阵的元素。

The adjacency matrix of an undirected graph must be a symmetric matrix (that is, the main diagonal from the upper left corner of the matrix to the lower right corner is the axis, and the elements in the upper right corner are all equal to the corresponding elements in the lower left corner). Therefore, only the elements of the upper (or lower) triangular matrix need be stored when the adjacency matrix is actually stored.

有向图邻接矩阵的表示如图 7.5 所示。

The representation of the adjacency matrix ofthe directed graph is shown in Figure 7.5.

主对角线上数值依然为 0，但因为是有向图，所以此矩阵并不对称。

The value on the main diagonal is still 0, but because it is adirected graph, this matrix is not symmetric.

由于邻接矩阵在稀疏图上效率很低（尤其是在节点数较多的图上，空间无法承受），所以一般只会在稠密图上使用邻接矩阵。

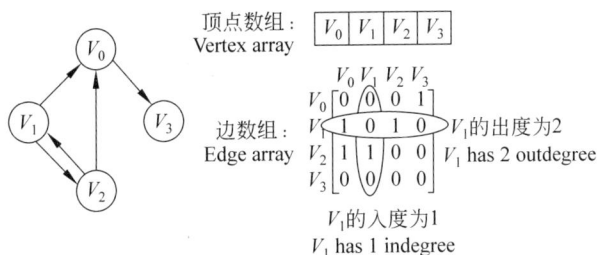

图 7.5　有向图邻接矩阵

Figure 7.5　Adjacency matrix of directed graph

Because adjacency matrix is inefficient on sparse graphs (especially on graphs with more vertexes, which the space cannot bear), they are generally only used on dense graphs.

7.2.4　链式前向星(Chain Forward Star)

链式前向星是一种静态链表存储,用边集数组和邻接表相结合,可以快速访问一个顶点的所有邻接点。

The chain forward star is a statically linked list storage, which combines an array of edge sets with an adjacency list to quickly access all the neighbors of a vertex.

链式前向星存储包括边数组和头节点数组两种结构。

The chain forward star storage includes two structures: edge array and head node array.

(1) 边数组。$edge[\ ]$,$edge[i]$表示第 i 条边。

Edge array. $edge[\ ]$, where $edge[i]$ represents the i edge.

(2) 头节点数组。$head[\ ]$,$head[i]$存储以 i 为起点的最后一条边的下标(在 $edge[\ ]$ 中的下标)。

Head node array. $head[\ ]$, where $head[i]$ stores the subscript of the last edge starting with i (subscript of $edge[\ ]$).

每一条边的结构如图 7.6 所示。

The structure of each edge is shown in Figure 7.6.

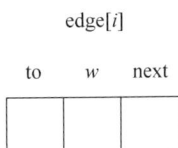

图 7.6　链式前向星边结构

Figure 7.6　Edge structure of the chain forward star

例如,一个无向图 G,如图 7.7 所示。按以下顺序输入每条边的两个端点,建立链式前向星,过程如下。

For example, an undirected graph G is shown in Figure 7.7. Enter the two vertexes of each edge in the following order to build a chain forward star. The process is as follows.

输入 1、2、5，创建一条边 1-2，权值为 5，创建第一条边 $edge[0]$，如图 7.8 所示。

Enter 1, 2, 5 to create an edge 1-2 with a weight of 5, and create the first edge $edge[0]$, as shown in Figure 7.8.

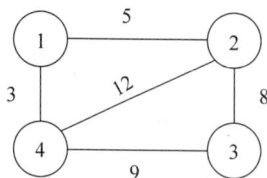

图 7.7　无向图 G
Figure 7.7　Undirected graph G

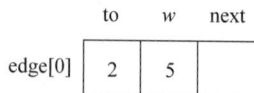

图 7.8　创建链式前向星的边
Figure 7.8　Creating the edge of the chain forward star

然后将该边链接到 1 号节点的头节点中。（初始时 $head[\ ]$ 数组全部初始化为−1），即 $edge[0].next=head[1]$；$head[1]=0$；表示 1 号节点关联的第一个条边为 0 号边，如图 7.9 所示。图中的虚线箭头仅表示它们之间的链接关系，不是指针。

This edge is then linked to the head node of vertex 1. (The initial $head[\ ]$ array is all initialized to −1), that is, $edge[0].next=head[1]$; $head[1]=0$; Indicating that the first edge associated with vertex 1 is edge 0, as shown in Figure 7.9. The dotted arrows in the diagram represent only the link relationship between them, not pointers.

因为是无向图，还需要添加它的反向边，2-1，权值为 5。创建第二条边 $edge[1]$，如图 7.10 所示。

Since it is an undirected graph, you also need to add its reverse edge, 2-1, with a weight of 5. Create a second edge, $edge[1]$, as shown in Figure 7.10.

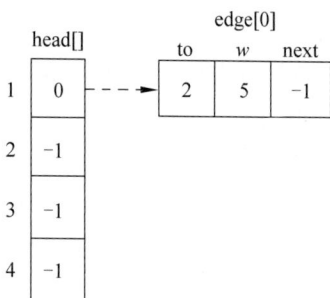

图 7.9　链接 1 号节点的边
Figure 7.9　Edge linking vertex 1

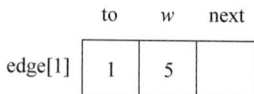

图 7.10　创建反向边
Figure 7.10　Creating a reverse edge

然后将该边链接到 2 号节点的头节点中，即 $edge[1].next=head[2]$；$head[2]=1$；表示 2 号节点关联的第一个条边为 1 号边，如图 7.11 所示。

Then link this edge to the head node of vertex 2, i.e. $edge[1].next=head[2]$; $head[2]=1$; The first edge representing the association of vertex 2 is edge 1, as shown in Figure 7.11.

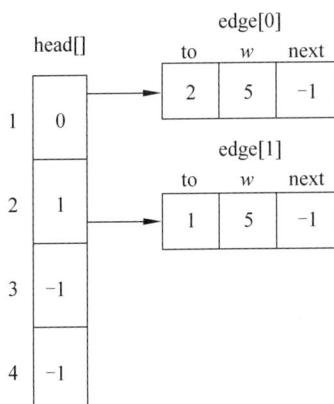

图 7.11　链接 2 号节点的边

Figure 7.11　Edge linking vertex 2

和邻接表一样,因为采用头插法进行链接,所以边输入顺序不同,创建的链式前向星也不同。对于无向图,每输入一条边,需要添加两条边。例如,输入第一条边 1、2、5,实际上添加了两条边,这两条边互为反向边。链式前向星具有边集数组和邻接表的功能,属于静态链表,不需要频繁地创建节点,应用十分灵活。

Like the adjacency list, because of the head insertion method for linking, the input order of the edge is different, and the chain forward stars are created differently. For undirected graphs, for each input edge, it needs to add two edges. For example, if input the first edge 1, 2, 5, it actually adds two edges, which are reverse edges to each other. The chain forward star has the functions of an edge set arrary and adjacency list. It is a static linked list and does not require frequent node creation. It is very flexible in application.

7.3　图的遍历(Traversing Graph)

图的遍历是指从图中的某一个节点出发,按照某种搜索方法沿着图中的边对图中的所有节点访问一次且仅访问一次。由于树是一种特殊的图,所以树的遍历实际上也可以看作一种特殊的图的遍历。图的遍历根据访问节点的顺序,主要分为广度优先搜索(Breadth First Search,BFS)、深度优先搜索(Depth First Search,DFS)和 A^* 搜索算法,本节主要讲解广度优先搜索和深度优先搜索算法。

The traversal of a graph means starting from a vertex in the graph and visiting all vertexes in the graph once and only once along the edges of the graph according to some search method. Since a tree is a special kind of graph, a tree traversal can be seen as a special kind of graph traversal as well. The traversal of the graph is mainly divided into Breadth First Search (BFS), Depth First Search (DFS) and A^* search algorithms according to the order of visiting vertexes. This section will focus on breadth first search and depth first search algorithms.

7.3.1 广度优先搜索（Breadth First Search，BFS）

广度优先搜索是一种用于图遍历的算法。它从图的起始节点开始，逐层地向外扩展，先访问当前节点的所有邻居节点，然后再逐层访问邻居的邻居节点，以此类推。该算法通常使用队列（Queue）来辅助实现。

Breadth first search is an algorithm for graph traversal. It starts at the starting vertex of the graph and spreads out layer by layer, first visiting all the neighbors of the current vertex, and then visiting the neighbors' layer by layer, and so on. The algorithm is usually implemented using a queue.

广度优先搜索的基本思想是从起始节点开始，将起始节点标记为已访问，并将其加入队列，再从队列中取出一个节点，访问该节点的所有未被访问过的邻居节点，并将它们标记为已访问并加入队列，然后重复步骤2，直到队列为空。

The basic idea of breadth first search is to start from the start vertex, mark the start vertex as visited, and add it to the queue. Then it will take a vertex from the queue, visit all the neighbor vertexes that have not been visited, and mark them as visited and added to the queue. After that, repeat step 2 until the queue is empty.

广度优先搜索会逐层遍历图，确保先访问离起始节点更近的节点；由于使用了队列，节点的访问顺序是按照其离起始节点的距离递增的；它可以用于求解最短路径问题，因为它首次访问到目标节点时，路径长度一定是最短的。

Breadth first search traverses the graph layer by layer, ensuring that vertexes closer to the starting vertex are visited first. Because of the use of queues, the access order of vertexes increases according to their distance from the starting vertex. It can be used to solve the shortest path problem because the path length must be the shortest when it accesses the target vertex for the first time.

例如，一个无向图 G 如图 7.12 所示。广度优先搜索算法先访问节点 a，然后依次访问与节点 a 相连的节点如 b、c，以此类推，其遍历的结果为 a、b、c、d、e、f、g、h。

For example, an undirected graph G is shown in Figure 7.12. The breadth first search algorithm first accesses vertex a, and sequential accesses vertexes connected to vertex a such as b, c, and so on. The traversal result is a, b, c, d, e, f, g, h.

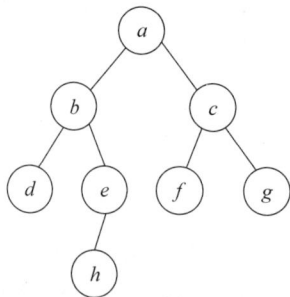

图 7.12　无向图 G

Figure 7.12　Undirected graph G

具体代码实现如下,运行结果如图 7.13 所示。

The specific code implementation is as follows, and the result is as shown in Figure 7.13.

```
def BFS(graph, q):                   #广度优先遍历, 基于队列(Breadth-first traversal, based on queue)
    queue = [ ]                      #建立队列(Create queue)
    queue.append(q)
    data = [ ]                       #记录已经遍历过的点(Record the vertexes that have been traversed)
    data.append(q)
    while queue:
        n = queue.pop(0)             #队列先进先出(Queue first in, first out)
        nodes = graph[n]
        for j in nodes:
            if j not in data:
                queue.append(j)
                data.append(j)
        print(n)
if _ _name_ _ = ='_ _main_ _':
    graph = {
        'a': ['b', 'c'],
        'b': ['d', 'e'],
        'c': ['f', 'g'],
        'e': ['h'],
        'd': [ ],
        'f': [ ],
        'g': [ ],
        'h': [ ],
    }
    print("广度优先遍历(Breadth-first traversal):")
    BFS(graph, 'a')                  #结果为a, b, c, d, e, f, g, h
```

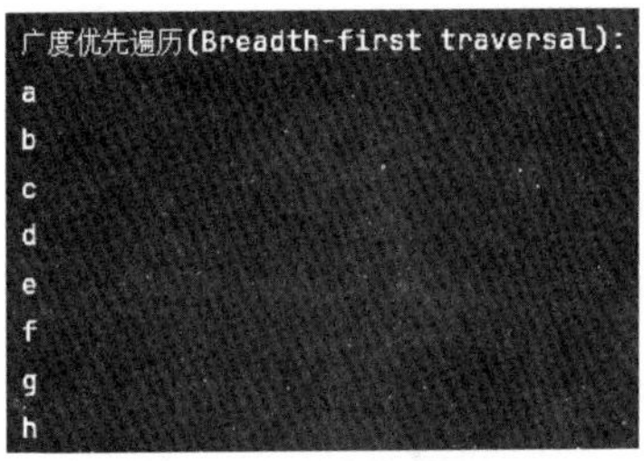

图 7.13　实现结果图

Figure 7.13　Implementation result diagram

7.3.2　深度优先搜索(Depth First Search, DFS)

深度优先搜索是可用于遍历树或者图的搜索算法,它从一个起始节点开始,沿着一条路径尽可能远地访问节点,直到到达不能继续前进的节点,然后返回上一层继续探索其他路径。这个过程是递归的,通过不断地深入节点的子节点,直到遍历完整个图。

Depth first search is a search algorithm that can be used to traverse a tree or graph, starting from a starting vertex, visiting vertexes as far as possible along a path until it reaches a vertex that can no longer proceed, and then returning to the previous layer to continue exploring other paths. This process is recursive, by going deeper and deeper into the child vertexes until the entire graph has been traversed.

深度优先搜索是图论中的经典算法,利用深度优先搜索算法可以产生目标图的相应拓扑排序表,利用拓扑排序表可以方便地解决很多相关的图论问题,如最短路径问题等。一般用栈数据结构来辅助实现此算法。根据深度优先搜索的特点,采用递归函数实现比较简单,但也可以不采用递归。

Depth first search is a classic algorithm in graph theory, which can generate the corresponding topological sorting table of the target graph. Topological sorting table can easily solve many related graph theory problems, such as the shortest path problem, etc. Generally, stack data structure is used to assist the implementation of this algorithm. According to the characteristics of depth first search, using recursive function is relatively simple, but recursion can also be avoided.

例如,一个无向图 G 如图 7.14 所示。深度优先搜索,先访问节点 a,再顺着节点 a 的一个节点往下遍历,以此类推,其遍历情况如图 7.15 所示,遍历结果为 a、b、d、e、h、c、f、g。

For example, an undirected graph G is shown in Figure 7.14. Depth first search, first visit vertex a, then traverse down along the vertex a, and so on. The traversal situation is shown in Figure 7.15, and the traversal result is a, b, d, e, h, c, f, g.

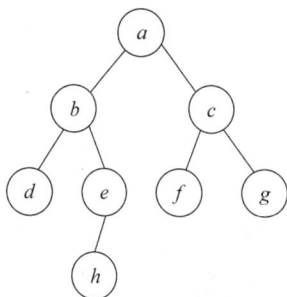

图 7.14　无向图 G
Figure 7.14　Undirected graph G

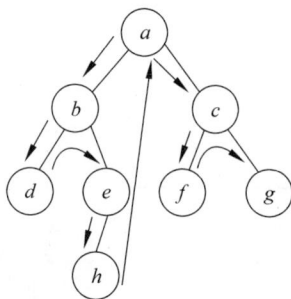

图 7.15　遍历顺序图
Figure 7.15　Traversal sequence graph

具体代码实现如下,运行结果如图 7.16 所示。

The specific code implementation is shown below, and the running results are shown in Figure 7.16.

```
def dfs(graph, start, visited=None):
    if visited is None:
        visited = set()
    print(start)
    visited.add(start)
    for neighbor in graph[start]:
        if neighbor not in visited:
            dfs(graph, neighbor, visited)
#测试例子
graph = {
```

续

```
        'a': ['b', 'c'],
        'b': ['a', 'd', 'e'],
        'c': ['a', 'f', 'g'],
        'd': ['b'],
        'e': ['b', 'h'],
        'f': ['c'],
        'g': ['c'],
        'h': ['e']
}
print("深度优先搜索(Depth-First-Search):")
start_node = 'a'
dfs(graph, start_node)
```

图 7.16　实现结果图

Figure 7.16　Implementation result diagram

7.4　Dijkstra 算法(Dijkstra Algorithm)

加权图是指每条边都带有权重的图。每个边的权重可以表示两个顶点之间的距离、成本或任何其他可以量化的指标。最短路径,是指两顶点之间经过的边上的权值之和最少的路径,并且路径上的第一个顶点是源点,最后一个顶点是终点。

A weighted graph is a graph with a weight attached to each vertex. The weight of each edge can represent the distance between two vertexes, the cost, or any other metric that can be quantified. A shortest path is the path with the least sum of weights on the edges between two vertexes, and the first vertex on the path is the source and the last vertex is the destination.

最短路径问题是图算法中的一个经典问题,有很多不同的算法可以解决,其中最著名的两个算法是迪杰斯特拉算法(*Dijkstra* Algorithm)和贝尔曼-福特算法(*Bellman-Ford* Algorithm)。

The shortest path problem is a classic problem in graph algorithms that can be solved by many different algorithms, the two most famous of which are *Dijkstra* algorithm and *Bellman-Ford* algorithm.

7.4.1　基本思想(Basic Idea)

Dijkstra 算法是一种用于解决单源最短路径问题的经典算法,由荷兰计算机科学家

E.W.Dijkstra 在 1956 年提出。其基本思想是从源节点开始,逐步探索到达图中其他节点的最短路径,直到找到所有节点的最短路径为止。

Dijkstra algorithm is a classical algorithm for solving the single-source shortest path problem, proposed by Dutch computer scientist *E.W.Dijkstra* in 1956. The basic idea is to start from the source vertex and gradually explore the shortest path to other vertexes in the graph until the shortest path of all vertexes is found.

以下是 *Dijkstra* 算法的基本思想和步骤。

Here are the basic ideas and steps of *Dijkstra* algorithm.

（1）初始化。将源节点的最短距离设置为 0,将所有其他节点的最短距离设置为无穷大（或一个足够大的值）,并将所有节点标记为未访问。

Initialization. Set the shortest distance of the source vertex to 0, set the shortest distance of all other vertexes to infinity（or a sufficiently large value）, and mark all vertexes as unvisited.

（2）选择最短路径节点。从未访问的节点中选择最短路径距离最小的节点作为当前节点。

The vertex with the smallest shortest path distance among the unvisited vertexes is selected as the current vertex.

（3）更新最短路径。对于当前节点的每个邻居节点,计算通过当前节点到达邻居节点的距离,如果通过当前节点到达邻居节点的距离比原始记录的距离小,则更新邻居节点的距离。

Update the shortest path. For each neighbor vertex of the current vertex, the distance to the neighbor vertex through the current vertex is calculated, and if the distance to the neighbor vertex through the current vertex is smaller than the original recorded distance, the distance of the neighbor vertex is updated.

（4）标记节点。将当前节点标记为已访问。

Mark the vertex. Mark the current vertex as accessed.

（5）重复（2）~（4）,直到所有节点都被标记为已访问,或者找到了目标节点的最短路径。

Repeat（2）to（4）until all vertexes are marked as visited or the shortest path to the destination vertex is found.

Dijkstra 算法使用了一种贪心策略,每次选择当前距离源节点最近的未访问节点,然后通过它更新到达其他节点的最短路径。该算法保证在无负权重边的情况下,找到源节点到所有其他节点的最短路径。

Dijkstra algorithm uses a greedy strategy, each time selects the current unvisited vertex that is closest to the source vertex, and then updating the shortest path to other vertexes through it. The algorithm guarantees to find the shortest path from the source vertex to all other vertexes without negative weight edges.

7.4.2　算法实例（Algorithm Example）

1. 问题描述（Problem Description）

在无向图 $G = (V, E)$ 中,找到由顶点 A 到其余各点的最短路径,无向图如图 7.17 所示。

In the undirected graph $G = (V, E)$, find the shortest path from vertex A to the remaining vertexes, as shown in Figure 7.17.

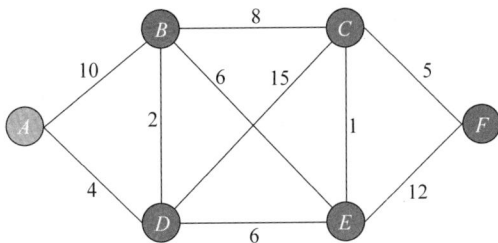

图 7.17 无向图 G

Figure 7.17 Undirected graph G

2. 问题分析(Problem Analysis)

S 集合表示已求出最短路径的节点集合,U 表示其余未确定最短路径的节点集合,与节点有直接接触的节点的距离为确定值,没有直接接触的节点为无穷大。具体访问顺序及访问节点如表 7.1 所示。

The S set represents the set of vertexes for which the shortest path has been found, and the U set represents the set of vertexes for which the shortest path has not been determined, where the distance between the vertexes that is in direct contact is a definite value, and the one that is not in direct contact is infinite. The specific access sequence and access vertexes are shown in Table 7.1.

表 7.1 访问节点情况

Table 7.1 Access vertexes

S 集合(S set)	U 集合(U set)
选 A 为起点,$S=\{A\}$,此时的最短路径 $A{\to}A=0$,以 A 为中心点,从 A 开始查找(Choose A as the starting vertex, $S=\{A\}$, then the shortest path $A{\to}A=0$, with A as the center point, start from A)	$U=\{B, C, D, E, F\}$,$A{\to}B=10$,$A{\to}D=4$,$A{\to}\{E, F, C\}=\infty$,$d(AD)$ 最短(Shortest)
选入 D,此时 $S=\{A, D\}$,此时最短路径 $A{\to}A=0$,$A{\to}D=4$,以 D 为中间点,从 $A{\to}D$ 这条最短路径进行查找(Select D, then $S=\{A, D\}$, then the shortest path $A{\to}A=0$, $A{\to}D=4$, take D as the middle point, search from the shortest path $A{\to}D$)	$U=\{B, C, E, F\}$,$A{\to}D{\to}B=6<10(A{\to}B)$,$A{\to}D{\to}E=10$,$A{\to}D{\to}C=19$,$A{\to}D{\to}F=\infty$,$d(ADB)$ 最短(Shortest)
选入 B,此时 $S=\{A, D, B\}$,此时最短路径 $A{\to}A=0$,$A{\to}D=4$,$A{\to}D{\to}B=6$,以 B 为中间点,从 $A{\to}D{\to}B$ 这条路径进行查找(Select B, then $S=\{A, D, B\}$, then the shortest path $A{\to}A=0$, $A{\to}D=4$, $A{\to}D{\to}B=6$, with B as the middle point, search from $A{\to}D{\to}B$ this path)	$U=\{C, E, F\}$,$A{\to}D{\to}B{\to}C=14<19(A{\to}D{\to}C)$,$A{\to}D{\to}B{\to}E=12>10(A{\to}D{\to}E)$,$A{\to}D{\to}B{\to}F=\infty$,$d(ADE)$ 最短(Shortest)

续表

S 集合（S set）	U 集合（U set）
选入 E，此时 S={A, D, B, E}，此时最短路径 A→A = 0，A→D=4，A→D→B=6，A→D→E=10，以 E 为中间点，从 A→D→E 这条路径进行查找（Select E, then S = {A, D, B, E}, then the shortest path A→A=0, A→D=4, A→D→B=6, A→D→E=10, with E as the middle point, search from the path A→D→E）	U={C, F}，A→D→E→C = 11<14，A→D→E→F=22，d(ADEC)最短（Shortest）
选入 C，此时 S={A, D, B, E, C}，此时最短路径 A→A=0，A→D=4，A→D→B=6，A→D→E=10，A→D→E→C=11，以 C 为中间点，从 A→D→E→C 这条路径进行查找（Select C, then S={A, D, B, E, C}, then the shortest path A→A=0, A→D=4, A→D→B=6, A→D→E = 10, A→D→E→C = 11, with C as the intermediate point, is searched from the path A→D→E→C）	U={F}，A→D→E→C→F=16，发现最短路径为（The shortest path is found）A→D→E→C→F
选入 F，此时 S={A, D, B, E, C, F}，此时最短路径 A→A=0，A→D=4，A→D→B=6，A→D→E=10，A→D→E→C=11，A→D→E→C→F=16，以 F 为中间点，从 A→D→E→C→F 这条路径进行查找（Select F, then S = {A, D, B, E, C, F}, then the shortest path A→A=0, A→D=4, A→D→B=6, A→D→E=10, A→D→E→C= 11, A→D→E→C→F=16, with F as the middle point, search from the path A→D→E→C→F）	集合为空，查找完毕（Set is empty. Search complete）

3. 问题解答（Answer）

问题分析可知，从 A 节点到其他各节点的最短路径如表 7.2 所示。

The analysis of the problem shows that the shortest path from vertex A to all other vertexes is shown in Table 7.2.

表 7.2　最短路径及距离
Table 7.2　Shortest path and distance

路径（Path）	距离（Distance）
A→A	0
A→D	4
A→D→B	6
A→D→E	10
A→D→E→C	11
A→D→E→C→F	16

根据各节点之间的路径和距离可以构建一个二维矩阵。

According to the paths and distances between vertexes, a two-dimensional matrix can be constructed.

	A	B	C	D	E	F
A	0	10	∞	4	∞	∞
B	10	0	8	2	6	∞
C	∞	8	0	15	1	5
D	4	2	15	0	6	∞
E	∞	6	1	6	0	12
F	∞	∞	5	∞	12	0

4. 程序实现(Program Implementation)

上面实例的实现代码如下,实现结果如图 7.18 所示。

The code for the above example is shown below, and the result is shown in Figure 7.18.

```
def dijkstra(matrix, start_node):
    MAX = float('inf')                              #定义 MAX 为无穷大(Define MAX as infinity)

    matrix_length = len(matrix)                     #矩阵一维数组的长度,即节点的个数
    used_node = [False] * matrix_length             #访问过的节点数组
    distance = [MAX] * matrix_length                #最短路径距离数组
    distance[start_node] = 0                         #初始化,将起始节点的最短路径修改成 0

    while used_node.count(False):                    #判断是否还有未访问的节点
        min_value = MAX
        min_value_index = -1

        #在最短路径节点中找到最小值,已经访问过的不再参与循环
        for index in range(matrix_length):
            if not used_node[index] and distance[index] < min_value:
                min_value = distance[index]
                min_value_index = index

        #将访问节点数组对应的值修改成 True,标志其已经访问过了
        used_node[min_value_index] = True

        #更新 distance 数组,只更新未访问节点的距离
        for index in range(matrix_length):
            #确保从当前节点到其他节点有路径且该节点未被访问
            if not used_node[index] and matrix[min_value_index][index] != MAX:
                distance[index] = min(distance[index], distance[min_value_index] +
                matrix[min_value_index][index])
    return distance

#示例矩阵
matrix_ = [
    [0, 10, float('inf'), 4, float('inf'), float('inf')],
    [10, 0, 8, 2, 6, float('inf')],
    [float('inf'), 8, 10, 15, 1, 5],
    [4, 2, 15, 0, 6, float('inf')],
    [float('inf'), 6, 1, 6, 0, 12],
```

续

$$\left[\text{float}('\text{inf}'), \text{float}('\text{inf}'), 5, \text{float}('\text{inf}'), 12, 0\right]$$
]

```
#运行 Dijkstra 算法
ret = dijkstra(matrix_, 0)
print("A 到 B, C, D, E, F 各点的最短距离为(The shortest distance from A to B, C, D, E, F is):")
print(ret)
```

```
A到B,C,D,E,F各点的最短距离为(The shortest distance from A to B,C,D,E,F is):
[0, 6, 11, 4, 10, 16]
```

图 7.18　实现结果图

Figure 7.18　Implementation result diagram

7.5　Bellman-Ford 算法（Bellman-Ford Algorithm）

Bellman-Ford 算法是一种用于解决单源最短路径问题的算法，可以处理带有负权重边的图，该算法由 *Richard Bellman* 和 *Leslie Ford* 在 1958 年提出，是一种基于动态规划的算法。

The *Bellman-Ford* algorithm is for solving the single-source shortest path problem that can handle graphs with weighted edges. The algorithm was proposed by *Richard Bellman* and *Leslie Ford* in 1958 and is based on dynamic programming.

7.5.1　基本思想（Basic Idea）

Bellman-Ford 算法的基本思想是通过对图中的所有边进行松弛操作（即尝试降低目标节点的最短路径估计值），来逐步逼近最短路径的正确解。它的主要步骤如下。

The basic idea of the *Bellman-Ford* algorithm is to gradually approach the correct solution of the shortest path by relaxing all the edges in the graph (that is, trying to reduce the shortest path estimate of the destination vertexes). Its main steps are as follows.

（1）初始化。将源节点的最短路径估计值设置为 0，将所有其他节点的最短路径估计值设置为无穷大。

Initialization. Set the shortest path estimate for the source vertex to 0, and set the shortest path estimate for all other vertexes to infinity.

（2）边的松弛操作。对图中的每条边进行松弛操作。对于每条边 (u,v)，如果从源节点 u 到节点 v 的路径长度比当前估计的最短路径长度小，则更新节点 v 的最短路径估计值为从源节点 u 经过边 (u,v) 到达节点 v 的路径长度。

Relaxation operation of edges. Relaxation is performed on each edge of the graph. For each edge (u, v), if the path length from source vertex u to vertex v is smaller than the current estimated shortest path length, then update the shortest path estimate for vertex v to be the path length from source vertex u through edge (u, v) to vertex v.

（3）重复松弛操作。重复进行($|V|-1$)次边的松弛操作,其中,($|V|$)是图中节点的数量。这样可以确保在没有负环存在的情况下,所有节点的最短路径估计值都将收敛到正确的值。

Repeat the relaxation operation. Repeat the edge relaxation operation $|V|-1$ times, where $|V|$ is the number of vertexes in the graph. This ensures that the shortest path estimates for all vertexes will converge to the correct value in the absence of negative rings.

（4）检测负环。如果在重复松弛操作之后,仍然存在可以被更新的最短路径估计值,则说明图中存在负环。负环是一个环路,其边的权重之和为负值。负环的存在会导致无法计算出正确的最短路径。

Detect negative rings. If there is still a shortest path estimate that can be updated after repeated relaxation, then there is a negative ring in the graph. A negative ring is a loop where the sum of the weights of its sides is negative. The existence of a negative ring makes it impossible to calculate the correct shortest path.

Bellman-Ford 算法的时间复杂度为 $O(mn)$,其中,m 是节点的数量,n 是边的数量。它的优点是可以处理带有负权重边的图,并且能够检测负环的存在。缺点是在一般情况下,它的时间复杂度较高,因此在处理大规模图时可能效率不高。

The time complexity of the *Bellman-Ford* algorithm is $O(mn)$, where m is the number of vertexes and n is the number of edges. It has the advantage of being able to handle graphs with negative weighted edges and being able to detect the presence of negative rings. The disadvantage is that in general, it has a high time complexity, so it may not be efficient when dealing with large-scale graphs.

7.5.2　算法实例（Algorithm Example）

1. 问题描述（Problem Description）

假设有向图 G 中顶点为 A、B、C、D、E,求顶点 A 到其他顶点的最短距离。有向图 G 如图 7.19 所示。

Assume that the vertexes in the directed graph G are A, B, C, D, E. Find the shortest distance between vertex A and other vertexes. Figure 7.19 shows the directed graph G.

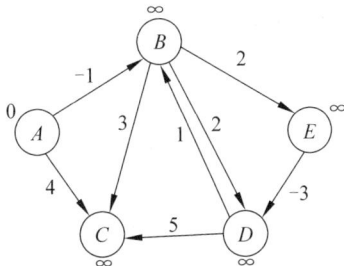

图 7.19　有向图 G

Figure 7.19　Directed graph G

2. 问题分析（Problem Analysis）

首先初始化 A 到其他顶点的距离为无穷大，同时初始化一个待处理顶点库为 $\{A\}$，各顶点之间距离如表 7.3 所示。

Firstly, the distance between A and other vertexes is initialized as infinity, and a library of vertexes to be processed is initialized as $\{A\}$. The distance between vertexes is shown in Table 7.3.

表 7.3 顶点 A 到各顶点之间距离
Table 7.3 Distance between vertex A and each vertex

A	B	C	D	E
0	∞	∞	∞	∞

顶点库：$\{A\}$。

Vertex library：$\{A\}$.

处理顶点库中的顶点 A。顶点 A 与 B、C 相邻，它到另外两个节点的距离分别是 -1,4，均小于 ∞，故更新距离表，并将顶点 B、C 加入顶点库中。更新后的距离表如表 7.4 所示。

Deal with vertex A in the vertex library. Vertex A is adjacent to B and C, and the distance from it to the other two vertexes is -1, 4, both are less than ∞, so the distance table is updated, and the vertexes B and C are added to the vertex library. Table 7.4 shows the updated distance table.

表 7.4 距离表（1）
Table 7.4 Distance table（1）

A	B	C	D	E
0	-1	4	∞	∞

顶点库 $\{B,C\}$。

Vertex library：$\{B, C\}$.

处理顶点库中的顶点 B。顶点 B 与 C、D、E 相邻，距离分别为 3、2、2。由于 A-B-C 这条路径比 A-C 这条路径耗时短，故更新顶点 C 的距离表，同时更新顶点 D 和 E 在距离表中的数据，并将顶点 D、E 加入顶点库。更新后的距离表如表 7.5 所示。

Deal with vertex B in the vertex library. Vertex B is adjacent to C, D, E at distances of 3, 2, 2, respectively. Because the path A-B-C takes less time than the path A-C, update the distance table of vertex C, update the data of vertexes D and E in the distance table, and add vertexes D and E to the vertex library. Table 7.5 lists the updated distances.

表 7.5 距离表（2）
Table 7.5 Distance table（2）

A	B	C	D	E
0	-1	2	1	1

顶点库 $\{C, D, E\}$ 。

Vertex library：$\{C, D, E\}$.

处理顶点库中的顶点 C。发现 C 没有通向其他顶点，更新顶点库为 $\{D, E\}$。

Deal with vertex C in the vertex library. Find that C does not lead to other vertexes, and update the vertex library to $\{D, E\}$.

处理顶点库的顶点 D。顶点 D 通向顶点 C，但是对于 C 来说，经过 D 的路径比不经过 D 的路径要费时，所以不做处理，更新顶点库为 $\{E\}$。

Deal with vertex D in the vertex library. Vertex D leads to vertex C, but a path through D takes more time for C than a path without D, so no processing is done and the vertex library is updated to $\{E\}$.

处理顶点库的顶点 E。E 通向顶点 D，且可以将顶点 D 的路径成本降低为 -2，故更新距离表，并将具有新路径成本的顶点 D 加入顶点库中。更新后的距离表如表 7.6 所示。

Deal with vertex E of the vertex library. E leads to vertex D, and the path cost of vertex D can be reduced to -2, so the distance table is updated, and vertex D with new path cost is added to the vertex library. Table 7.6 lists the updated distances.

表 7.6　距离表（3）
Table 7.6　Distance table（3）

A	B	C	D	E
0	-1	2	-2	1

顶点库 $\{D\}$ 。

Vertex library：$\{D\}$.

处理顶点库中的顶点 D。D 通向 B 和 C，但是不能将 B 和 C 的路径成本进一步降低了，故路径表不更新，同时顶点库中没有其他顶点了，结束整个过程。

Deal with vertex D of the vertex library. D leads to B and C, but the path cost of B and C cannot be further reduced, so the path table is not updated, and there are no other vertexes in the vertex library, ending the whole process.

3. 问题解答（Answer）

由问题分析可知，遍历完所有节点后，顶点 A 到所有顶点的最短距离如表 7.7 所示。

According to the problem analysis, after traversing all vertexes, the shortest distance between vertex A and all vertexes is shown in Table 7.7.

表 7.7　顶点 A 到各顶点的最短距离
Table 7.7　Minimum distance from vertex A to each vertex

A	B	C	D	E
0	-1	2	-2	2

4. 代码实现（Program Implementation）

在上面的例子中，使用数字 0~4 代表顶点 A~E，具体代码实现如下，实现结果如图 7.20

所示。

In the above example, the vertices A through E are represented by the numbers 0 to 4. The specific code is shown below, and the result is shown in Figure 7.20.

```
class Graph:
    def __init__(self, vertexes):
        self.vertexes = vertexes
        self.graph = []
    def add_edge(self, u, v, w):
        self.graph.append([u, v, w])
    def bellman_ford(self, start):
        #初始化最短路径估计值(Initializes the shortest path estimate)
        distance = [float('inf')] * self.vertexes
        distance[start] = 0
        #进行|V| - 1次边的松弛操作(Perform the edge relaxation operation |V| -1)
        for _ in range(self.vertexes - 1):
            for u, v, w in self.graph:
                if distance[u] !=float('inf') and distance[u] + w < distance[v]:
                    distance[v] = distance[u] + w
        #检查负环(Check negative loop)
        for u, v, w in self.graph:
            if distance[u] !=float('inf') and distance[u] + w < distance[v]:
                print("Graph contains negative weight cycle")
                return
        #打印最短路径结果(Print the shortest path result)
        for i in range(self.vertexes):
            print(f"Shortest distance from {start} to {i} is {distance[i]}")
            print(f"从 {start} 到 {i} 的最短路径是 {distance[i]}")
#创建一个图实例(Create a diagram instance)
g = Graph(5)
#添加边
g.add_edge(0, 1, -1)
g.add_edge(0, 2, 4)
g.add_edge(1, 2, 3)
g.add_edge(1, 3, 2)
g.add_edge(1, 4, 2)
g.add_edge(3, 2, 5)
g.add_edge(3, 1, 1)
g.add_edge(4, 3, -3)
#运行 Bellman-Ford 算法,从节点 0 开始(Run the Bellman-Ford algorithm, starting at node 0)
g.bellman_ford(0)
```

```
Shortest distance from 0 to 0 is 0
从 0 到 0 的最短路径是 0
Shortest distance from 0 to 1 is -1
从 0 到 1 的最短路径是 -1
Shortest distance from 0 to 2 is 2
从 0 到 2 的最短路径是 2
Shortest distance from 0 to 3 is -2
从 0 到 3 的最短路径是 -2
Shortest distance from 0 to 4 is 1
从 0 到 4 的最短路径是 1
```

图 7.20　实现结果图

Figure 7.20　Implementation result diagram

7.6 Floyd-Warshall 算法(Floyd-Warshall Algorithm)

Floyd-Warshall 算法是一种用于解决图中所有节点对之间最短路径的动态规划算法。它适用于有向图或无向图,能够处理带有负权重边的图,但不能处理存在负权重环路的图。

Floyd-Warshall algorithm is a dynamic programming for solving the shortest path between all pairs of vertexes in a graph. It is suitable for directed or undirected graphs and can deal with graphs with negative weighted edges, but not graphs with negative weighted loops.

7.6.1 基本思想(Basic Idea)

算法的基本思想是维护一个二维矩阵,其中,元素 $dist[i][j]$ 表示从节点 i 到节点 j 的最短路径长度。算法通过逐步更新这个矩阵,使得矩阵中的每个元素都表示相应节点对之间的最短路径。

The basic idea of the algorithm is to maintain a two-dimensional matrix where element $dist[i][j]$ represents the shortest path length from vertex i to vertex j. The algorithm updates the matrix step by step so that each element in the matrix represents the shortest path between the corresponding pair of vertexes.

Floyd-Warshall 算法的步骤如下。

The steps of the *Floyd-Warshall* algorithm are as follows.

(1) 初始化矩阵。对于图中的每一对节点 i 和 j,如果存在边 (i, j),则将 $dist[i][j]$ 设为这条边的权重;否则,将 $dist[i][j]$ 设为无穷大。对角线上的元素 $dist[i][i]$ 设为 0。

Initialize the matrix. For each pair of vertexes i and j in the graph, if there is an edge (i, j), $dist[i][j]$ is set as the weight of this edge; Otherwise, set $dist[i][j]$ to infinity. The element $dist[i][i]$ on the diagonal is set to 0.

(2) 动态规划更新。对于每一个中间节点 k,在当前的 $dist$ 矩阵上尝试通过节点 k 来改善 i 到 j 的路径。如果路径 $dist[i][k]+dist[k][j]$ 的长度小于当前已知的 $dist[i][j]$,则更新 $dist[i][j]$ 为新的更短路径长度。

Dynamic programming updates. For each intermediate vertex k, try to improve the path from i to j by vertex k on the current $dist$ matrix. If the length of path $dist[i][k]+dist[k][j]$ is less than that of currently known $dist[i][j]$, then $dist[i][j]$ is updated to a new shorter path length.

(3) 重复更新。对于每一个可能的中间节点 k,重复上述动态规划更新步骤。经过该步骤的迭代后,$dist$ 矩阵将包含所有节点对之间的最短路径长度。

Repeat the update. Repeat the dynamic programming update step for each possible intermediate vertex k. After iterating through this step, the $dist$ matrix will contain the shortest path length between all vertex pairs.

7.6.2　算法实例（Algorithm Example）

1. 问题描述（Problem Description）

现在用 A、B、C 分别表示三个城市，边的权重表示从一个城市到另一个城市的路径长度。给定的邻接矩阵表示城市之间的路径长度，其中，∞ 表示两城市之间没有直接路径。现在要找出每个城市之间的最短距离。简化实例如图 7.21 所示。

Now the three cities are represented by A, B, C respectively and the weight of the edges represent the path length from one city to the other. The given adjacency matrix represents the path length between cities, where ∞ means that there is no direct path between two cities. Now find the shortest distance between each city. A simplified example is shown in Figure 7.21.

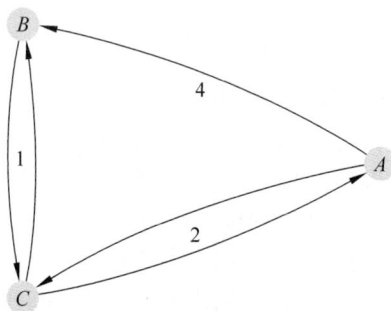

图 7.21　示例图

Figure 7.21　Example diagram

2. 问题分析（Problem Analysis）

构造各城市之间距离矩阵（Construct the distance matrix between cities）

$$\begin{array}{c|ccc} & A & B & C \\ \hline A & 0 & 4 & 2 \\ B & \infty & 0 & 1 \\ C & 2 & 1 & 0 \end{array}$$

，构造初始的前驱矩阵（Construct the initial precursor matrix）

$$\begin{array}{c|ccc} & A & B & C \\ \hline A & -1 & -1 & -1 \\ B & -1 & -1 & -1 \\ C & -1 & -1 & -1 \end{array}$$

（-1 表示无前驱节点）（-1 indicates that there is no precursor vertex）。

借助 A 间接到达其他顶点。分析：借助 A，A 可以直接到达 B、C，所以第一行保持不变；B 没有到 A 的路径，所以 B 这一行不变；C 借助 A 到 B 为 $2+4=6$ 大于 C 直接到 B 的权值 1，所以 C 这一行不变，对角始终为 0。

Other vertexes are reached indirectly with the help of A. Analysis: with the help of A, A can directly reach B and C, so the first row remains unchanged; B doesn't have path to A, so B stays the same; C is $2+4=6$ with the help of A to B, which is greater than the weight 1 of C directly to B, so the row C does not change, and the diagonal is always 0.

在这一轮中距离矩阵不变,仍为

$$\begin{array}{c|ccc} & A & B & C \\ \hline A & 0 & 4 & 2 \\ B & \infty & 0 & 1 \\ C & 2 & 1 & 0 \end{array}$$

,没有找到更短的路径,所以前驱矩阵也保持不变。

In this round the distance matrix does not change, no shorter path is found, so the precursor matrix remains the same.

借助 B 间接到达其他顶点。分析: 借助 B,A 到 B 是 4,A 借助 B 到 C 为 4+1=5>2(A 直接到 C)因此不变;B 到 A 没有路,因此仍然是无穷,B 到 C 是 1;C 借助 B 到 A 不通,所以是无穷,无穷大于 2,因此 2 不变,C 到 B 是 1。

Other vertexes are reached indirectly with the help of B. Analysis: with the help of B, A to B is 4, A with the help of B to C is 4+1=5>2 (A directly to C) so unchanged. There's no way from B to A, so it's still infinite, and B to C is 1. C doesn't get to A through B, so it's infinity, and infinity is greater than 2, so 2 doesn't change, and C goes to B is 1.

在这一轮中距离矩阵不变仍为

$$\begin{array}{c|ccc} & A & B & C \\ \hline A & 0 & 4 & 2 \\ B & \infty & 0 & 1 \\ C & 2 & 1 & 0 \end{array}$$

,没有找到更短的路径,所以前驱矩阵也保持不变。

In this round the distance matrix does not change, no shorter path is found, so the precursor matrix remains the same.

借助 C 间接到达其他顶点。分析: 借助 C,A 借助 C 到 B 为 2+1=3<4(A 直接到 B)因此 3 替换 4,A 借助 C 到 C 仍是 2;B 借助 C 到 A 为 1+2=3 小于无穷(B 到 A),因此 3 替换掉无穷,B 到 C 是 1。

Indirectly reaching other vertexes with C. Analysis: with C, A with C to B is 2+1= 3<4 (A goes directly to B) so 3 replaces 4, and A with C to C is still 2; B by virtue of C to A being 1+2=3 is less than infinity (B to A), so 3 replaces infinity and B to C is 1.

在这一轮中距离矩阵变为

$$\begin{array}{c|ccc} & A & B & C \\ \hline A & 0 & 3 & 2 \\ B & 3 & 0 & 1 \\ C & 2 & 1 & 0 \end{array}$$

,找到了更短的路径,所以前驱矩阵变

为

$$\begin{array}{c|ccc} & A & B & C \\ \hline A & -1 & 2 & -1 \\ B & 2 & -1 & -1 \\ C & -1 & -1 & -1 \end{array}$$

。

In this round the distance matrix changes, and find a shorter path, so the precursor matrix changes.

3. 问题解答(Answer)

综上,得到最终的距离矩阵

$$\begin{array}{|c|ccc|}\hline & A & B & C \\ A & 0 & 3 & 2 \\ B & 3 & 0 & 1 \\ C & 2 & 1 & 0 \\ \hline\end{array}$$

和前驱节点矩阵

$$\begin{array}{|c|ccc|}\hline & A & B & C \\ A & -1 & 2 & -1 \\ B & 2 & -1 & -1 \\ C & -1 & -1 & -1 \\ \hline\end{array}$$

。

In summary, the final distance matrix and precursor vertex matrix are obtained.

4. 程序实现(Program Implementation)

上面实例的具体代码实现如下,实现结果如图 7.22 所示。

The specific code implementation of the above example is shown below, and the result is shown in Figure 7.22.

```
import numpy as np
#定义无穷大(Define infinity)
INF = float('inf')
#定义图的邻接矩阵(Define the adjacency matrix of the graph)
graph = np.array([[0, 4, 2],
                  [INF, 0, 1],
                  [2, 1, 0]])
#获取节点个数(Get the number of nodes)
n = len(graph)
#初始化距离矩阵和前驱节点矩阵(Initialize the distance matrix and the precursor node matrix)
distance_matrix = graph.copy()
predecessor_matrix = np.full((n, n), -1)
# Floyd-Warshall 算法(Floyd-Warshall algorithm)
for k in range(n):
    for i in range(n):
        for j in range(n):
            if distance_matrix[i][j] > distance_matrix[i][k] + distance_matrix[k][j]:
                distance_matrix[i][j] = distance_matrix[i][k] + distance_matrix[k][j]
                predecessor_matrix[i][j] = k
#打印最终的距离矩阵和前驱节点矩阵(Print the final distance matrix and precursor node matrix)
print("最终距离矩阵(Final distance matrix):")
print(distance_matrix)
print("\n前驱节点矩阵(Precursor node matrix):")
print(predecessor_matrix)
```

```
最终距离矩阵(Final distance matrix):
[[0. 3. 2.]
 [3. 0. 1.]
 [2. 1. 0.]]

前驱节点矩阵(Precursor node matrix):
[[-1  2 -1]
 [ 2 -1 -1]
 [-1 -1 -1]]
```

图 7.22 实现结果图

Figure 7.22 Implementation result diagram

习题 7(Exercises Seven)

1. 什么是图的邻接矩阵和邻接表？它们分别适用于什么样的场景？

What are the adjacency matrices and adjacency list of graphs？What kind of scenarios do they apply to？

2. 图的遍历算法有哪些？它们之间有何区别？

What are the traversal methods of the graph？What's the difference between them？

3. *Dijkstra* 算法的基本思想是什么？说出它的优缺点。

What is the basic idea of *Dijkstra* algorithm？Name its advantages and disadvantages.

4. 广度优先搜索算法描述如下，请填写空缺部分的语句。

The breadth first search algorithm is described below. Please fill in the blanks.

```
def dfs(graph, start, visited=None):
    if visited is None:
        _____
    print(start)
    visited.add(start)
    for neighbor in graph[start]:
        _____:
            dfs(graph, neighbor, visited)
```

5. 给定一个加权有向图，如图 7.23 所示，找到从起点 *A* 到所有其他节点的最短路径及其长度。请用 *Python* 代码实现。

Given a weighted directed graph, as shown in Figure 7.23, find the shortest path and its length from the starting vertex *A* to all other vertexes. Implement this in *Python* code.

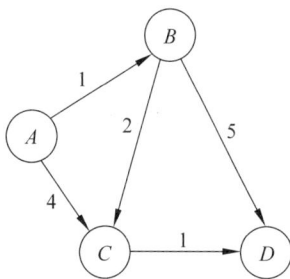

图 7.23 加权有向图

Figure 7.23 Weighted directed graph

第8章

概率分析和随机算法

Chapter 8　Probability Analysis and Stochastic Algorithm

概率分析是根据使用概率预测分析不确定因素和风险因素对项目经济效果的影响的一种定量分析方法。其实质是研究和计算各种影响因素的变化范围，以及在此范围内出现的概率和期望值。由于概率的原因所引起的实际价值与估计价值或预期价值之间的差异，通常称为风险性，因此概率分析也可称为风险分析。在项目评价中所用的概率是指各种基本变量(如投资、成本、收益等)出现的频率。其分析结果的可靠性很大程度上取决于每个变量概率值判断的准确性。

Probability analysis is a quantitative analysis method based on the use of probability prediction to analyze the impact of uncertain factors and risk factors on project economic results. Its essence is to study and calculate the variation range of various influence factors, as well as the probability and expected value of occurrence within this range. The difference between the actual value and the estimated value or the expected value caused by the probability is usually called risk, so the probability analysis can also be called risk analysis. The probability used in project evaluation refers to the frequency of occurrence of various basic variables (such as investment, cost, benefit, etc.). The reliability of its analysis results largely depends on the accuracy of the judgment of the probability value of each variable.

相较于确定性算法在执行过程中面临的多重选择情况，随机性选择通常能够比最优选择更加省时省力，因此随机算法可以在很大程度上降低算法的复杂度。

Compared with the multiple choices faced by the deterministic algorithm in the execution process, the stochastic choice can usually save more time and effort than the optimal choice, so the stochastic algorithm can largely reduce the algorithm complexity.

随机算法以概率分析为基础，基本特征是对所求解问题的同一实例用同一随机算法求解两次，可能得到完全不同的结果，并且其求解所需时间和结果可能会有很大的差异。由于随机的特性，随机算法结果往往是不可再现的，算法分析需要结合概率论、统计力和数论的知识。

Stochastic algorithms are based on probability analysis, and their basic characteristics are that if the same stochastic algorithm is used to solve the same instance of the problem twice, the results may be completely different, and the time required for solving the problem and the results may be very different. Due to the characteristics of randomness, the results of random algorithms are often not reproducible, and the analysis of algorithms needs to combine the knowledge of probability theory, statistical power, and number theory.

一般可以将随机算法分为数值随机化算法、蒙特卡洛算法、拉斯维加斯算法和舍伍德算法。

Generally, stochastic algorithms can be classified into numerical randomization algorithm, *Monte Carlo* algorithm, *Las Vegas* algorithm and *Sherwood* algorithm.

8.1 概率分析(Probability Analysis)

概率分析又称为风险分析,是通过研究各种不确定性因素发生不同变动幅度的概率分布及其对项目经济效益指标的影响,对项目可行性和风险性以及方案优劣做出判断的一种不确定性分析法。概率分析常用于对大中型重要若干项目的评估和决策之中。通过计算项目目标值(如净现值)的期望值及目标值大于或等于零的累积概率来测定项目风险大小,为投资者决策提供依据。

Probability analysis, also known as risk analysis, is an uncertainty analysis method to judge the feasibility and risk of the project and the pros of the program by studying the probability distribution of different changes of various uncertainty factors and their influence on the project economic benefit index. Probability analysis is often used in the evaluation and decision-making of large and medium-sized important projects. By calculating the expected value of the project target value (such as net present value) and the cumulative probability that the target value is greater than or equal to zero, the project risk is determined, which provides the basis for investors to make decisions.

8.1.1 概率分析方法(Probability Analysis Method)

概率分析通常将数据的期望值和标准差作为主要的结果指标。进行概率分析具体的方法主要有期望值法、效用函数法和模拟分析法等。

Probabilistic analysis usually takes the expected value and standard deviation of the data as the main outcome indicators. The specific methods of probability analysis mainly include expectation value method, utility function method and simulation analysis method.

1. 期望值法(Expectation Value Method)

期望值法在项目评估中应用最为广泛,是通过计算项目净现值的期望值和净现值大于或等于零时的累计概率,来比较方案优劣、确定项目可行性和风险程度的方法。

Expectation Value Method is the most popular application in project evaluation, it is a method to compare the pros and cons of the project, and determine the feasibility and risk degree of the project by calculating the expected value of the project and the cumulative probability when the net present value of the project is greater than or equal to zero.

2. 效用函数法（Utility Function Method）

所谓效用，是对总目标的效能价值或贡献大小的一种测度。在风险决策的情况下，可用效用来量化决策者对待风险的态度。通过效用这一指标，可将某些难以量化、有质的差别的事物（事件）给予量化，将要考虑的因素折合为效用值，得出各方案的综合效用值，再进行决策。

The so-called utility is a measure of the effectiveness value or contribution to the total goal. In the case of risk decision, utility can be used to quantify the decision maker's attitude towards risk. Through the index of utility, some things (events) that are difficult to quantify and have qualitative differences can be quantified, the factors to be considered can be converted into utility values, and the comprehensive utility value of each scheme can be obtained, and then the decision can be made.

效用函数反映决策者对待风险的态度。不同的决策者在不同的情况下，其效用函数是不同的。

The utility function reflects the maker's attitude towards the risk. Different decision-makers have different utility functions in different situations.

3. 模拟分析法（Simulation Analysis Method）

模拟分析法就是利用计算机模拟技术，对项目的不确定因素进行模拟，通过抽取服从项目不确定因素分布的随机数，计算分析项目经济效果评价指标，从而得出项目经济效果评价指标的概率分布，以提供项目不确定因素对项目经济指标影响的全面情况。

Simulation analysis method is the use of computer simulation technology to simulate the uncertain factors of the project, by extracting random numbers subject to the distribution of project uncertainty factors, calculate and analyze the project economic effect evaluation index, so as to obtain the probability distribution of the project economic effect evaluation index, in order to provide a comprehensive situation of the impact of project uncertainty factors on the project economic index.

8.1.2 分析步骤（Analysis Procedure）

期望值法是通过计算项目净现值的期望值和净现值大于或等于零时的累计概率，来比较方案优劣、确定项目可行性和风险程度的方法。一般步骤如下。

The expectation value method is a method to compare the pros and cons of the project, and determine the feasibility and risk degree of the project by calculating the expected value and the cumulative probability of the net present value of the project is greater than or equal to zero. The general steps are as follows.

（1）选用净现值作为分析对象，并分析选定与之有关的主要不确定性因素。

Choose net present value as the analysis object, and analyze and select the main uncertainty factors related to it.

（2）按照穷举互斥原则，确定各不确定性因素可能发生的状态或变化范围。

According to the principle of exhaustion and mutual exclusion, determine the possible state or change range of each uncertainty factor.

（3）分别估算各不确定性因素每种情况下发生的概率。

Estimate the probability of each uncertainty factor in each case.

（4）计算第 t 年各可能发生情况下的净现值。

Calculate the net present value of each possible scenario in year t.

$$E(NPV_t) = \sum_{i=1}^{n} X_{it} P_{it} \tag{8-1}$$

其中，X_{it} 为第 t 年第 i 种情况下的净现值；P_{it} 为第 t 年第 i 种情况发生的概率，n 为发生的状态或变化范围数。

Where, X_{it} is the net present value under the i case in year t; P_{it} is the probability of the i case occurring in year t, and n is the number of states or ranges of changes occurring.

整个项目寿命周期净现值的期望值的计算公式如下。

The expected value of the net present value over the life cycle of the project is calculated as follows.

$$E(NPV) = \sum_{i=1}^{m} \frac{E(NPV_t)}{(1+i)^t} \tag{8-2}$$

其中，i 为折旧率，m 为项目寿命周期长度。项目净现值期望值大于零，则项目可行，否则，不可行。

Where, i is the depreciation rate and m is the life cycle length of project. If the expected net present value of the project is greater than zero, the project is feasible; otherwise, it is not feasible.

（5）各年净现值标准差的计算公式如下。

The standard deviation of net present value for each year is calculated as follows.

$$\delta_t = \sqrt{\sum_{i=1}^{n} [X_{it} - E(NPV_t)]^2 P_{ut}} \tag{8-3}$$

整个项目寿命周期的标准差计算公式如下。

The standard deviation of the entire project life cycle is calculated as follows.

$$\delta = \sqrt{\sum_{i=1}^{m} \frac{\delta_t^2}{(1+i)^t}} \tag{8-4}$$

整个项目寿命周期的标准差系数如下。

Standard deviation coefficient for the entire project life cycle is as follows.

$$V = \frac{\delta}{E(NPV)} \times 100\% \tag{8-5}$$

一般地，V 越小，项目的相对风险就越小，反之，项目的相对风险就越大。

In general, the smaller the V, the smaller the relative risk of the project, and vice versa, the greater the relative risk of the project.

（6）计算净现值大于或等于零时的累计概率。累计概率值越大，项目所承担的风险就越小。

Calculate the cumulative probability when the net present value is greater than or equal to zero. The greater the cumulative probability value, the less risk the project takes.

（7）分析结果做综合评价，说明项目是否可行及承担风险性大小。

The analysis results are comprehensively evaluated to show whether the project is feasible and the degree of risk borne.

8.1.3　实例演示（Example Demonstration）

某投资者以 25 万元购买了一个商铺两年的经营权，第一年净现金流量可能为 22 万元、18 万元、14 万元，概率分别为 0.2、0.6、0.2；第二年净现金流量可能为 28 万元、22 万元、16 万元，概率分别为 0.15、0.7、0.15。若折旧率为 10%，问该商铺的投资是否可行。

An investor buys a two-year operation right of a shop with 250,000 yuan, the net cash flow in the first year may be 220,000 yuan, 180,000 yuan, 140,000 yuan, with probabilities of 0.2, 0.6, 0.2 respectively. The second year net cash flow may be 280,000 yuan, 220,000 yuan, 160,000 yuan, the probability is 0.15, 0.7, 0.15. If the depreciation rate is 10%, ask whether the investment of the shop is feasible.

数学解法。（Mathematical Solution）

$$E(NPV_1) = 22 \times 0.2 + 18 \times 0.6 + 14 \times 0.2 = 18$$

$$E(NPV_2) = 28 \times 0.15 + 22 \times 0.7 + 16 \times 0.5 = 22$$

$$E(NPV) = \frac{E(NPV_1)}{(1+i)} + \frac{E(NPV_2)}{(1+i)^2} - 25 = 9.54$$

$$\delta_1 = 2.530 \tag{8-6}$$

$$\delta_2 = 3.286$$

$$\delta = 3.840$$

$$V = \frac{\delta}{E(NPV)} \times 100\% = 40.25\%$$

计算机解法运行结果如图 8.1 所示，与数学计算结果符合，*Python* 代码如下。

The operation result of the computer solution is shown in Figure 8.1, which is consistent with the mathematical calculation result. The *Python* code is as follows.

```python
import random
import math
import numpy as np
def NPV1():    #定义第一年随机收益函数(Define the first year random payoff function)
    pp = random.random()
    if pp <= 0.2:
        NP1 = 22
    elif pp <= 0.8:
        NP1 = 18
    else:
```

续

```
        NP1 = 14
    return NP1
def NPV2():  #定义第二年随机收益函数(Define the random return function of the second year)
    pp = random.random()
    if pp <=0.15:
        NP2 = 28
    elif pp <=0.85:
        NP2 = 22
    else:
        NP2 = 16
    return NP2
if __name__ == "__main__":  #主函数(Main function)
    invest = 25
    repn = 10000  #定义重复实验次数(Define the number of repeated experiments)
    rep1 = []
    rep2 = []
    rep = []
    i = 1.1
    for t in range(0, repn):
        rep1.append(NPV1())
        rep2.append(NPV2())
        rep.append(NPV1()/i+NPV2()/i/i)
    E_NPV1 = np.mean(rep1)
    thet1 = np.std(rep1)
    E_NPV2 = np.mean(rep2)
    thet2 = np.std(rep2)
    E_NPV = np.mean(rep) - invest
    thet = math.sqrt(thet1 **2/i+thet2 **2/i/i)
print("E(NPV1)=", E_NPV1)
print("标准偏差1(Standard deviation 1) = ", thet1)
print("E(NPV2)=", E_NPV2)
print("标准偏差2(Standard deviation 2) = ", thet2)
print("E(NPV)=", E_NPV)
print("标准偏差(Standard deviation ) = ", thet)
print("V=", thet/E_NPV * 100, "%")
```

```
E(NPV1)= 18.0088
标准偏差1(Standard deviation 1) = 2.5354925675300253
E(NPV2)= 21.9304
标准偏差2(Standard deviation 2) = 3.2921658281441415
E(NPV)= 9.507107438016533
标准偏差(Standard deviation ) = 3.8472863309606544
V= 40.467475055307816 %
```

图 8.1 投资收益随机实验运行结果

Figure 8.1　Result of the randomized experiment on investment return

8.2　随机方法(Stochastic Method)

8.2.1　随机抽样(Random Sampling)

在根据样本资料推论总体时,可用概率的方式客观地测量推论值的可靠程度,从而使这种推论建立在科学的基础上。因此，随机抽样在社会调查和社会研究中应用较广泛。常用的随机抽样方法主要有纯随机抽样、分层抽样、系统抽样、整群抽样、多阶段抽样等。

When inferring the total from sample data, the reliability of the inferring value can be measured objectively in the way of probability, so that such inferences are based on science. Therefore, random sampling is widely used in social investigation and social research. The commonly used random sampling methods include pure random sampling, stratified sampling, systematic sampling, cluster sampling, multi-stage sampling and so on.

随机抽样的基本特点有随机原则抽样、部分调查推断总体、事先计算误差并适当控制。由此，随机抽样的最主要优点是能够根据样本统计估计总体，同时计算误差，以得到可靠的总体结论。但是相对地，抽样调查相较于直接计算，在时间成本和空间成本上不可避免地增加。

The basic characteristics of random sampling are as follows: random principle sampling, partial investigation to infer the whole, calculation of errors in advance and appropriate control. Thus, the most important advantage of random sampling is the ability to estimate the whole from the sample statistics and simultaneously calculate the error to obtain reliable whole conclusions. However, compared with direct calculation, sampling survey inevitably increases in time cost and space cost.

1. 单纯随机抽样（Simple Random Sampling）

单纯随机抽样又称为简单随机抽样，是最基本的抽样方法，根据抽样后是否放回分为重复抽样和不重复抽样。纯随机抽样方法有抽签法、随机数字表法等。虽然简单随机抽样简单有效，但是在总体太大等各类情况下，难以得到准确有效的样本，故在大规模社会调查中很少采用纯随机抽样。

Simple random sampling is the most basic sampling method. It is divided into repeated sampling and non-repeated sampling according to whether it is put back after sampling. Pure random sampling methods include lottery method, random number table method and so on. Although simple random sampling is simple and effective, it is difficult to obtain accurate and effective samples in various situations such as the population is too large, so pure random sampling is rarely used in large-scale social surveys.

2. 分层抽样（Stratified Sampling）

先按照一定规则特征，将总体分为若干子层并各自单独抽样，最后合起来。各层样本数的确定有分层定比、奈曼法、非比例分配法。分层的原则是增加层内的同质性和层间的异质性。分层随机抽样在实际抽样调查中广泛使用，在同样样本容量的情况下，它比纯随机抽样的精度高，管理方便且效率高。

Firstly, according to certain rule characteristics, the total is divided into several sub-layers and sampled separately, and finally combined. The sample number of each layer can be determined by stratified ratio, *Neiman* method and non-proportional distribution method. The principle of layering is to increase homogeneity within layers and heterogeneity between layers. Stratified random sampling is widely used in actual sampling survey. It has higher precision, convenient management, and higher validity than pure random sampling under the same sample size.

3. 系统抽样(Systematic Sampling)

系统抽样又称为等距抽样,是纯随机抽样的变种,用总体单位数与样本容量相除,以确定分组的组内数量和抽样间隔。在第一组中随机抽取,并以间隔数在总体中逐个提取,作为样本。

Systematic sampling, also known as isometric sampling, is a variant of pure random sampling in which the number of total units is divided by the sample size to determine the number of groups and sampling intervals. Random sampling in the first group, and the number of intervals in the total one by one, as a sample.

4. 整群抽样(Cluster Sampling)

整群抽样又称为聚类抽样,先按照某种标准将总体分为若干个群体,然后从中随机抽取样本群体,抽中的群体中所有单位都作为样本。整体抽样要求群间异质性小,群内异质性大,只需列出入样群的单位,因此可节约大量财力、人力。

Cluster sampling divides the total into several groups according to a certain standard, and then randomly selects sample groups from which all units are taken as samples. Overall sampling requires small inter-group heterogeneity and large intra-group heterogeneity, and only the units in the sample group need to be listed, so it can save a lot of financial resources and manpower.

5. 多阶段抽样(Multi-Stage Sampling)

多阶段抽样又称为多级抽样,将抽样过程分为几个阶段,结合使用上述方法中的两种或数种。当研究总体广泛且分散时,多采用多阶段抽样,以降低调查费用。但由于每级抽样都会产生误差,经过多级抽样产生的样本,误差也相应增大。

Multi-stage sampling divides the sampling process into several stages, using two or more of the above methods in combination. When the research is extensive and scattered, multi-stage sampling is used to reduce the investigation cost. However, since each stage of sampling will produce errors, the errors of the samples generated by multi-stage sampling will increase correspondingly.

8.2.2 四种常见的随机算法(Four Common Stochastic Algorithms)

随机算法的原理与随机抽样同源,以概率为核心,按照规则重复实验,获取样本,并嵌入具体的实例或确定性算法中,改善算法性能。

The principle of stochastic algorithm is the same as random sampling. With probability as the core, experiments are repeated according to rules to obtain samples and embed them into specific instances or deterministic algorithms to improve the performance of the algorithm.

1. 数值随机化算法（Numerical Randomization Algorithm）

数值随机化算法常用于数值问题的求解，所得到的解常常都是近似解，且近似解的精度随计算时间的增加而不断提高。在许多情况下，待求解的问题在原理上可能不存在精确解，或者说精确解存在但无法在可行时间内求得，因此使用数值随机化算法可得到非常不错的解。

Numerical randomization algorithm is often used to solve numerical problems, and the obtained solutions are often approximate solutions, and the accuracy of approximate solutions increases with the increase of computing time. In many cases, the problem to be solved may not have an exact solution in principle, or the exact solution exists but cannot be obtained in feasible time, so this algorithm can be used to get a very good solution.

2. 蒙特卡洛算法（Monte Carlo Algorithm）

蒙特卡洛算法是计算数学中的一种计算方法，它的基本特点是以概率与统计学中的理论和方法为基础，以是否适合于在计算机上使用为重要标志。蒙特卡洛是摩纳哥的一个著名城市，以赌博闻名于世，这与蒙特卡洛算法的特点相契合。

Monte Carlo algorithm is a calculation method in computational mathematics, its basic characteristics are based on the theories and methods in probability and statistics, and whether it is suitable for use in computers is an important symbol. *Monte Carlo* is a famous city in Monaco, which is famous for gambling, which fits with the characteristics of *Monte Carlo* algorithm.

对于许多问题来说，近似解是毫无意义的，而蒙特卡洛算法是用于求解问题的准确解，其基本特点是能够求得问题的一个解，但是未必是正确的。求得正确解的概率与算法执行所用的时间相关。但是一般情况下不能有效地确定求得的解是否正确。

For many problems, the approximate solution is meaningless, and the *Monte Carlo* algorithm is used to solve the exact solution of the problem. Its basic characteristic is that it can find a solution to the problem, but it may not be correct. The probability of finding the correct solution is related to the time taken by the algorithm to execute. But in general, it cannot be effectively determined whether the obtained solution is correct or not.

3. 拉斯维加斯算法（Las Vegas Algorithm）

拉斯维加斯算法不会得到不正确的解，但是也可能找不到解。拉斯维加斯算法得到正确解的概率也随着算法执行时间的增加而提高。对于任一实例，只要用同一个拉斯维加斯算法反复求解足够多的次数，就可以使求解失败的概率任意小（无限接近于0）。

The *Las Vegas* algorithm will not get an incorrect solution, but it may not find a solution either. The probability of the *Las Vegas* algorithm getting the correct solution also increases with the increase of the execution time of the algorithm. For any instance, as long as the same *Las Vegas* algorithm is repeated enough times, the probability of failure can be arbitrarily small (infinitely close to 0).

4. 舍伍德算法（Sherwood Algorithm）

一个确定性算法在最坏情况下的时间复杂度与其他所有情况下的平均复杂度有较大差异时，可以在其中引入随机性来降低最坏情况出现的概率，进而消除或减少好坏实例之间的差异，这样的随机算法称为舍伍德算法。

When the time complexity of a deterministic algorithm in the worst case is quite different from the average complexity of all other cases, randomness can be introduced to reduce the probability of the worst case, and then eliminate or reduce the difference between good and bad instances. Such a random algorithm is called *Sherwood* algorithm.

舍伍德算法不会改变对应确定性算法的求解结果，每次运行都能够得到问题的正确解，其核心是为了避免算法最坏情况的发生以降低概率。因此，舍伍德算法不改变原有算法的平均性能，而是设法保证以更高的概率获得算法的平均计算性能。

Sherwood algorithm does not change the solution result of the corresponding deterministic algorithm, and can get the correct solution of the problem every time it runs. The core of Sherwood algorithm is to avoid the worst case of the algorithm and reduce the probability. Therefore, the *Sherwood* algorithm does not change the average performance of the original algorithm, but tries to ensure that the average computing performance of the algorithm is obtained with a higher probability.

8.2.3 随机数发生器（Random Number Generator）

1. 伪随机数（Pseudo Random Number）

随机数在概率算法设计中扮演着十分重要的角色。在现实计算机上无法产生真正的随机数，因此，在概率算法中使用的随机数都是伪随机数。线性同余法是产生伪随机数的最常用的方法，其产生的随机序列 $a_1, a_2, \cdots, a_n$ 满足下面的公式。

Random numbers play an important role in the design of probabilistic algorithms. Real random numbers cannot be generated on real computers, so the random numbers used in probabilistic algorithms are pseudorandom numbers. Linear congruence method is the most used method to generate pseudorandom numbers, and the random sequence $a_1, a_2, \cdots, a_n$ generated by it satisfies the following formula.

$$\begin{cases} a_0 = d \\ a_n = (ba_{n-1} + c)\bmod m, \ n = 1,2,\cdots \end{cases} \tag{8-7}$$

其中，d 称为随机序列种子，$b \geqslant 0$；c 为增量，$c \geqslant 0$；m 为模数，应当满足 $m>0$。b、c 和 m 取值越大，且 b 与 m 的互质可以使随机性能更好。需要注意的是，对于确定 b、c 和 m 的一个随机数发生器，相同的随机种子会产生相同的序列，这就是伪随机数。

Where d is a random sequence seed, $b \geqslant 0$; c is increment, $c \geqslant 0$; m is the modulus and must meet $m>0$. The larger the values of b, c, and m, and the prime of b and m can make the random performance better. It should be noted that for a random number generator that determines b, c, and m, the same random seed will produce the same sequence, which is the pseudorandom number.

2. 程序实现（Program Implementation）

用 *Python* 实现随机数生成器。每次计算时，利用线性同余计算出新的种子 *d*，作为产生下一个随机数的种子。程序运行结果如图 8.2 所示。

Implementing a random number generator in *Python*. At each calculation, a new seed *d* is calculated using linear congruence as the seed to produce the next random number. The result of the program is shown in Figure 8.2.

```
import time
class RandomNumber：
    m = 65536
    b = 1194211693
    c = 12345
    _ _d = 0 #d 为当前种子(d is the current seed)
    def _ _init_ _(self, s=0)：            #默认值 0 表示由系统自动产生种子(The default value 0
                                          #indicates that the system automatically generates seeds)
        if(s ==0)：
            self.d = time.time( )
        else：
            self.d = s
    def random1(self, n = 10)：           #产生 0：n-1 之间的随机整数(Generates a random integer between 0：n-1)
        self.d = int(self.b * self.d + self.c)   #线性同余法计算新的种子(New seeds are computed by
                                          #linear congruence method)
        return (self.d >> 16) % n
    def fRandom(self)：                   #产生[0, 1)之间的随机实数(Produces a random real number between [0, 1))
        return self.random1(self.m)/self.m
a = RandomNumber( )
b = a.random1(10)
a.fRandom( )
print("b=", b, "a.fRandom( )= =", a.fRandom( ))
```

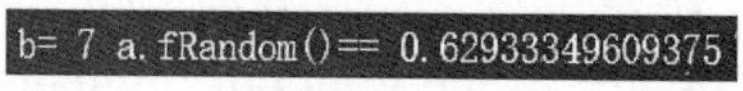

b= 7 a. fRandom()== 0. 62933349609375

图 8.2　随机数生成器的运行结果

Figure 8.2　Result of the random number generator

注意在 *Python* 实现其他的算法时，*random* 类包中已经集成了平均分布、正态分布等常用的随机数生成函数，可以直接调用而无须自行编写。

Note that when *Python* implements other algorithms, the *random* class package has integrated common random number generation functions such as average distribution and Gaussian distribution, it can be called directly without writing it yourself.

8.3　数值随机化算法（Numerical Randomization Algorithm）

8.3.1　计算圆周率（Calculate Circular Constant）

将 *n* 个点随机投入一个边长为 2*r* 的二维正方形空间内，假设任一点的落点概率是相同

的,若此时该正方形内切圆(半径 r)中落点的数目为 k,如图 8.3 所示,当点足够多时,则有

$$\frac{k}{n} = \frac{(2r)^2}{\pi r^2},\ \text{即}\ \pi = \frac{4k}{n}。$$

If n points are randomly put into a two-dimensional square space with a side length of $2r$, assuming that the probability of landing at any point is the same, if the number of landing points in the tangent circle (radius r) of the square is k, as shown in Figure 8.3, when there are enough points, there is $\frac{k}{n} = \frac{(2r)^2}{\pi r^2}$, then $\pi = \frac{4k}{n}$.

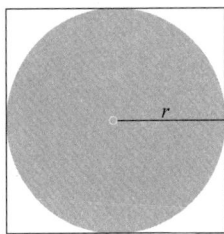

图 8.3 正方形内切圆

Figure 8.3 Inner circle of a square

设计使用随机投点计算,在二维平面的横、纵坐标分别生成范围内的随机数表示即可。重复若干次随机点的实验,统计落点情况计算。

Uses random drop point calculation to generate random number representation in the range of horizontal and vertical coordinates of two-dimensional plane respectively. The experiment of random points is repeated several times to calculate the statistical landing point.

用 *Python* 实现需要导入 *random* 类包,这是可以直接使用的随机数生成的函数。定义投点模拟实验和主控制函数,最后输出结果。实验结果如图 8.4 所示,程序代码如下。

Python implementations need to import the *random* package, which is a random number generation function that can be used directly. Define the drop point simulation experiment and the main control function, and finally output the result. The experimental result is shown in Figure 8.4, and the program code is as follows.

```
import random
def Darts(n):
    k = 0  #记录落入四分之一圆内的点数(Record the points that fall into the quarter park)
    for i in range(n):
        x = random.random()    #产生一个[0,1)之间的实数,赋给 x(Produces a real number between [0,1),
                               #assigned to x)
        y = random.random()    #产生一个[0,1)之间的实数,赋给 y(Produces a real number between [0,1),
                               #assigned to y)
        if x **2 + y **2 <= 1:
            k += 1
    return 4 * k / n
if __name__ == "__main__":
    pi = Darts(100000)
    print("pi=" + str(pi))
```

`pi=3.14368`

图 8.4　计算圆周率程序运行结果

Figure 8.4　Result of pi calculation program

8.3.2　计算定积分（Compute Definite Integral）

设函数 $f(x)$ 是定义域为 $[0,1]$ 的连续函数,且函数值为 $0 \leqslant f(x) \leqslant 1$,需要计算定积分 $I = \int_0^1 f(x)\,\mathrm{d}x$,即为图 8.5 中阴影部分的面积。

Let the function $f(x)$ be a continuous function whose domain is $[0,1]$ and value is $0 \leqslant f(x) \leqslant 1$, and need to compute the definite integral $I = \int_0^1 f(x)\,\mathrm{d}x$, which is the area of the shaded part in Figure 8.5.

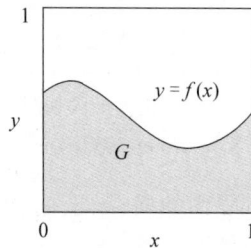

图 8.5　定积分面积关系

Figure 8.5　Relation of constant integral area

采用随机投点的测定策略,落点在阴影区域的概率如下。

Using the random drop point measurement strategy, the probability of the drop point in the shadow region is as follows.

$$p = \int_0^1 \int_0^{f(x)} \mathrm{d}y\,\mathrm{d}x = \int_0^1 f(x)\,\mathrm{d}x = I \tag{8-8}$$

因此,向正方形区域内投 n 个点,并统计阴影区域的落点 m,那么定积分计算结果为 $I \approx p = m/n$。

Therefore, casting n points into the square area and counting the drop point m of the shaded area, then the definite integral is calculated as $I \approx p = m/n$。

用横、纵坐标分别生成范围内的随机数来投点,需要导入 *random* 依赖包以生成随机数,以及 *math* 包用于数学计算。定义函数 $f(x)$ 和随机投点模拟实验,并设置入口 *main*( ) 函数,调用定义函数,输出打印结果。代码如下所示。程序结果如图 8.6 所示。

Random numbers in the range of horizontal and vertical coordinates are generated to drop points, and the *random* dependency package needs to be imported to generate random numbers, and the *math* package is used for mathematical calculations. Define function $f(x)$ and random drop point simulation experiment, and set the entry *main*( ) function, call the defined function, and output the printed result. The program result is shown in Figure 8.6, and the code is shown below.

```
import random
import math
def f(x):
    return abs(math.sin(x))
def Definite_integral(n):
    k = 0#记录落入四分之一圆内的点数(Record the number of points that fall into the quarter circle)
    for i in range(n):
        x = random.random()    #产生一个[0, 1]之间的实数,赋给x(Produces a real number between [0, 1],
                               #assigned to x)
        y = random.random()    #产生一个[0, 1]之间的实数,赋给y(Produces a real number between [0, 1],
                               #assigned to y)
        if y<f(x):
            k += 1
    return k/n
if __name__ == "__main__":
    s = Definite_integral(10000)
    print("I=" + str(s))
```

I=0.4534

图 8.6　定积分计算结果

Figure 8.6　Result of definite integral

8.3.3　解非线性方程组(Solving Nonlinear Equations)

1. 问题描述(Problem Description)

解非线性方程组最常用的解决方法有线性化方法和求函数极小值方法,但是容易在计算中出现各类问题而使方法失效。虽然概率算法耗时较多,但是其设计思想简单,易于实现,能够适应一定的精度要求,在实际使用中是比较有效的。

The most used methods for solving nonlinear equations are linearization method and function minimum value method, but it is easy to cause all kinds of problems in the calculation and make the method invalid. Although the probabilistic algorithm is more time-consuming, its design idea is simple, easy to implement, and can meet certain accuracy requirements, so it is more effective in practical use.

设有一组实变量 $x_1, x_2, \cdots, x_n$,其非线性实函数 $f_i(i=1,2,\cdots,n)$ 构成一组非线性方程组,如式(8-9),求一组指定范围内的解 $x_1^*, x_2^*, \cdots, x_n^*$。

There is a set of real variables x_1, x_2, $\cdots$, x_n, whose nonlinear real functions $f_i(i=1, 2, \cdots, n)$ constitute a set of nonlinear equations, as shown in formula (8-9). Please find a set of solutions within a specified range x_1^*, x_2^*, $\cdots$, x_n^*.

$$\begin{cases} f_1(x_1,x_2,\cdots,x_n) = 0 \\ f_2(x_1,x_2,\cdots,x_n) = 0 \\ \cdots \\ f_n(x_1,x_2,\cdots,x_n) = 0 \end{cases} \tag{8-9}$$

构造函数 $\Phi(x) = \sum_{i=1}^{n} f_i^2(x)$，其中，$x = (x_1, x_2, \cdots, x_n)$，则函数 $\Phi(x)$ 的零点即为所求非线性方程组的一组解。

Constructor function $\Phi(x) = \sum_{i=1}^{n} f_i^2(x)$, among $x = (x_1, x_2, \cdots, x_n)$ then the zero of the function $\Phi(x)$ is a set of solutions to the nonlinear equations.

2. 简单随机模拟算法（Simple Stochastic Simulation Algorithm）

在指定求根区域内，选定一个初始根值 x_0，然后按照预先选定的分布策略，如以 x_0 为中心的正态分布、均匀分布、三角分布等，逐个选取随机点 x。多次选取随机点，代入函数 $\Phi(x)$，并把满足精度要求的点作为近似解。

Within the specified root finding area, an initial root value x_0 is selected, and then select random points x one by one according to the pre-selected distribution strategy, such as normal distribution, uniform distribution or triangular distribution centered on x_0. Select random points several times, substitute into the function $\Phi(x)$ and take the points that meet the precision requirements as the approximate solution.

这种方法简单直观，但是工作量较大。

This method is simple and intuitive, but it requires a lot of work.

3. 随机搜索（Random Search）

在指定的求根区域 D 内，选定一个随机点 x_0 作为搜索的出发点。搜索规则为，设当前搜索点为 x_j，首先计算出搜索方向 r 和搜索步长 a，由此得到随机搜索增量 Δx_j，则下一步的随机搜索点为 $x_{j+1} = x_j + \Delta x_j$。当 $\Phi(x_{j+1}) < \varepsilon$ 时，取到近似解，否则进行下一步搜索。

Within the specified root finding region D, a random point x_0 is selected as the starting point of the search. The search rule is, let the current search point x_j, be the first to calculate the search direction r and search step a, to obtaining the random search increment Δx_j, then the next random search point is $x_{j+1} = x_j + \Delta x_j$. When $\Phi(x_{j+1}) < \varepsilon$, get the approximate solution, otherwise proceed to the next search.

4. 代码示例（Example Code）

简单的示例运行结果如图 8.7 所示，代码如下。

The result of a simple example implementation is shown in Figure 8.7, with the following code.

```python
import random
import math
#定义要解的非线性方程组(Define the nonlinear system of equations to be solved)
def equations(x, y):
    eq1 = x **2 + y **2 -25
    eq2 = x **2 -y -1
    return eq1, eq2
#简单随机模拟算法(Simple stochastic simulation algorithm)
def simple_random_simulation():
    x = random.uniform(-10, 10)
    y = random.uniform(-10, 10)
```

续

```
    T = 1.0
    T_min = 0.00001
    alpha = 0.9

    while T > T_min:
        i = 1
        while i <= 100:
            eq1, eq2 = equations(x, y)
            E = abs(eq1) + abs(eq2)
            x_new = x + random.uniform(-0.1, 0.1)
            y_new = y + random.uniform(-0.1, 0.1)
            eq1_new, eq2_new = equations(x_new, y_new)
            E_new = abs(eq1_new) + abs(eq2_new)
            if E_new < E:
                x = x_new
                y = y_new
            else:
                if random.random() < math.exp((E - E_new) / T):
                    x = x_new
                    y = y_new
            i += 1
        T = T * alpha
    return x, y
if __name__ == "__main__":
    result_x, result_y = simple_random_simulation()
    print("solution: x = ", result_x, "y = ", result_y)
```

```
solution: x = 2.330725196424911 y = 4.42345874367702
```

图 8.7　解非线性方程组运行结果

Figure 8.7　Results of solving nonlinear equations

8.4　蒙特卡洛算法(Monte Carlo Algorithm)

8.4.1　基本思想(Basic Idea)

设 p 是一个 $0.5\sim1$ 的实数,若蒙特卡洛算法对于问题的任一实例能够得到的正确解的概率不小于 p,则称算法是 p 正确的。对于同一实例,蒙特卡洛算法不会给出两个不同的正确解,即算法的解是一致的。因此,对于一个一致的 p 正确的蒙特卡洛算法,要提高获得正确解的概率,只需要执行算法若干次,从中选择出现频率最高的解即可。

Let p be a real number between 0.5 and 1, and the *Monte Carlo* algorithm is said to be p-correct if the probability that it can obtain the correct solution for any instance of the problem is not less than p. For the same instance, the *Monte Carlo* algorithm will not give two different correct solutions, that is, the algorithm is consistent. Therefore, for a uniformly p-correct *Monte Carlo* algorithm, to increase the probability of obtaining the correct solution, it is only necessary to execute the algorithm several times and select the solution with the highest frequency.

为使得正确解的概率不低于 $1-\varepsilon(0<\varepsilon\leqslant1-p)$，至少需要调用 x 次，即有 $p+(1-p)p+(1-p)^2p+\cdots+(1-p)^{x-1}p\geqslant1-\varepsilon$，可以得出 $1-(1-p)^x\geqslant1-\varepsilon$，则 $x\geqslant\log_{1-p}\varepsilon$，即 $x\geqslant\left\lceil\dfrac{\log_2\varepsilon}{\log_2(1-p)}\right\rceil$。理论上，无论 ε 取值多小，都能够用多次调用的方法获得结果。

In order to make the probability of the correct solution not less than $1-\varepsilon(0<\varepsilon\leqslant1-p)$, it's need to be called at least x times, that is, there is $p+(1-p)p+(1-p)^2p+\cdots+(1-p)^{x-1}p\geqslant1-\varepsilon$, can be obtained $1-(1-p)^x\geqslant1-\varepsilon$, then $x\geqslant\log_{1-p}\varepsilon$, i.e. $x\geqslant\left\lceil\dfrac{\log_2\varepsilon}{\log_2(1-p)}\right\rceil$. In theory, no matter how small the ε value is, the result can be obtained with multiple calls.

8.4.2　主元素问题（Principal Element Problem）

设 $T[1{:}n]$ 是一个含有 n 个元素的数组。当元素 x 在数组中的出现次数超过一半时，即 $|\{i\,|\,T[i]=x\}|>n/2$，则称元素 x 为数组 T 的主元素。则 T 数组中最多只有一个主元素。

Let $T[1{:}n]$ be an array of n elements. When element x occurs more than half of the time in the array, that is $|\{i\,|\,T[i]=x\}|>n/2$, element x is said to be the principal element of array T. There is at most one principal element in the T-array.

算法设计主要分为蒙特卡洛算法和重复调用控制。蒙特卡洛算法在给定的数组中随机选择一个元素，并逐个对比所有元素，统计"相等"的次数是否超过一半，输出结果的真假。重复调用控制算法根据预设的概率阈值，在有限次的调用过程中能否达到要求。

The algorithm design is mainly divided into *Monte Carlo* algorithm and repeated call control. The algorithm randomly selects an element in a given array, and compares all elements one by one, counting whether the number of " equal" is more than half, and the output result is true or false. According to the preset probability threshold, the repeated call control algorithm can reach the requirement in the process of finite calls.

现在用 *Python* 实现算法，首先调用 *random* 包用于产生随机数，*math* 包用于数学计算使用。然后定义蒙特卡洛算法函数和重复调用算法函数（其中包含蒙特卡洛算法的调用）。主程序中设计输入实例和可信度要求，调用重复控制算法，打印输出结果。程序结果如图 8.8 所示，代码算法如下。

To implement the algorithm in *Python*, the *random* package is first called to generate random numbers, and the *math* package is used for mathematical calculations. Then define the *Monte Carlo* algorithm function and the repeated call algorithm function (which contains the *Monte Carlo* algorithm call). In the main program, the input example and reliability requirements are designed, the repetitive control algorithm is called, and printed output results. The program results are shown in Figure 8.8, and the algorithm code is as follows.

```
import random
import math
def majority(T):                #判定主元素的蒙特卡洛算法(Monte Carlo algorithm for determining principal elements)
    global p
    n = len(T)
    i=random.randint(0, n-1)        #产生 1~n 的随机下标(Generates random subscripts between 1 and n)
    x=T[i]                          #随机选择元素(Randomly selected element)
    k=0
    for j in range(n):
        if T[j]==x:
            k +=1
    p = k/n
    return p>0.5 #当 p>0.5 时,T 含有主元素(When p≥0.5, T contains the main element)
def majorityMC(T, threshold):
    #重复调用算法 majority(Call the majority algorithm repeatedly)
    result1 = majority(T)
    if result1:
        return True
    else:
        k=int(math.ceil(math.log2(threshold)/math.log2(1-p)))
        for i in range(1, k):
            if (majority(T)):
                return True
        return False

if __name__=="__main__":
    T = [1, 5, 5, 6, 3, 2, 5, 5, 5, 6, 2, 5, 5, 5, 5, 5, 5, 5, 5]
    T1 = [1, 5, 6, 6, 3, 2, 5, 6, 5, 6, 2, 6, 5, 5, 6, 5, 6, 5, 6]
    p = 0
    resultT = majorityMC(T, 0.01)
    resultT1 = majorityMC(T1, 0.01)
    print("T 中是否有主元素？结果为:", resultT)
    print("T1 中是否有主元素？结果为:", resultT1)
    print("Is there a principal element in T? result:", resultT)
    print("Is there a principal element in T1? result:", resultT1)
```

```
T中是否有主元素？，结果为： True
T1中是否有主元素？，结果为： False
Is there a principal element in T？，result： True
Is there a principal element in T1？，result： False
```

图 8.8　主元素问题运行结果

Figure 8.8　Results of principal element problem

8.4.3　素数测试(Prime Number Test)

素数在密码学中有很大的作用,而素数测试又是素数研究中的一个重要课题。

Prime numbers play an important role in cryptography, and prime number testing is an important subject in the study of prime numbers.

1. 试除法（Trial Division Method）

根据素数的定义，最简便的方式是试除法，用 $2,3,\cdots,\sqrt{n}$ 去除 n，判断能否被整除。如果能，即存在除了 1 和它本身的因素，则该数为合数；否则，为素数。

According to the definition of prime numbers, the easiest way is to try division, divide n by 2, 3, $\cdots$, $\sqrt{n}$, to see if it can be divided exactly. If it can, that is, there are factors other than 1 and itself, then the number is composite; otherwise, it is a prime.

该方法在确定素数或合数时，能够同时确定合数因子分解。但是，试除法的时效取决于 n，当 n 很大且没有较小的因子时，效率很低。

This method can also determine the factorization of composite numbers when determining prime or composite. However, the timeliness of trial division depends on n, which is inefficient when n is large and there are no smaller factors.

2. Wilson 定理（Wilson Theorem）

对于给定的正整数 n，判断 n 是素数的充要条件是 $(n-1)! \equiv -1 \pmod{n}$。

For a given positive integer n, the necessary and sufficient condition for determining that n is a prime number is $(n-1)! \equiv -1 \pmod{n}$.

根据此原理设计的算法，需要逐次计算 n 的 mod 关系，计算量较大，无法实现对较大素数的测试。因此考虑引入随机化思想，但是正确率非常低。

The algorithm designed according to this principle needs to calculate the mod relation of n one by one, and the calculation is too large to realize the test of large prime numbers. Therefore, the idea of randomization was considered, but the accuracy rate was very low.

3. 费尔马小定理（Fermat Theorem）

如果 p 是一个素数且 a 是整数，则有 $a^p \equiv a \pmod{p}$，其中，若 a 不能被 p 整除，则 $a^{p-1} \equiv 1 \pmod{p}$。利用其逆否定理，对于任意整数 a，若 a 不能被 p 整除且不满足 $a^{p-1} \equiv 1 \pmod{p}$，则 p 不是一个素数。

If p is a prime number and a is an integer, then there is $a^p \equiv a \pmod{p}$, If a is not divisible by p, then $a^{p-1} \equiv 1 \pmod{p}$, Using its inverse negation theorem, for any integer a, if a is not divisible by p and is not satisfied $a^{p-1} \equiv 1 \pmod{p}$, the p is not a prime number.

通过计算 $d = a^{p-1} \bmod p$ 来判断整数 p 的素数性。当 $d \neq 1$ 时，p 一定不是素数；当 $d=1$ 时，p 可能是素数，但也存在合数的可能性。随机选取 $1 \sim p$ 中的整数 a，利用条件判断 p 的素数性。

To determine the primality of the integer p by calculation $d = a^{p-1} \bmod p$。If $d \neq 1$ then p must not be a prime. When $d = 1$, p may be prime, but there is also the possibility of composite numbers. The integer a between 1 and p is randomly selected, and the condition is used to judge the prime of p.

4. 二次探测定理（Quadratic Detection Theorem）

有一些合数也能够满足费尔马小定理，因此设计一个二次探测定理，用于避免满足条件的合数被判定为素数的情况。

Some composite numbers can also satisfy *Fermat* theorem, so a quadratic detection theorem is designed to avoid the situation that the composite numbers satisfying the condition are judged to be prime.

如果 p 是一个素数，x 是整数且 $0<x<p$，则方程 $x^2\equiv1(\bmod\ p)$ 等价于 $(x^2-1)\equiv0(\bmod\ p)$，因此 $(x-1)(x+1)\equiv0(\bmod\ p)$，故 p 必须整除 $x-1$ 或 $x+1$，$x=1$ 或 $x=p-1$。

If p is a prime number and x is an integer, and $0<x<p$, then the equation $x^2\equiv1(\bmod\ p)$ is equivalent to $(x^2-1)\equiv0(\bmod\ p)$, therefore $(x-1)(x+1)\equiv0(\bmod\ p)$, p must be divisible by $x-1$ or $x+1$, $x=1$ or $x=p-1$.

可以在利用费尔马小定理计算过程中，增加对整数 p 的二次试探，并在此基础上设计 *Miller_Rabin* 素数测试算法，再通过多次调用降低错误概率。

In the process of calculating with *Fermat* theorem, it can be increased the quadratic test of integer p, and design the *Miller_Rabin* prime number test algorithm on this basis, and then reduce the error probability by calling it many times.

5. 程序综合测试（Program Synthesis Test）

在 *Python* 程序中，将上述求解方法都进行测试。程序运行结果如图 8.9 所示，代码如下。

In a *Python* program, test all the above solutions. The program running result is shown in Figure 8.9, and the code is as follows.

```python
#采用4种方法进行素数测试(Four methods are used to test prime numbers)
#试除法(Trying division method)
import math
import random
def Prime(n):
    m = math.floor(math.sqrt(n))
    for i in range(2, m+1):
        if n % i==0:
            return False
    return True
#Wilson 定理(Wilson Theorem)
def WilsonP(n):
    fac_mod = 1
    for i in range(2, n):
        fac_mod = (fac_mod * i) % n
    if fac_mod ==n-1:
        return True
    else:
        return False
#费马定理
def fermat_prime(n):
    power = n - 1
    d = 1
    a = random.randint(2, n)
    while(power > 1):
        if(power % 2 ==1):
```

续

```
                    d = d * a % n
            power = power              //2 #整除（Exact division）
            a = a * a % n
        if(a * d % n ==1):
            return True
        else:
            return False
#二次探测定理（Quadratic Detection Theorem）
def Secondary_detection(n):
    result = True
    for x in range(2, n-1):
        if(x **2 % n ==1):
            result = False
            break
    return result
def power(n):
    global a
    power = n - 1
    d = 1
    while(power > 1):
        if(power % 2 ==1):
            d = d * a % n
        power = power//2
        result = a * a % n
        if result ==1 and a !=1 and a !=n-1:
            return False
        a = result
    if(a * d % n ==1):
        return True
    else:
        return False
#二次探测定理+费马定理（Quadratic Detection Theorem + Fermat Theorem）
def Miller_Rabin1(n, k):
    global a
    for i in range(k):
        a = random.randint(2, n-1)
        result = power(n)
        if not result:
            return False
    return True
if __name__ =="__main__":
    n = 12346
    print("试除法测试结果为:", Prime(n))
    print("Wilson 定理测试结果为:", WilsonP(n))
    print("费马定理测试结果为:", fermat_prime(n))
    print("二次探测定理测试结果为:", Secondary_detection(n))
    print("二次探测定理+费马定理测试结果为:", Miller_Rabin1(n, 10))
    print("trial divisor:", Prime(n))
    print("Wilson :", WilsonP(n))
    print("Fermat:", fermat_prime(n))
    print("quadratic probing:", Secondary_detection(n))
    print("quadratic probing+Fermat:", Miller_Rabin1(n, 10))
```

```
试除法测试结果为： False
Wilson定理测试结果为： False
费马定理测试结果为： False
二次探测定理测试结果为： True
二次探测定理+费马定理测试结果为： False
trial divisor: False
Wilson : False
Fermat: False
quadratic probing: True
quadratic probing+Fermat: False
```

图 8.9 素数测试程序运行结果

Figure 8.9 Results of prime number test program

8.5 拉斯维加斯算法(Las Vegas Algorithm)

8.5.1 算法原理(Principle of Algorithm)

拉斯维加斯算法所做的随机性决策,有可能导致算法找不到所需的解,因此通常使用一个 *bool* 型函数来表示拉斯维加斯型算法,当算法找到一个解时返回 *True*,否则返回 *False*。

The random decision made by the *Las Vegas* algorithm may cause the algorithm to fail to find the desired solution, so a *bool* function is usually used to represent the *Las Vegas* algorithm, which returns *True* when the algorithm finds a solution and *False* otherwise.

一种典型的拉斯维加斯算法调用形式为 $success = LV(x, y)$,LV 表示算法名称,x 为输入参数。当 $success$ 结果是 *True* 时,y 返回问题的解;当 $success$ 返回结果为 *False* 时,表示未能找到问题的一个解,此时可对同一实例再次独立地调用相同的算法。

A typical *Las Vegas* algorithm call is of the form $success = LV(x, y)$, where LV represents the algorithm name and x is the input parameter. When $success$ is *True*, y returns the solution to the problem. When $success$ returns *False*, a solution to the problem could not be found, and the same algorithm can be invoked again, independently, on the same instance.

设 $p(x)$ 是对输入 x 调用拉斯维加斯算法获得问题的一个解的概率。一个正确的拉斯维加斯算法应该对所有输入 x 均有 $p(x)>0$,在更强的意义下,要求存在一个常数 $\delta>0$,使得对问题的每一个实例 x 均有 $p(x)\geqslant\delta$。

Let $p(x)$ be the probability of calling the *Las Vegas* algorithm on input x to obtain a solution to the problem. A correct *Las Vegas* algorithm should have $p(x)>0$ for all inputs x, and in a stronger sense, requires a constant $\delta>0$ so that for every instance x of the problem $p(x)\geqslant\delta$。

设 $s(x)$ 和 $e(x)$ 分别是算法对于具体实例 x 求解成功或失败所需的平均时间,只要有足够的时间,对于任何的实例 x,总能找到问题的一个解。做 n 次实验,成功次数是 $np(x)$,不成功的次数为 $n(1-p(x))$,实验总时间为 $np(x)s(x)+n(1-p(x))e(x)$,则找到一个解

所需的平均时间为 $t(x) = \dfrac{np(x)s(x) + n(1-p(x))e(x)}{np(x)} = s(x) + \dfrac{1-p(x)}{p(x)}e(x)$。

Let $s(x)$ and $e(x)$ be the average time required for the algorithm to succeed or fail for a specific instance x, and given enough time, a solution to the problem can always be found for any instance x. Do n experiments, the number of successful times is $np(x)$, the number of unsuccessful times is $n(1-p(x))$, the total experiment time is $np(x)s(x) + n(1-p(x))e(x)$, then the average time required to find a solution is $t(x) = \dfrac{np(x)s(x) + n(1-p(x))e(x)}{np(x)} = s(x) + \dfrac{1-p(x)}{p(x)}e(x)$.

8.5.2　整数因子分解（Integer Factorization）

整数因子分解是指将大于 1 的整数 n 分解为如 $n = p_1^{m_1} p_2^{m_2} \cdots p_k^{m_k}$ 形式，其中，$p_1, p_2, \cdots,$ p_k 是 k 个逐个增大的素数，$m_1, m_2, \cdots, m_k$ 是 k 个正整数。

Integer factorization refers to the factorization of an integer n greater than 1 into a form of $n = p_1^{m_1} p_2^{m_2} \cdots p_k^{m_k}$ where there are k increasing prime numbers p_1, p_2, $\cdots$, p_k, which k are the positive integers m_1, m_2, $\cdots$, m_k。

如果 n 是一个合数，则 n 必有一个非平凡因子 x，$1 < x < n$，x 可以整除 n。整数的因子分割问题就是求 n 的一个非平凡因子。

If n is a composite number, then n must have a nontrivial factor x, $1 < x < n$, and x divides into n. The problem of partition of integers is to find a nontrivial factor of n.

进一步分析可知，通过对 $1 \sim x$ 的数进行试除，从而得到 $1 \sim x^2$ 的任一整数的因子分割。根据这一特点，*Pollard* 提出了用于实现因子分割的拉斯维加斯算法。

Further analysis shows that by trial division of the numbers between 1 and x, we can get the factor partition of any integer between 1 and x^2. According to this feature, *Pollard* proposed a *Las Vegas* algorithm for factor segmentation.

（1）产生 $0 \sim n-1$ 范围内的一个随机数 x，令 $y = x$。

Generate a random number x in the range of $0 \sim n-1$, so that $y = x$.

（2）按照 $x_i = (x_{i-1}^2 - 1) \bmod n$，$i = 2, 3, 4, \cdots$ 产生一系列的 x。

Produces a series of x according to $x_i = (x_{i-1}^2 - 1) \bmod n$, $i = 2, 3, 4, \cdots$.

（3）对于 $k = 2^j (j = 0, 1, 2, \cdots)$ 以及 $2^j < i \leqslant 2^{j+1}$，计算 $x_i - x_k$ 与 n 的最大公因子 $d = \gcd(x_i - x_k, n)$，如果 $1 < d < n$，就实现对 n 的一次分割，输出 d。

For $k = 2^j (j = 0, 1, 2, \cdots)$ and $2^j < i \leqslant 2^{j+1}$, calculate the largest common factor $d = \gcd(x_i - x_k, n)$ of $x_i - x_k$ and n, if $1 < d < n$, then to achieve a partition of n, output d.

整数的因子分解问题可以用确定性算法或随机化算法实现，程序运行结果如图 8.10 所示，程序代码如下。

The factorization problem of integers can be realized by deterministic algorithm or randomized algorithm. The result of the program is shown in Figure 8.10, and the program code is as follows.

```python
#整数因子分割——拉斯维加斯算法(Integer factor segmentation-Las Vegas algorithm)
import random
import math
#因子分割的确定性算法(Deterministic algorithm for factor segmentation)
def split(n):
    k = math.floor(math.sqrt(n))
    for i in range(2, k+1):
        if (n % i ==0):
            return i
    return 1
#求 a 和 b 的最大公约数(Find the greatest common divisor of a and b)
def gcd(a, b):
    if b ==0:
        return a
    else:
        return gcd(b, a % b)
#因子分割的 Pollard 算法(Pollard algorithm for factor segmentation)
def pollard(n):
    x = random.randint(0, n-1)#随机整数(Random integer)
    y = x
    k = 2
    i = 0
    while (i<=64):#64 为最大迭代次数(64 indicates the maximum number of iterations)
        i +=1
        x = (x * x - 1) % n
        d = gcd(y-x, n)#求 n 的非平凡因子 xk-x(Find the nontrivial factor xk-x for n)
        if d > 1 and d < n:
            return d
        if y ==x:
            return n
        if (i ==k):
            y = x
            k *=2
if __name__ =="__main__":
    n = 1000
    a1 = split(n)
    a = pollard(n)
    print("确定性算法分割的 n 的一个因子为:", a1)
    print("Pollard 算法分割的 n 的一个因子为:", a)
    print("确定性算法:", a1)
    print("Pollard algorithm:", a)
```

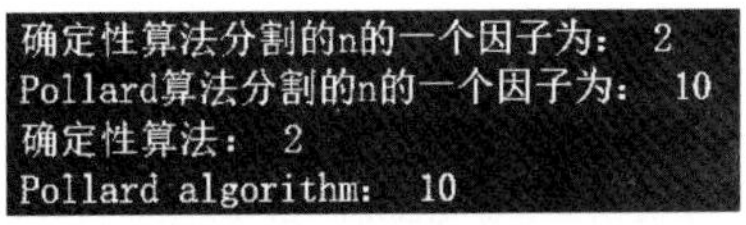

```
确定性算法分割的n的一个因子为: 2
Pollard算法分割的n的一个因子为: 10
确定性算法: 2
Pollard algorithm: 10
```

图 8.10 整数因子运行结果

Figure 8.10 Result of integer factor running

8.5.3　n 皇后问题（n-Queen Problem）

这类问题要求将国际象棋中的 n 个皇后棋子放在 n×n 棋盘的不同行、不同列、不同斜线的位置，找出相应的放置方案。这个问题仅要求任意两个皇后的位置满足要求，并没有其他放置规律。因此，可以随机选择棋盘上的一个位置，只要和其他皇后的位置不冲突即可。

This problem requires that n queens in chess be placed in different rows, different columns, and different slash positions on the n×n board, and find out the corresponding placement scheme. This problem only requires the position of any two queens to meet the requirements, and there is no other placement rule. Therefore, a position on the board can be chosen at random, if it does not conflict with the positions of other queens.

用 n 维向量 $X=(x_1,x_2,\cdots,x_n)$ 表示 n 个皇后在棋盘中的位置，x_i 表示第 i 个皇后在 i 行 x_i 列，i、j 两个皇后不同列表示为 $x_i \neq x_j$，不在同一斜线表示为 $|i-j| \neq |x_i-x_j|$。

The position of n queens on the chessboard is represented by n-dimensional vector $X=(x_1, x_2, \cdots, x_n)$, x_i indicating that the i queen is in the i row and x_i column, and the different columns of i and j queens are represented by $x_i \neq x_j$, and the different slash line are represented by $|i-j| \neq |x_i-x_j|$.

该算法需要设计三个函数。首先需要判断当前位置能否放置，然后在当前所有能够放置的位置中随机放置，最后反复调用求解 n 皇后的放置问题。

The algorithm requires the design of three functions. First, it needs to determine whether the current position can be placed, then randomly place in all the current positions that can be placed, and finally repeatedly call to solve the placement problem of n queens.

算法的程序实现结果如图 8.11 所示，算法代码如下。

The program implementation result of the algorithm is shown in Figure 8.11, and the algorithm code is as follows.

```
import random
def Place(k):                #判断能否在第 k 行放置第 k 个皇后，皇后从 0 开始编号(Determines whether the k
                             #queen can be placed in row k, numbered from 0)
    global x
    for j in range(k):       #当前皇后和前面已经放置好的皇后是否同一斜线、是否同一列(Whether the current
                             #queen and the previous queens are in the same slash line and column)
        if((abs(k-j)==abs(x[j]-x[k])) or (x[j]==x[k])):
            return False
    return True
def queensLV(n):
    global x
    k=0
    count = 1                #记录当前要放置的第 k 个皇后在第 k 行的有效位置(Records the current effective
                             #position of the k queen to be placed in row k)
    while((k<n) and (count>0)):
        count=0
        y=[]
```

续

```
        for i in range(n):
            x[k] =i
            if(Place(k)):
                y.append(i)      #第 k 个皇后在第 k 行的有效位置存于 y 数组(The effective position of the Kth
                                 #queen in line k is stored in the y array)
                count += 1
    #从有效位置中随机选取一个位置放置第 k 个皇后(A position is randomly selected from the valid positions
    #to place the k queen)
        if(count>0):
            x[k] =y[random.randint(0, count-1)]
            k += 1
    return (count>0)#count>0      #表示放置成功(Indicates successful placement)
def queensLV1(n):
    global x
    k = 0
    maxcount = 20                #最大尝试次数(Maximum attempts)
    while((k < n) and (count > 0)):
        count = 0
        y = []
        for i in range(n):
            x[k] =i
            if(Place(k)):
                y.append(i)      #第 k 个皇后在第 k 行的有效位置存于 y 数组(The effective position of the k
                                 #queen in line k is stored in the y array)
                count += 1
    #从有效位置中随机选取一个位置放置第 k 个皇后(A position is randomly selected from the valid positions
    #to place the k queen)
        if(count>0):
            x[k] =y[random.randint(0, count-1)]
            k += 1
    return (count>0)             #count>0 表示放置成功(If count>0, the storage is successfully placed)
def nQueen(n):
    success = queensLV(n)
    while(not success):
        success = queensLV(n)
    return success
if __name__ == "__main__":
    n = 8
    x = [-1 for i in range(n)]
    success = nQueen(n)
    if success:
        print("n 皇后问题的一种放置方案为:", x)
        print("n-queen:", x)
```

```
n皇后问题的一种放置方案为:  [7, 1, 3, 0, 6, 4, 2, 5]
n-queen:  [7, 1, 3, 0, 6, 4, 2, 5]
```

图 8.11 n 皇后问题程序运行结果

Figure 8.11 Result of n queen problem program

8.6 舍伍德算法（Sherwood Algorithm）

分析算法在平均情况下的计算复杂性时，通常假定算法的输入数据服从某一特定的概率分布。例如，在输入数据是均匀分布时，快速排序算法所需的平均时间是 $O(n\log n)$。而当其输入已"几乎"排好序时，这个时间界就不再成立。在这种情况下，通常可以采用舍伍德算法来消除算法所需计算时间与输入实例间的这种差异。

When analyzing the computational complexity of an algorithm in the average case, it is usually assumed that the input data of the algorithm obey a certain probability distribution. For example, when the input data is evenly distributed, the average time required by the quicksort algorithm is $O(n\log n)$. When the input is "almost" sorted, this time horizon no longer holds. In this case, Sherwood's algorithm can often be used to eliminate this difference between the computation time required by the algorithm and the input instance.

8.6.1 随机快速排序（Random Quicksort）

快速排序算法始终选择待排序列的第一个元素作为基准元素进行划分，但是最坏情况的时间复杂性与平均情况差别较大，因此在基准元素的选择上需要做优化。将随机性引入基准元素选择，此时的快速排序算法就是舍伍德算法，可以以高概率获得平均计算性能。

The quicksort algorithm always selects the first element of the sequence to be sorted as the base element, but the time complexity of the worst case is much different from that of the average case, so the selection of the base element needs to be optimized. When randomness is introduced into the selection of base elements, the quicksort algorithm is the *Sherwood* algorithm, which can obtain the average computing performance with high probability.

在序列中随机选择一个位置的元素，将其与第一个位置的元素互换，然后作为基准元素，从序列的两端交替向中间扫描，划分序列，然后执行快速排序算法。用 *Python* 实现的程序结果如图 8.12 所示，算法代码如下。

Randomly select an element in one position in the sequence, interchange it with the element in the first position, and then used as a reference element, alternately scan from the two ends of the sequence to the middle, divide the sequence, and then perform a quick sort algorithm. The result of the program implemented in *Python* is shown in Figure 8.12, and the algorithm code is as follows.

```
#随机快速排序(Random quicksort)
import random
#随机选取基准元素进行划分(Base elements are randomly selected for partitioning)
def RandPartition(R, left, right):    #R:待排序元素, left:起始索引, right:结束索引(R: elements to be sorted,
                                      #left: start index, right: end index)
        i = left
```

续

```
        j = right
        randi = random.randint(left, right)      #随机选取一个位置(Pick a random location)
        R[left], R[randi] = R[randi], R[left]     #将随机选择的位置和第一个位置元素互换(Swap the randomly
                                                  #selected position with the first position element)
        pivot = R[left]                           #用序列的第一个元素作为基准元素(Use the first element of the
                                                  #sequence as the base element)
        while(i<j):                               #从序列的两端交替向中间扫描,直至 i 等于 j 为止(Scan
                                                  #alternately from both ends of the sequence to the middle until
                                                  #i equals j)
            while(i<j and R[j]>=pivot):           #pivot 相当于在位置 i 上(pivot is equivalent to position i)
                j -= 1                            #从右向左扫描,查找第 1 个小于 pivot 的元素(Scan from right to
                                                  #left to find the first element less than pivot)
            if(i<j):                              #表示找到了小于基准元素的元素(indicates that an element
                                                  #smaller than the reference element was found)
                R[i], R[j] = R[j], R[i]           #交换 R[i]和 R[j],交换后 i 执行加 1 操作(Switch R[i] and R[j],
                                                  #after the switch i perform the plus 1 operation)
                i += 1
            while(i<j   and R[i]<=pivot):         #从左向右扫描,查找第 1 个大于 pivot 的元素(Scan from left to
                i += 1                            #right to find the first element larger than pivot)
            if(i<j):                              #表示找到了大于基准元素的元素(indicates that an element larger
                                                  #than the reference element is found)
                R[i], R[j] = R[j], R[i]           #交换 R[i]和 R[j],交换后 j 执行减 1 操作(Switch R[i] and R[j],
                                                  #after the switch j performs minus 1 operation)
                j -= 1
        return j

#快速排序函数(Quick sort function)
def quickSort(R, left, right):
    if left < right:
        j = RandPartition(R, left, right)
        quickSort(R, left, j-1)
        quickSort(R, j+1, right)

if _ _name_ _ == "_ _main_ _":
    arr = [54, 26, 93, 17, 77, 88, 5, 44, 55, 20, 3]
    n = len(arr)
    quickSort(arr, 0, n-1)
    print (arr)
```

[3, 5, 17, 20, 26, 44, 54, 55, 77, 88, 93]

图 8.12 随机快速排序运行结果

Figure 8.12 Result of random quicksort running

8.6.2 线性时间选择(Linear Time Selection)

线性时间选择的分治算法随基准元素的选择较为复杂。首先分组,然后取每一组的中位数,再取中位数的中位数,最后以第二次的中位数对 n 个元素进行划分。根据舍伍德算法的思想,可以在基准元素的选择上引入随机性,将线性时间选择算法改造为舍伍德算法。

The divide-and-conquer algorithm of linear time selection is more complex with the selection of base elements. First grouping, then taking the median of each group, then taking the median of the medians, and finally divide n elements by the median of the second time. According to the idea of Sherwood algorithm, randomness can be introduced into the selection of base elements, and the linear time selection algorithm can be transformed into *Sherwood* algorithm.

采用随机划分的方法将给定元素划分为两部分，计算包括基准元素在内的较小部分元素个数 count，如果 count<k，则在较大部分的元素中递归找第 k-count 小；如果 count>k，则在较小部分元素个数中递归找到第 k 小的数。算法程序实现的结果如图 8.13 所示，程序代码如下。

The given element is divided into two parts by random partition method, and the number *count* of the smaller part including the base element is calculated. If *count<k*, the *k-count* smaller is found in the larger part of the element recursively. If *count>k*, the k smallest number is recursively found among the smaller number of elements. The result of the algorithm program implementation is shown in Figure 8.13, and the program code is as follows.

```
#随机选择第 k 小元素（Select the kth smallest element at random）
import random
def RandPartition(R, left, right):        #R：待查找元素序列,left：起始索引,right：结束索引（R: Sequence
                                          #of elements to be searched, left: start index, right: end index）
    i = left
    j = right
    randi = random.randint(left, right)
    R[left], R[randi] = R[randi], R[left]
    pivot=R[left]                         #用序列的第一个元素作为基准元素（Use the first element of the
                                          #sequence as the base element）
    while(i<j):                           #从序列的两端交替向中间扫描,直至 i 等于 j 为止（Scan alternately
                                          #from both ends of the sequence to the middle until i equals j）
        while(i<j and R[j]>=pivot):       #pivot 相当于在位置 i 上（pivot is equivalent to position i）
            j -=1                         #从右向左扫描,查找第 1 个小于 pivot 的元素（Scan from right to left
                                          #to find the first element less than pivot）
        if(i<j):                          #表示找到了小于基准元素的元素（indicates that an element smaller
                                          #than the base element was found）
            R[i], R[j] = R[j], R[i]       #交换 R[i]和 R[j],交换后 i 执行加 1 操作（Switch R[i] and R[j].
                                          #After the switch, i adds 1）
            i +=1
        while(i<j  and R[i]<=pivot):      #从左向右扫描,查找第 1 个大于 pivot 的元素（Scan from left to right
                                          #to find the first element larger than pivot）
            i +=1
        if(i<j):                          #表示找到了大于基准元素的元素（indicates that an element larger
                                          #than the reference element was found）
            R[i], R[j] = R[j], R[i]       #交换 R[i]和 R[j],交换后 j 执行减 1 操作（Switch R[i] and R[j], and
                                          #j will subtract 1 after the switch）
            j -=1
    return j
#随机选择第 k 小元素（Select the kth smallest element at random）
def select(R, left, right, k):
```

续

```
    if left >=right：
        return R[left]
    j = RandPartition(R, left, right)
    count = j - left + 1
    if count ==k：
        return R[j]
    elif count < k：
        returnselect(R, j + 1, right, k - count)
    else：
        returnselect(R, left, j, k)
if _ _name_ _ ==" _ _main_ _"：
    arr = [54, 26, 93, 17, 77, 5, 44, 55, 20]
    n = len(arr)
    k = int(input("请输入正确的k/Please enter the correct k"))
    while( k<0 or k>n)：
        print("Error")
        k = int(input("请输入正确的k/Please enter the correct k"))
    min_k = select(arr, 0, n-1, k)
    print ("R 中第", k, "小的数为：", min_k)
    print ("The", k, "TH smallest number in R is：", min_k)
```

```
请输入正确的k/Please enter the correct k   -2
Error
请输入正确的k/Please enter the correct k   3
R中第 3 小的数为： 20
The 3 TH smallest number in R is： 20
```

图 8.13   线性时间选择问题程序运行结果

Figure 8.13   Result of linear time selection problem program

习题 8(Exercises Eight)

1. 请简述造成伪随机数的原因。

Please briefly describe the reason for the pseudorandom number.

2. 请简述数值随机算法、蒙特卡洛算法、拉斯维加斯算法和舍伍德算法的适用情况。

Please briefly describe the application ofstochastic algorithm, *Monte Carlo* algorithm, *Las Vegas* algorithm and *Sherwood* algorithm.

3. 请另外选择两个函数以及定义域,用数值随机化算法计算各自的定积分。

Please select the other two functions and the domain, and use the Stochastic algorithm to calculate their definite integrals.

4. 请写一个六面筛子程序。

Please write a six-sided sieve program.

5. 有一枚质地均匀的硬币,请写一个抛硬币程序,并测试记录抛 20 次、50 次、100 次、1000 次和 10 000 次时,正反面的次数和概率。

There is a fair coin with uniform texture, please write a coin toss program and test the number and probability of the positive and negative flips for 20, 50, 100, 1000 and 10 000 times.

参 考 文 献

[1]　王秋芬,赵刚彬. 算法设计与分析[M]. 北京:清华大学出版社,2023.

[2]　霍红卫. 算法设计与分析[M]. 2 版. 西安:西安电子科技大学出版社,2021.

[3]　王红梅. 算法设计与分析[M]. 3 版. 北京:清华大学出版社,2023.

[4]　程振波,李曲,王春平. 算法设计与分析(Python)[M]. 北京:清华大学出版社,2018.

[5]　江红,余青松. Python 程序设计与算法基础教程微课版[M]. 2 版. 北京:清华大学出版社,2018.

[6]　王晓东. 算法设计与分析[M]. 4 版. 北京:清华大学出版社,2018.

[7]　郭宪,方勇纯. 深入浅出强化学习原理入门[M]. 北京:电子工业出版社,2018.

[8]　王磊,王晓东. 机器学习算法导论[M]. 北京:清华大学出版社,2019.

[9]　阿布舍克·维贾亚瓦吉亚. Python 机器学习[M]. 宋格格,译. 北京:人民邮电出版社,2019.

[10]　Andreas C. Müller Sarah Guido. Python 机器学习基础教程[M]. 张亮,译. 北京:人民邮电出版社,2018.

[11]　斋藤康毅. 深度学习入门:基于 Python 的理论与实现[M]. 陆宇杰,译. 北京:人民邮电出版社,2018.

[12]　伊藤真. 用 Python 动手学机器学习[M]. 郑明智,司磊,译. 北京:人民邮电出版社,2021.

[13]　杰弗瑞·希顿. 人工智能算法(卷 1):基础算法[M]. 李尔超,译. 北京:人民邮电出版社,2024.

[14]　杰弗瑞·希顿. 人工智能算法(卷 3):深度学习和神经网络[M]. 王海鹏,译. 北京:人民邮电出版社,2024.

[15]　结城浩. 程序员的数学[M]. 管杰,译. 北京:人民邮电出版社,2012.

[16]　周志华. 机器学习[M]. 北京:清华大学出版社,2016.

[17]　Allen B. Downey. 统计思维:程序员数学之概率统计[M]. 张建锋,陈钢,译. 北京:人民邮电出版社,2013.